DATE DUE

ROBOTICS AND
FLEXIBLE MANUFACTURING TECHNOLOGIES

ROBOTICS AND FLEXIBLE MANUFACTURING TECHNOLOGIES

Assessment, Impacts and Forecast

by

Robert U. Ayres and Steven M. Miller

Department of Engineering and Public Policy
Carnegie-Mellon University
Pittsburgh, Pennsylvania

James Just, Keith King, Michael Osheroff
George Berke, Peter Spidaliere, Tran Ngoc

DHR, Incorporated
McLean, Virginia

np **NOYES PUBLICATIONS**
Park Ridge, New Jersey, U.S.A.

Published in the United States of America by
Noyes Publications
Mill Road, Park Ridge, New Jersey 07656

10 9 8 7 6 5 4 3 2 1

Library of Congress Cataloging-in-Publication Data

Ayres, Robert U.
    Robotics and flexible manufacturing technologies.

    Bibliography: v. 2, p.
    Includes index.
    Contents: pt. 1. Carnegie-Mellon assessment --
pt. 2. DHR assessment.
    1. Robotics. 2. Flexible manufacturing systems.
3. Technology assessment. I. Miller, Steven M.
II. Title.
TJ211.A96 1985   670.42   85-15312
ISBN 0-8155-1043-8

Foreword

A critical assessment of current robotics and flexible manufacturing technologies, their impacts on the industrial base, and a forecast of future functional capabilities and emerging applications areas are presented in this book.

Nearly three times as many robots were installed in industry in 1984 as in 1979, and the prospects for the future can only increase, probably by orders of magnitude. This valuable up-to-date and critical assessment indicates that robotics and flexible manufacturing technologies will be a major driving force in the industrial world.

Part I of the book reviews the role of tools, machines and controls in our society; the extensions of capabilities of commercially available robots, integration of sensory information processing with robotic devices, and the integration of robots, machine tools, parts handling and transport devices and computers into flexible manufacturing systems (FMSs). Analyses are given on changes in unit costs and production labor requirements which will accompany the more widespread use of industrial robots and FMSs.

Part II reviews key worldwide R&D activities and discusses the principal thrusts and trends in robotics development. It provides a technological forecast addressing future directions of robotics producers and end-users.

The information in this book is from:

> *An Exploratory Assessment of Second Generation Robotics and Sensor-Based Systems* prepared by Robert U. Ayres and Steven M. Miller of the Department of Engineering and Public Policy, Carnegie-Mellon University for the National Science Foundation, September 1984.

Robotic Technology: An Assessment and Forecast prepared by James Just, Keith King, Michael Osheroff, George Berke, Peter Spidaliere and Tran Ngoc of DHR, Incorporated for the Aerospace Industrial Modernization Office of the Air Force Systems Command, July 1984.

The table of contents is organized in such a way as to serve as a subject index and provides easy access to the information contained in the book.

NOTICE

Contents and Subject Index

PART I

CARNEGIE-MELLON ASSESSMENT

The information in Part I is from *An Exploratory Assessment of Second Generation Robotics and Sensor-Based Systems* prepared by Robert U. Ayres and Steven M. Miller of the Department of Engineering and Public Policy, Carnegie-Mellon University for the National Science Foundation, September 1984.

1. Introduction

The first systematic assessment of the potential impacts of robots on employment and productivity was the report, "Technology Assessment: The Impact of Robots," submitted to the National Science Foundation by the Eikonix Corporation in 1979. The assessment primarily focused on robots in manufacturing, but gave some attention to household applications and to applications in the nuclear industry. The technology assessment was a useful examination of what was then known about the technology, what could be surmised about its future, and the implications for society, for industry, and for the government.

The Eikonix report stated that robot capabilities were, and would remain for at least ten to fifteen years, much too limited for the needs of most manufacturers; and that as a result, diffusion of the technology would be slow and gradual. The assessment concluded that "it is highly unlikely that diffusion of robotics will be rapid. . .The technical constraints against a vast change in the scope of robotic capabilities is one reason for the slow diffusion rate. . . there is just no evidence to suggest that the impacts will be anything but very minor, at least up to the turn of the century."[1]

NSF has sponsored this new assessment for three reasons. First, progress in the development of robotics technology has accelerated since 1979, particularly in the areas of sensory information processing. Second, the earlier study did not give much attention to parallel and concurrent developments in manufacturing technology such as flexible manufacturing systems and design for automated fabrication and assembly. Third, the U.S. now has con-

---

[1] page 231.

siderably more experience with robotic use and impacts. There are nearly three times as many robots installed in U.S. industry in 1984 as in 1979.

This updated assessment is comprised of two major components. The first is a set of four papers by the principal investigators, Robert Ayres and Steven Miller. The first paper by Ayres is an overview concerned mainly with articulating a set of broad issues. The second paper, by Miller, is a survey of recent developments in flexible manufacturing technology and the specific role of robotics and sensory information processing. The third and fourth papers were adapted from Miller's recent (1983) Ph.D. thesis. They provide detailed background information and analysis of the coming impacts of robotics in the metalworking industries specifically.

The second major component is the "proceedings" of a computer conference held to review and expand upon the ideas in the four Ayres and Miller papers. The computer conference was organized by the Center for Social and Economic Issues of the Industrial Technology Institute, Ann Arbor, Michigan. Over a two month period, 18 researchers in the U.S. and one in Sweden discussed a variety of issues related to the social and economic impacts of flexible automation via the computer conference network. Abstracts of the four papers, and an outline of the proceedings of the computer conference follow.

The first paper "The Robotics Society" discusses the role of tools, machines and controls in human society, and the conditions under which humans as machine controllers can expect to be replaced in the future by computers linked to sensors. It concludes that, during the coming two or three decades, most "operative" jobs on the factory floor could well be eliminated. On the other hand, relatively few non-factory jobs will be taken over by robots.

A critical benefit of this new form of computer-integrated automation is that it will greatly increase the flexibility of large scale manufacturing and shift the basis for product competition away from price and towards quality and performance.[2] Potential socio-economic implications for the present generation of industrial workers, the industrial unions, and the specialized manufacturing communities, cannot be dismissed lightly. These impacts deserve further study in detail together with possible education and other responses from the public sector.[3]

In the longer run, the paper argues that specialized robots will find many other uses, but that a general purpose "Humanoid" robot is most unlikely to be developed. The reason is, briefly, that a generalized "human" appearance and capability could only be achieved by sacrificing the special skills that would give robots advantages in particular applications. Moreover, even without special skills, it is most unlikely that humans could design a truly general purpose machine that could closely approach the human level of physical performance, sensitivity and intelligence simultaneously across the whole range. Hence, the impact of robotics on human society in the next century will be mostly of an indirect kind; viz. through productivity and employment. The robotic society of the 21st century will in no way resemble the traditional science-fiction view. Robots are and will remain machines that extend human

---

[2] See Ayres, Robert U., The Next Industrial Revolution: Reviving Industry Through Innovation, Ballinger Publishing Co., Cambridge, MA, 1984.

---

[3] See Ayres, Robert U. and Steven M. Miller, Robotics: Applications and Social Implications, Ballinger Publishing Co., MA, 1983. Also Ayres and Miller, Robotics and the Conservation of Human Resources, Technology and Society, Vol. 4(3), winter, 1982, pages 181–97 and Ayres and Miller, Robotic Realities: Near-Term Prospects and Problems, Annals of the American Academy of Political and Social Sciences, Vol. 470, November, 1983, pages 28–55.

capabilities. They will eventually serve people directly in some capacities, as they now serve other machines. In the long run, the major social benefit may be that it will no longer be necessary for a class of human workers to operate, load, unload, and watch over production machines, thus--in effect-- being paced by and "serving" the machines.

The second paper, "Recent Developments in Robotics and Flexible Manufacturing Systems," reviews some of the recent developments in robotics and flexible manufacturing technology. Attention is focused on three broad classes of developments:

1. Extensions of the capabilities of commercially available robots.

2. Integration of sensory information processing with robotic devices, and

3. Integration of robots, machine tools, parts handling and transport devices, parts feeders and assembly machines, "smart sensors" and computers into flexible manufacturing systems.

Topics covered that relate to the first class of development include increases in the number of robot models designed for precision placement and increases in the use of computer controlled robots. Topics that relate to the second class include capabilities and limitations of machine vision systems for controlling robot positioning and manipulation in parts acquisition, tool handling, and assembly. Topics relating to the third class include:

- The integration of multiple FMS for machining in one plant

- The use of FMS in manufacturing processes other than machining

- The addition of artificial intelligence and sensor-based systems to FMS

In the third paper, "Impacts of Robotic and Flexible Manufacturing Technologies on Manufacturing Costs and Employment," the issues considered include the extent to which unit costs and production labor requirements might be reduced if there is more widespread use for industrial robots and flexible manufacturing systems. These issues are analyzed from two different perspectives. The technological focus of the first perspective is narrowly confined to the use of robotic manipulators. It is assumed that robotic manipulators will be "retrofitted" into existing production facilities without making major changes in the organization of production within the factory, other than modifying individual work stations so that robots can replace one (or perhaps several) operators. The critical variable in this perspective is an estimate of the percent of the production worker jobs that will be replaced by robots. Reductions in unit cost are calculated by assuming that a given percentage of labor costs is reduced. The question of whether decreases in production labor requirements could be offset by an increase in demand stimulated by a reduction in price is also discussed.

The focus of the second perspective is much broader than the first, and is concerned with the impacts of integrating robots with other types of computer assisted manufacturing (CAM) technologies into flexible manufacturing systems. It is assumed that a factory using general purpose machines to produce specialized products in batches can be reorganized and integrated so the machines are fully utilized and used more efficiently. One critical variable in this perspective is an estimate of the potential increase in output that could be realized if all of the time in a year available for production were utilized. The other critical variable is an estimate of the unit cost and of the labor requirements in a fully utilized batch production plant. Based on an analysis of a cross section of metalworking industries, a relationship is specified between the level of output and the level of unit cost. Reductions in unit cost for a given increase in output are derived from this relationship. Reductions in unit labor requirements are calculated in a similar manner.

In the fourth paper, "Custom Batch, and Mass Production in the Metal-working Industries," 101 metalworking industries (at the four digit SIC level) are classified as either being dominated by custom/small batch production, mid-batch production, large batch production, or mass production. Given the assignments of industries to a mode of production, the distribution of value added, output, and production worker employment by mode of production is calculated. This analysis corroborates the claim that approximately 75 percent of the value added in the metalworking sector is accounted for by products which are batch produced.

The significance of the claim that most industries in metalworking produce batches of specialized products can only be appreciated by considering the difference in unit cost between batch and mass production analyzed in the previous paper. Our analysis supports previously published estimates that the unit cost using the most efficient mass production techniques would be 100–500 times lower than if the products were produced in a "one-of-the-kind" mode (and 10–30 times lower than if it were batch produced). Considering that 1) most of the value added in metalworking is accounted for by batch production, and 2) products which are produced are much more expensive than products mass produced in large volumes, much of the value added within the metalworking sector can be viewed as a type of penalty cost that has been unavoidable because of the inherent inefficiencies of custom and batch production relative to mass production. This is the foundation for many of the arguments citing the need to accelerate the development and use of "robotic" and other types of flexible production technologies which are applicable to batch production.

Comments of the researchers participating in the computer conference are grouped into the following 15 categories:

- Robot/FMS Technology
- Cost/Benefits of Robots/FMS
- Importance of Flexibility
- Structure of Industries Providing and Using Flexible Automation
- Alternative Forms of Social Organization of Production
- Factors Limiting the Rate of Technological Change
- Impacts of Flexible Automation on Skills
- Implications for Education and Training
- Implications for Plant Location
- Implications for Economic Growth and Job Displacement
- Implications for Industrial Policy
- Comments on Forecasting Methodology
- Afterthoughts on Teleconferencing
- Other Comments and Criticisms, not elsewhere classified

As an addendum to the assessment we include a short paper by Gerald Ross, formerly of the Industrial Technology Institute, on the need for flexible management systems to operate flexible manufacturing systems. The paper was motivated, in large part, by the discussions in the computer conference.

2. The Robotics Society

Robert U. Ayres

Department of Engineering and Public Policy
Carnegie-Mellon University
Pittsburgh, PA 15213

March 1984

What Kind of Jobs Will Robots Be Good at?

Worldwide there are fewer than 30,000 robots only 6,000 of which are in the U.S. These robots are stiff, ponderous mechanical arms, with an elbow, a 'wrist' and two fingers. They are immobile and insensate. They do simple tasks like spot welding, spray painting, or loading and unloading a die-casting machine, strictly from memory. There is one robot for every 40,000 citizens of this country, one for every 1,000 workers in the manufacturing industries and perhaps one for every 400 workers doing the dull, routine, repetitive jobs that robots do best. These are facts.

Perceptions are also facts of a sort. Media coverage of "robotics" has been frenzied for the past three years, at least. There have probably been more published articles and interviews on the subject each year than there are robots sold. Hundreds of small firms have sprung up to manufacture or sell robots to a supposedly insatiable market. Investors seek hot new robotics stocks—though seekers far outnumber finders. Pundits see robotics and flexible automation as the salvation of an aging sclerotic U.S. industrial base that is being soundly thrashed in the marketplace by the dynamic Japanese. Upscale consumers buy robot toys and seem to be waiting eagerly for the household (or personal) robot to take over dreary chores like washing windows, vacuuming floors and taking out the garbage. Amid all this excitement, understandably, many workers fear for their jobs.

To be sure, the robot C3PO in "Star Wars" is a fantasy—for now. But is today's science fiction tomorrow's reality? Arthur C. Clark virtually invented the communication satellite in a science fiction story published a decade before the reality. Isaac Asimov, whose novels give him a claim, of sorts, to the title "father of robotics" is beginning again, after a 30 year hiatus, to publish tales of hypothetical future human societies in which humans and robots mingle, coexist and interact as distinct species of intelligent beings. Asimov has lots of company in writing on this theme. Is Asimov describing our own future a generation or so ahead?

Curiously, most of the recent science fiction writers seem to imagine an inevitable technological evolution toward "humanoid" robots—similar to C3PO—with arms, hands, legs, eyes, ears and brains ("positronic" or not) similar to those of humans. True, in modern science fiction (not including Karel Capek's apocalyptic RUR, written in 1920) the robots are generally physically and mentally inferior (though stronger and more durable) than their human counterparts. In science fiction, of course, the literary convention is that humans always have to maintain or regain dominance in the denouements.

What about the real world? Obviously literary conventions are irrelevant and the good guys don't always come out on top. Do robots in the real world have the potential of taking over most—if not all—human jobs eventually? If this should occur what kind of an economic system would result? Would intelligent robots become citizens, receive income, pay union dues, pay taxes, etc.? If so will the humans be mostly unemployed? Will our descendants be living on welfare? Or will robots become the principle form of human wealth as domestic animals and slaves once were?

Or are we endowing robots with too much potential? After all, we are the products of five hundred million years of biological evolution, not to mention five million years of social evolution. Granted that machines are stronger and tougher than human limbs, while computers can carry out specialized numerical manipulations much faster than humans. Granted that computers can acquire knowledge (from other computers) faster than humans can. These advantages obviously account for the important role machines and computers already play in the economic system. But humans also have some inborn abilities that will be very hard—if not impossible—for machines to imitate.

The important point is that neither humans nor machines can be very good at everything at the same time. The qualities that make a good jockey are utterly wrong for a football player or a basketball player. Poets and artists do not make good accountants. Surgeons do not box and boxers don't make good surgeons. Similarly animals or plants that become especially well adapted to one environment generally become ill-suited to others. Humans, among all the species of animals, are notable for being the least specialized, i.e., for being flexible and adaptable. To take one example, the human hand is an outstanding general purpose manipulator, partly because it lacks the specialized sharp claws of a predator, the spatulate claws of a digger, the web of a swimmer, the feathered wing of a flyer or the hoof of an ungulate. Humans have equally unspecialized teeth, lacking tusks, fangs or special cud-chewing molars.

A similar point can be made about human reliance on vision. Many ground living animals rely primarily on smell, of course. Smell is far more specialized, it is an excellent way of tracking prey or finding a mate even in the dark, but is relatively ineffective for warning against enemies—which can approach from downwind—and quite useless for facilitating manipulative tasks. Moreover, it is impossible to walk with your head up to see while simultaneously sniffing the ground. Nature has had to choose between alternatives. Bats and the most advanced marine animals rely on acute sense of hearing, but only because the former fly at night and live in caves, while water conducts sound much better than light. Binocular color vision seems to be the most general purpose sensory system for land animals functioning in daylight. But to have good vision in daylight an animal cannot also have the best possible night vision, whence nocturnal animals generally find dark crannies to hide in during the day. Only humans operate both in daylight and in darkness.

How do machines surpass humans? Mechanically, they have several advantages. First, they can be made very strong (if enough power is available). Second, their metal skins can make them relatively impervious to surface damage. Third, their inherent rigidity makes it possible to repeat a sequence of motions many times with minimal wear and high precision. Fourth, they can perform rotational motions (which human joints do not permit). Fifth, they can continue to function for long periods, as long as power is available, subject only to wear, corrosion and metal fatigue, but not to tiredness or boredom.

Note that all these mechanical advantages arise from one core fact: most machines are made of dense, hard, rigid materials—mainly metal. But machines made from metal also suffer from some significant disadvantages arising from the same central fact. One is that when most machines break down, they tend to be totally nonfunctional until they are fixed, including the period during which they are being worked on. Clearly, broken machines do not fix themselves (i.e., they do not heal, as living organisms can often do). Another disadvantage of machines is that their strength and rigidity often makes them somewhat inappropriate for tasks involving close contacts with brittle or easily-injured materials.

Let us turn the question around and ask: how do humans surpass machines? Thinking about it this way quickly reveals a very important aspect of many tasks, as carried out by human workers. It is simply that humans seldom memorize a sequence of motions completely. In practice, visual or tactile senses are relied upon constantly to control fine motions (as opposed to gross movement). Moreover, in practice it seems that this sensory feedback is really essential, even in many routine tasks. Machines programmed to repeat a sequence of motions from memory alone cannot reliably perform insertion tasks required in parts assembly, for instance. Engaging a nut on a threaded bolt is an example of a task that is very hard for a machine lacking tactile (or force feedback) sensory capability. Often the nut will jam and the machine, not knowing any better, will strip the thread or cause even worse damage. (This is why auto mechanics often use mechanical bolt tighteners only after they have properly engaged the nut on the bolt by hand.)

Once the nature of the difficulty is clear, most people can immediately think of many everyday tasks that would offer comparable difficulties for insensate machines. Briefly, machines (in their present state of development) find it very difficult to discriminate shades, colors, shapes and textures, to manipulate soft and floppy objects, or to function in an environment that is changing rapidly and unpredictably. This is because of the inability of machines to process and interpret sensory information and to modify their actions accordingly.

Humans, by contrast, clearly find these things comparatively easy. On the other hand humans find it difficult to respond to information that is not directly accessible to eyes, ears or touch. For instance, the inability of humans to sense radar signals directly makes it very likely that computers (linked to radar systems) will totally replace human air traffic controllers in the not very distant future. Humans also find it difficult to discriminate or manipulate very small (microscopic) objects; or to manipulate very large, very heavy, very hot or very cold objects; or to work in space, under water, or in a very noisy or toxic environment. Finally, and this is a very important fact indeed, humans find it very difficult to perform the same task over and over again without getting bored and fatigued and making mistakes.

Summarizing what we know or can reasonably infer about the inherent capabilities of humans and robots (i.e., machines) leads to a map something like Figure 1. Moving from right to left tasks become more difficult for humans. Moving down, tasks become increasingly difficult for machines, assuming they are made of metal and that machine-senses utilize known types of sensors and strategies for sensory information processing and decision making. Clearly, if totally new physical principles are discovered and applied (e.g., "positronic brains") the relationships suggested in Figure 1 may change. But for the next few decades, at least, this picture seems likely to remain valid.

Generally speaking, the top left-hand corner of Figure 1 includes tasks that are fairly easy for both humans and machines; the lower right corner includes tasks that would be difficult for either, and at present are the exclusive domain of humans, while the opposite (upper right) corner includes tasks that are likely to be easier for machines. Tasks in a given row should be roughly equally difficult for a robot, while tasks in a given column should be roughly equally difficult for a human.

As a rough generalization, increasing sensory feedback capabilities for robots will enable them to gradually take over tasks starting from the top right corner and working down toward the bottom left; this "rule" is obviously subject to economic and other constraints. For example, while most aircraft landings and takeoffs are quite routine, humans are likely to be more competent at handling emergencies for a long time to come. Similar considerations apply to medicine, dentistry and surgery.

Needless to say, tasks that are intrinsically nonroutine, e.g., firefighting or mechanical repairs, are also less likely to justify the use of sophisticated robots than tasks that are frequently repeated. The economics of substitution of machines for humans are also strongly influenced by the conditions under which a human would have to work, since human workers are naturally reluctant to take or keep boring unpleasant and dangerous jobs. To attract workers, employers must pay higher wages. This favors earlier use of robots for such jobs.

From the above discussion it is clear that sensory feedback is an essential prerequisite for many tasks, as they are normally performed. Outside the factory environment, in fact, high level sensory capabilities are required for vitually all tasks, with rare and minor exceptions. Within the factory environment, however, it is sometimes possible to engineer a high degree of predictability into the

FIGURE 1

INCREASINGLY DIFFICULT FOR HUMANS ⟶

Pick and place a medium sized oriented metal part Spotweld a repetitive pattern Sandblast a wall Spray paint a simple surface Drive a train on tracks Arcweld along a seam	Assemble a wooden cabinet, electric motor or pump, repetitively Pick and place a heavy metal part Solder very tiny wire connectors Spray paint a complex surface	Pick and place a very heavy metal part in a hot noisy (toxic) environment Operate a fire extinguisher inside a burning building Laser brain surgery (ex diagnosis) Land a spacecraft, good weather
Pick and place a randomly oriented part from a bin (medium size) Pick unripe fruits/veg. (size alone) Wash windows, selectively Wash dishes, glassware individually Inspect eggs in a hatchery	Inspect a PC board for faults Build a brick wall (3-D) Cut coal from a face Operate a farm tractor Assemble a wire harness. Finish (e.g. lacquer) a cabinet, to order Land a small plane (day, good weather, no traffic) Cut and assemble a suit Identify counterfeit paper money.	Assemble a mechanical watch Control air traffic at a busy airport Weld a broken waterpipe, from inside Inspect a VLSI chip for faults
Pick and place very floppy objects, e.g. thread, yearn, wire Cut and arrange flowers Harvest ripe soft fruits/veg. by color texture Separate crabmeat from shells/clean shrimp Inspect seedlings in a nursery for quality Plant rice seedlings (or similar)	Drive a truck or bus through traffic Deliver a baby (normal) and inspect for faults Dental hygiene Repair lace 'Invisible mend' a garment	Land an airliner at night bad weather Identify a counterfeit "old master" Repair a damaged "old master" High speed auto chase through city traffic Diagnose a medical condition Heart or liver transplant

INCREASINGLY DIFFICULT FOR "INTELLIGENT" ROBOTS ⟶

task by careful design and engineering of the production system as a whole. Further potential for reducing the need for sensory information arises from the flexibility of design: a given functional product (say an electric mixer) may be manufactured from many alternative materials, in a variety of different configurations. It can be assembled by a variety of sequences of operations. Nowadays a product designer can choose designs that are particularly well-adapted for machines, as compared to humans. Thus if one version of the product involves one or more tasks that are intrinsically very difficult for a machine (such as the assembly of a wiring harness) it may be possible to eliminate that task by redesigning the product to utilize a PC board instead. Another way to minimize assembly difficulty is to minimize the number of hard-to-handle parts, especially by eliminating screws, bolts, washers and nuts.[1]

Until the Year 2000

During the next two decades by far the major uses of robots (and machine vision) will undoubtedly be in factories. Viewed from an historical perspective a long term trend in manufacturing technology can be seen. It is the increasing integration of factory operations under centralized control (Table 1). In practice individual machines are increasingly linked together into clusters and clusters of clusters. In the past these linkages have been strictly mechanical (via synchronous indexing transfer lines) but in the future computers offer an alternative, as illustrated by the schematic diagram Figure 2. In effect, computers will gradually take over the monitoring coordination and scheduling functions that human workers have hitherto performed. In this context, the role of robots is primarily materials handling, i.e., passing workpieces from machine to inspection device and on to the next machine. Robots also can and do handle tools per se in situations where workpieces are large or tools are small, and it is more convenient to move the tool to the workpiece than conversely. Welding and spray painting are common examples, but robots can also manipulate portable routers, electric drills, jigsaws and glue guns.

A mechanically integrated transfer line system is enormously expensive to design and build. It is usually one-of-a-kind and almost totally inflexible in terms of adapting to design changes. On the other hand, an electronically integrated system is inherently flexible, in that it can accommodate design change by straightforward reprogramming. It therefore achieves high production rates without requiring standardization and bridges the traditional gap (to a degree) between mass production unit costs and batch production unit costs. While it is not yet unequivocally clear that electronically integrated flexible technologies are economically competitive for very large scale production, because flexibility requires greater complexity, it is clear that integration can bring significant saving to batch producers. Thus costs of production of capital goods (e.g., machinery) can definitely be expected to decline somewhat relative to other, es-

---

[1]See G. Boothroyd and P. Dewhurst "Design for Assembly" Department of Mechanical Engineering, University of Massachusetts, 1983.

TABLE 1

FIVE GENERATIONS OF AUTOMATION (I)

	Pre-Manual Control	First (1300) Fixed Mechanical Stored Program (Clockwork)	Second (1800) Variable Sequence Mechanical Program (Punched Card/Tape)
Source of instructions for machine manually	Human operator	Machine/designer builder	Off-line programmer/ operator records sequences of instruc- tions manually
Mode of storage (How is message stored?)	NA	Built-in e.g. as patterns of cams gears	Serial: coded patterns of holes punched in cards/tapes
Interface with Controller (How is message received?)	Mechanical linkage to power source	Mechanical: machine is self- controlled by direct mech. links to drive shaft or power source	Mechanical: machine is controlled by mech. linkage actuated by cards via peg-in-hole mechanism
Sensors providing feedback	NA	NA	NA
Communication with higher level controller?	NA	NA	NA

TABLE 1 con't

FIVE GENERATIONS OF AUTOMATION (II)

Third (1950) Variable Sequence Electro-Mechanical (Analog/Digital)		Fourth (1975) Variable Sequence Digital (CNC) (Computer Control)	Fifth (1990?) Adaptive Intelligent (AC) A.I. (System Integration)	
On-line operator "teaches" machine manually	Off-line programmer prepares instructions	Generated by computer based on machine level stored program instructions modified by feedback.	Generated by computer based on high-level language instructions, modified by feedback	
Serial: as mech. (analog) record e.g. on wax or vinyl disc	Serial as purely electrical impulses e.g. on magnetic tape/wire	In computer memory as program with branching possibilities	In computer memory as program with interpretive/adaptive capability	
Electro-mechanical: controlled by valves switches, etc. that are activated by tranducers - in turn controlled by playback of recording		Electronic: machine reproduces motions computed by program based on feedback	Electronic: (as in CNC) machine adjusts to cumulative changes in state.	
NA		Narrow-spectrum analog digital (converted e.g. to digital) optical e.g. voltm./ encoders strain.gauge	Wide-spectrum analog or digital computer descriptions visual, tactile, requiring computer processing	
NA		NA	Optional primary program down- loaded from higher level	Essential, because microprocessor at machine level must pass visual and tactile info to higher levels to coordinate	High level controller has learning ability

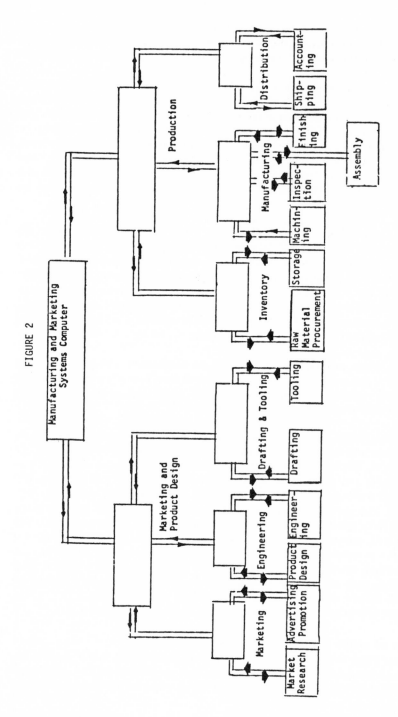

FIGURE 2

In a completely computer-aided design/computer-aided manufacturing operation there will be hierarchies of computers. Thus, the information and control loop from any one point in the operation to any other point will be easily facilitated.

pecially labor, costs. The long term shift towards capital intensive techniques, which stalled temporarily in the 1970s due to the sharp rise in energy costs, can be expected to resume at an accelerated pace in the late 1980s and 1990s.

Another factor that will affect the batch production sectors even more dramatically is the integration of human design effort with computer-aided-design (CAD) and computer-aided-manufacturing (CAM). For many capital goods that are produced in relatively small quantities (aircraft would be an excellent example) design costs can be quite large per unit produced. CAD offers the potential for directly cutting design costs dramatically, by half or even more in some instances, as the computer takes over many of the repetitive computations that are required in engineering design, as well as virtually eliminating the need for draftsmen.

Moreover, CAD offers additional advantages. First, the resulting designs are less likely to include flaws of the kind that have plagued the auto industry in recent years as the opportunity for human error is reduced. Second, many more variations of a given design theme can be tested using the computer. Hence the performance of the resulting design can be improved significantly in many cases. This can be extremely important in optimizing turbine blades, airfoil surfaces, ship's hulls, heat exchangers and so on. It can be equally important in designing products for ease of assembly and maintenance.

Third, the design process can be speeded up enormously. Once CAD is fully utilized, new product designs can be generated in a fraction of the time that was formerly required, thus shortening the time required to bring a new product (or a new model of an older product) to market.

Fourth, and finally, CAD automatically generates computerized parts lists, specifications and blueprints that can be communicated electronically to the manufacturing facility. The ultimate objective of manufacturing engineers is to integrate CAD and CAM, that is, to translate computerized part specifications directly into instructions for the machine tools. This is already virtually state-of-the-art for one special category of manufacturing: cutting or engraving flat parts (for later assembly) such as sheet metal, paper board, plywood, printed circuit boards, computer chips, or fabric. However CAD cannot yet be integrated directly (by computer) with CAM for the more general case of parts requiring several cutting, forming, joining or finishing stages. The reason is, simply, that CAD defines only the final dimensions of the parts whereas a fully integrated CAD/CAM system requires detailed specifications for workpieces in all intermediate stages of processing with sequences from each possible starting point.

This kind of model is still a fairly distant dream, at least in its most general form. Most likely it would be approached in stages, over a period of decades beginning with specialized models for particular categories of simplified part shapes (e.g., axles, engine blocks, turbine blades or metal cans) for which a great deal of empirical experience exists and the need for extrapolation beyond known cases is accordingly reduced.

Above and beyond the favorable impact on reducing the cost of capital goods, CAD/CAM will have another kind of impact that is qualitatively much more important. In bringing the unit cost of products made in batches nearer the cost now achieved only by products that are manufactured in extremely large numbers, while retaining (even increasing) the traditional flexibility of job shops, manufacturers will have far less incentive than they do today to adopt a competitive strategy of product standardization and mass production. In fact, U.S. based firms that have adopted this strategy in the past are currently having the most difficulty competing with newer mass production facilities located in the Far East, where workers are more disciplined, more highly motivated, more productive and considerably less well-paid than in the U.S.

It is increasingly clear that the only viable long-term competitive strategy for U.S.-based manufacturing firms is to shift quickly to a flexible batch-production mode and at the same time to accelerate sharply their traditional rate of product innovation. Quality and performance must displace price as the major objective of product design. Only by this strategy can the competitive advantages currently enjoyed by Japanese manufacturers be successfully countered. It is because of rapid and unpredictable innovation that the Japanese have failed to capture any significant fraction of the exploding personal computer market in the U.S. Obviously, a competitive strategy based on raid innovation for U.S. capital-goods producers depends on rapid near-term adoption of CAD/CAM and robotics.

How will a newly built factory in the year 2000 differ from one built in the recent past? Taking into consideration all of the points mentioned above, the differences will be quite profound.

First the local availability of cheap labor will not be a major consideration because the plant will employ very few people even during the day shift. Second, it will operate virtually unmanned at night and over weekends. Third, computer control will sharply reduce the need for on-site storage of raw materials and semifinished parts. Fourth, for the reasons mentioned above, the new plant will be relatively small in size in relation to its total output, as compared to present day plants of comparable capacity. On the other hand, for its size it will generate much more goods traffic than a present day plant.

The typical enormous 1960s vintage plant spread over several hundred acres, surrounded by vast parking lots and employing armies of hourly-paid shift-workers will be obsolete. The flexible factory of the year 2000 will still be located close to major rail, truck, or air transportation facilities, but it will not itself require a great deal of land nor many workers. It may, therefore, be closer to the city center than most factories built in the 1960s, probably near the major freeway linking the city center with its airport. Its location will be determined primarily by transportation costs and juxtaposition to major markets, and secondarily to facilitate interaction with product engineering and design. (In fact, the two may often be colocated.) The big change is that flexible automated factories will not be located in remote rural areas or foreign countries simply to minimize labor costs.

Thus the impetus that has shifted so much production away from major U.S. population centers in the Northeast and Midwest to the "sunbelt" will eventually exhaust itself. Sometime around the year 2000 the pendulum will begin to swing in the other direction as the congestion, increasing water shortages and declining quality-of-life in the sunbelt reduces its inherent attractiveness.

What of the fourteen million or so hourly-paid blue collar workers in the U.S.? The forces of international economic competition from the Far East, not the threat of competition by robots and "steel collar" workers, are already rapidly reducing their ranks. With each downswing of the economic cycle hundreds of thousands of highly paid steelworkers, auto workers and their unionized brethren are now being laid off. With each upswing a much smaller number are rehired. Employers are closing their oldest, most unionized plants, and shifting production to the South or overseas. Unions are becoming weaker and the general public (outside of a few states where union influence is still strong) is not impressed by their arguments for protection, if it means higher prices for consumers.

Unless this situation changes dramatically (which I do not expect) the major industrial unions such as the UAW and USW will be severely weakened by the year 2000. Many of the younger factory workers are already recognizing the inherent insecurity of their present jobs and are beginning to seek education and training to fit them for other careers. The older workers by and large will continue until they retire (early, in many cases), but they will not be replaced. The blue-collar culture of cities like Pittsburgh, Detroit and Chicago is already beginning to die out.

There will, predictably, be some casualties. There are a number of smaller cities and milltowns with only one major plant and many of these plants are already obsolete and likely to be shut down and not renovated or replaced. Some of the older (20 year plus) workers (especially steelworkers) are protected by their union contracts and will simply retire. Others are less fortunate and some will lose homes (which have become unsalable) and be impoverished, unless significant government sponsored relief efforts are undertaken to help them relocate and retrain. Many will be forced by circumstances to take jobs paying much lower wages than they have been accustomed to receiving.

The most plausible method of providing assistance for displaced industrial workers in the smaller cities and milltowns is to embark on some major (but limited) regional infrastructure renewal projects. Former steel workers and auto workers could be employed, with minimal retraining, in construction jobs. The challenge will be to identify and refinance appropriate large-scale construction projects, manage them well and ensure that the newly created jobs are not monopolized and that displaced industrial workers have a fair chance at the resulting job openings. This almost certainly requires a public-private "social contract" similar in concept to the Municipal Assistance Corporation that pulled New York City back from the brink of bankruptcy a decade ago.

After the Year 2000

Although the natural "domain of robots" is gradually moving from the upper right toward the lower left of Figure 1, it is not by any means certain that this shift will ever extend beyond the main diagonal. One might conclude differently if technological progress were a truly autonomous phenomenon, independent of socio-economic driving forces. In effect, this trend, if it exists, implies robot and computers are a kind of evolving species of life (or pseudo-life). This is a fascinating conjecture, entirely suitable for literary exploration. But it makes very little real-world economic sense as I hope to demonstrate. Why? The answer cuts to the core of human-machine relationships.

In brief, humans represent an extreme degree of flexibility and adaptability to change. We humans have, so to speak, specialized in being unspecialized. We can learn to do almost anything, in time, though we do not learn particularly efficiently. It is hardly surprising that our in-born capabilities in any given function are often easily surpassed by special purpose machines (or even other animals). It follows that we humans tend to use tools and machines to extend our limited inherent capabilities—but it also follows that these tools and machines are invariably much more specialized than we ourselves are.

A point that sometimes causes confusion needs to be explained. Obviously some machines and factories are much more specialized than others. An automobile engine plant exemplifies one extreme: it is totally specialized to a single product design. On the other hand, a conventional general purpose machine shop is relatively unspecialized, meaning that it can be used to custom-produce a wide variety of different sizes and shapes of metal product. True, the auto engine plant produces engines much more cheaply than the machine shop (provided at least a few million identical copies are to be made) but only by sacrificing the ability to make anything else. The tradeoff between flexibility and efficiency is quite a general one in nature, and in engineering. Mass production economies are achieved by exploiting this relationship, accepting rigidity (inflexibility) as the price of maximum efficiency. The disadvantage associated with a commitment to mass production is, of course, the inability to respond quickly to changing markets.

But the fact that some machines (such as robots) can be much less specialized than other machines must not obscure the fact that even the least specialized machine cannot begin to compare with humans in this dimension. There is no such thing, really, as a general purpose machine. One speaks of general purpose shops or general purpose computers, to be sure, but this is a semantic trap for the unwary. True, some computers (e.g., analog devices) are more specialized than others, but the range from most flexible to least flexible is narrow. By analogy, one might say that a 12 string guitar is more flexible than a 6 string guitar, but both are far less flexible than a full orchestra. The comparison between machines and humans is like the comparison between guitars and orchestras. The orchestra is not simply a collection of instruments: it alters its composition and capabilities for each piece of music. It may utilize a guitar for one piece, a harp for another and an organ for a third. So it is with humans. Not only are humans capable of a wide variety of actions but they uniquely have the ability to utilize specialized tools (i.e., machines) to solve particular problems efficiently, as needed, without sacrificing their own inherent flexibility.

Seen in this light, humans and machines are complementary, not competitive. While it might conceivably be possible for robots to be given humanoid form (2 arms, 2 legs, eyes, voice, etc.) and to be given electronic brains capable of imitating some human thought processes, there would be no economic reason for ever doing so. The result would be unnecessarily complex, unreliable, not very good at any specified task, and enormously costly. The essential point is that no single robot would even be required to do the whole range of things that a human can do, even in the unlikely event such a robot could do everything better. To say it another way, individual robots will always be used for (comparatively) special purposes. Hence, robots will always be specialized, by design. Even now there are materials-handling robots, painting robots, welding robots, assembly robots, inspection robots and mail delivery robots. Why give robot arms more degrees of freedom than the task requires? Why make a robot bigger, stronger, faster or smarter than necessary?

In fact, industrial robots are rapidly differentiating into many subspecies, with a wide range of diverse capabilities (size and weight, speed, precision, number of arms, number of fingers, number of joints, sensory capabilities, etc.). The physical resemblance of a welding robot to an assembly robot is already slight and likely to become still less in the future. Mining robots, undersea exploration and repair robots and space construction robots will be even more specialized. In brief, as regards industrial robots the evolutionary trend is clearly away from the general-purpose humanoid form, not toward it.

Before considering the long run implications of this trend, it should perhaps be tested in another context. After all, as I have pointed out, the factory environment is inherently more controllable than the general human environment. It is therefore less demanding of high quality sensory capabilities and intelligence. Is it possible that the ability to operate in other more diverse and variable environments (such as mines, highways, farms, households or battlefields) would justify the development of a humanoid robot?

The answer seems clear, as soon as the question is posed this way. Each of these potential uses for a robot is itself specialized. An undersea robot would have to have very different capabilities and characteristics than a construction robot or a household robot, or a robot soldier. Thus, the undersea robot would be an autonomous unmanned submarine with a manipulator arm (or two) and some sensory devices (based more on sonar than on vision) adapted to the marine environment. The mining robot would probably be an armored tracked vehicle specialized to a specific task such as drilling holes and inserting roof bolts. The farming robot would be an autonomous intelligent tractor capable of orienting itself at all times with respect to the boundaries of the field, and towing various pieces of equipment. (It would probably have to call for help if it got stuck in the mud.)

The robot soldier would also probably be an armored, autonomous vehicle (like the mining machine), but with no particular top or bottom. It might come in several variants for different kinds of terrain, ranging from spheres, air-cushion vehicles, or worm-like crawlers for flat ground or marsh (similar to tanks) to insect-like multilegged forms capable of maneuvering over rough ground and through forests.

The robot truck or bus driver would, in fact, be a robot vehicle of some sort, without a driver. The problem of designing a humanoid robot to drive safely in traffic is not worth solving, if it could be solved. When driverless vehicles are ultimately developed, they will probably be controlled by a network of sensors and computers implanted in the road system. The passenger will specify a final destination on a display map and the vehicle will communicate this data to a central regional traffic monitoring computer, which will route the vehicle and turn over control to sectoral control computers.

The future construction robot would be similarly specialized. There may someday be autonomous bricklaying robots, concrete-pouring robots, riveting robots, welding robots, painting robots, and so on. But none of these machines will be humanoid and none will have capabilities beyond those actually required for the task. A robot for construction of solar power satellites in orbit would, of course, be specialized for different tasks than a ground based construction robot, but unquestionably it would be specialized.

Finally, what of the domestic robot? In this case, it could be argued that robots might be more "acceptable" if they looked somewhat like human servants. I concede this question to the marketing experts, but it seems rather doubtful that customers would demand a humanoid appearance if that added greatly to the cost while detracting from performance. The first household robot seems very likely to be a semiautomated vacuum cleaner that (in robot mode) will move automatically around a room, avoiding obstructions but otherwise at random. It would naturally look somewhat like a conventional tank-type vacuum cleaner on wheels but with a shorter arm ending in an air intake designed to pass back and forth over the floor. It would probably utilize an ultrasonic directional range finder (similar to the devices found in self-focussing cameras) and force-feedback (tactile) sensors on the sides and tip of the vacuum-arm, which would wag back and forth automatically, somewhat like a dog's tail. In manual mode, the unit would have an extendable hose for emptying ash trays, cleaning venetian blinds, sofa cushions, etc.

The robot-vacuum cleaner would require one powered steerable driving wheel in addition to the angular and vertical degrees-of-freedom of the vacuum arm. A fairly sophisticated microprocessor would be required to interpret the various possible signals from these sensors and operate the joint controls. (For example, if the ultrasonic ranging device does not sense an obstruction along its direction of motion, but if the force-feedback sensor at the tip of the arm does meet an obstruction near floor level, the chances are it is the edge of a carpet.) A likely strategy on encountering a ground level obstruction would be to stop, lift the arm slightly, move forward a short distance, and if no obstruction is encountered, lower the arm again until it is the desired distance above the floor, and continue. Clearly, a program to cover all likely contingencies would be quite extensive. Probably a laboratory prototype could be designed and built today, but it would cost a million dollars or so and would be at the very edge of the state-of-the-art. Conceivably, such a unit could be commercialized in a decade if a large firm made a major effort, but the final product would necessarily be quite expensive even in large-scale production. How many people would pay $1,000 for a semiautomated vacuum cleaner?

Other types of household robots are obviously feasible in principle. One of the simplest (in concept) is an automated drink-server. The device would be an arm capable of pouring drinks from a designated set of bottles (in a fixed rack) adding ice and mixers on demand. It could be programmed by means of an ultrasonic remote control device similar to a TV control or by a light pointer, or conceivably by voice instruction.

A simple button-operated drink server was demonstrated by Unimation Inc., several years ago and several other robot manufacturers have their own demonstration versions. A practical commercial version, designed for serving at cocktail parties, would probably have to be mobile (like a robot mail delivery cart). Most likely it would slowly follow a specified route, fixed by a wire in the floor, announcing its presence verbally by "beeping" or by a programmable phrase such as "Would you care for a drink or a canape?" It would also specify the appropriate response, for instance, "If you would like a drink, please touch the white button," and would proceed to offer its menu of possibilities.

This sort of robot could be manufactured today, in moderate quantity, probably for a few thousand dollars each. On really large-scale production the price might come down to $500 to $1,000. But it is by no means clear how many people would pay that sort of price for an automated drink-server of such marked limitations.

Evidently, there would be a much greater potential market for a mobile household robot with more generalized "pickup," "putaway" and "cleanup" capabilities. Suppose we consider the requirements for such a robot. In "pickup" mode it should be able to pick up and store (for later "put away") all objects not belonging in a given room. This obviously requires the robot to possess a detailed internal map (in 3 dimensions) of each room, and its contents, stored in its memory. It must also have a high quality vision and tactile system to compare the actual disposition of objects in the room with the "correct" disposition. It must be able to accurately distinguish and ignore humans and pets from inanimate "objects," possibly by means of a CO_2 sensor combined with an infrared sensor. Apart from these feats of memory and recognition, the robot must have a rational navigation strategy based on a knowledge of its own position and orientation in the room, relative to all potential obstructions.

For each object encountered in a search pattern (say an empty glass) the robot must then identify it precisely if it belongs in the room, or else estimate its size, weight and attributes. It must decide if the object is rigid or soft (how soft) and, if rigid, its likely breakability and correct orientation with respect to the vertical. Possibly the robot will have to devise some experiments, utilizing tactile senses on its finger(s) to refine its estimates of these attributes. (Is it rotationally symmetric around any axis? Does it have a "flat" surface orthogonal to the rotational axis? If placed on the flat end would it be stable against small perturbations? Does it contain a liquid?)

Having made these decisions, the robot must devise a "pickup" strategy. Such a strategy comprises many elements, including the position and orientation of the gripper as it approaches the object, the amount of gripping force to be applied; how to detect slippage; what adjustment to be made in the event of slippage; the trajectory of the pickup arm/gripper and object; choice of interim storage location (e.g., a rack, slot or shelf location) and the position/orientation of the object in its place of interim storage. Trivial as these decisions are to a human, they are far beyond the current state-of-the-art of so-called "intelligent" computers. On the other hand, there seems to be no fundamental reason why computers, even small ones, should not eventually be able to solve such problems fairly efficiently.

Evidently the "putaway" mode would require much the same set of capabilities as the "pickup" mode, except that the robot must cope with the fact that objects found out of place in one room may belong in another. Thus the robot, having completed "pickup" in a given room would immediately put away those objects belonging in the same room by reference to its internal map of that room. To deal with objects belonging elsewhere it would need a more complex strategy. It would probably then call up from mass storage maps of adjoining rooms, comparing the unidentified found items with items known to belong there. The robot would then have to go to each room and see if the found object corresponds to a missing object or it would have to identify each found object, find it on a master list, and look up its correct location.

For a "cleanup," the room would need to be in the ordered (picked-up) state, with no objects out of place. Thus, "cleanup" must follow "pickup." The robot would then begin by vacuuming both carpeted and noncarpeted areas, either following a path stored in memory or computed from a map based on the observed locations of the furniture. The first method would be far faster and cheaper, since path-optimization calculations are known to require extraordinary amounts of computer capacity. It would then perform necessary floor scrubbing and waxing tasks, if desired, on noncarpeted parts of the floor, again following memorized motions and paths using the room map as a guide.

The most effective motions for scrubbing, waxing or polishing would probably be taught to the robot in manual, or "teach," mode by a human, much as a robot arc-welder is taught by an expert human welder.

As described above, a "cleanup" robot would need various tool attachments, such as rotating brushes, one or more tanks for storing liquids, and a vacuum system capable of collecting liquid droplets as well as air. To be truly automated, it would also need to have the ability to empty and refill its storage tanks, to wash out its internal plumbing with the help of suitable nonflammable solvents (that can be separately recycled) and to change its own attachments.

It is likely, on closer consideration, that the "pickup/putaway" would not be combined in a single unit with the "cleanup" robot, though it might be combined with a "serve" robot. The cleanup unit would require several tanks, a vacuum pump, or two, and hoses, as well as a number of tool attachments, all of which would be useless and in the way for other modes of operation. On the other hand the pickup/putaway/serve unit requires a flexible arm and several fingers, or perhaps two arms, and a flexible internal storage space for a variety of objects. It remains to be seen whether other functions can be combined efficiently.

What is quite clear, however, is that future robots will be specialized to particular tasks or families of tasks, and will not be endowed with a high level of general intelligence. For instance, the programming needed to permit a robot pickup/serve unit to function effectively would require an ability to solve certain limited classes of problems. These include: What class of object is x? Where does x belong? How should object x be gripped to avoid breakage? How should object x be oriented? How should internal storage space be loaded to maintain balance? What is the best route from point A to point B to avoid obstructions? These problems are intrinsically very difficult and will require far more computer power and more sophisticated software than is presently available. There is no doubt that such a robot would be "intelligent," as the term is used by computer scientists. But robots will not resemble humans in any respect (except perhaps with regard to arm-wrist-hand configuration) and their intelligence will be narrowly limited to the specialized purposes for which they are designed.

The Social Questions

When the future of robotics is discussed nowadays outside of scientific and engineering circles, two kinds of questions arise with particular urgency. The first of these burning issues is: what are the implications of microprocessors and robotics for the future of work, in terms of the future relationship of humans and machines in the workplace? The second major question (the exact formulation of which depends on the answer to the first): What do the coming changes in the workplace portend for social and economic institutions? As an instance of the first type of question, some social scientists have worried about whether humans will increasingly have to work "with" robots on future assembly lines and, if so, whether this would increase the monetary pressures of machine pacing, and the resulting alienation of human workers. As an instance of the second type of question, some social scientists have worried about the effects of a decline in the number of highly paid, economically secure, politically conservative semiskilled factory workers and their social role as a kind of bridge (or ladder) between the large numbers of low-paid unskilled entry-level workers and the small numbers of elite upper level professional workers and executives. In short, some have even foreseen the increasing substitution of robots for factory workers as a somewhat malign social force undermining the social basis of our political democracy.

As a matter of fact, these basic questions have been discussed for at least a century in the context of increasing automation and they will not be easily settled. I can only hope to offer a very modest contribution to the debate. What follows is strictly my personal opinion based on the facts and arguments presented earlier in this paper.

With regard to the future relationship of humans and the workplace, I believe that robots will certainly be widely utilized in the future, but that the number of existing jobs actually displaced by robots in the next half century or so will be relatively moderate. As stated already, by far the largest job loss will occur in the manufacturing industries, where, by 2035 or so, up to 90% of all jobs on the factory floor will be eliminated by automation. The total number of such jobs in the U.S. today is only about 14 million, however. In the nonmanufacturing sectors of the economy there will also eventually be a large number of robot hall sweepers, window cleaners, garbage collectors, street cleaners, bridge painters, farm tractors, pothole fillers, brick layers, gasoline (or methanol) pumps, fast-food dispensers, ticket-sellers, and so forth. But the net displacement from all of these specialized applications will probably be less than that in the manu-facturing sectors alone.

Many robots on the other hand, will find jobs where humans no longer wish to work (e.g., in mines or in other people's homes) or where humans cannot com-fortably or safely work at all (e.g., under water, in burning buildings, on the moon, in space, in radioactive or toxic environments, in the arctic, etc.). In fact, specialized robots will create major new industries that also create new em-ployment opportunities for humans.

To summarize the job situation, machines will eventually take over most of the repetitive, boring, demeaning jobs in our society, especially the jobs where workers interact mainly with machines or materials, not people. But robots will not take over the jobs that require high quality sensory capability, rapid re-sponses to a changing environment, or effective interaction with other humans. In fact, most jobs are already in the latter category rather than the former. Thus by 2035 most factories will be largely unmanned except for security guards, maintenance personnel and engineers. But most other places of work such as schools, hospitals, offices, shops and restaurants will continue to be operated by humans. Similarly, mining, construction, transportation, communications and agriculture will continue to use human workers, in most of the same functions that such workers currently perform. Handicrafts and popular arts are likely to be a major growth sector as factory employment declines.

I think in short, that the problem of worker alienation by assembly-line type jobs will be eliminated in the simplest possible way: by 2035 humans will con-tinue to design, install, modify and maintain and repair machines, but in the advanced industrial societies they will no longer "feed" them or directly control them. The future human worker will often be assisted by machines. But the unnatural situation where humans, in effect, had to assist machines (because the latter were too primitive) will have disappeared.

With regard to the second of the two great social questions, I have already noted that there will inevitably be some casualties of technological progress in auto-mation in the coming two decades. A human society should be prepared to ameliorate the economic impact of such job displacements as do occur, and to assist the workers to retrain and relocate if necessary.

But what of the impact on our social structure? Will the blue collar middle-class shrink or disappear? Will the opportunities for upgrading and promotion from entry-level positions to "skilled" jobs via on-the-job training be sharply restricted, resulting in an elite defined by educational opportunity? I think that, in fact, there is some danger of this happening in the short run, because the income hierarchy in the future will be based on different criteria than in the re-cent past.

The kinds of skills that will justify hiring humans rather than buying machines in the future are primarily either inborn (hence available at minimum wage) or they are learned at home or in school but not, by and large, on-the-job. Eyesight, hearing and motor-coordination are examples of inborn characteristics in humans that would not be significantly improved by extensive on-the-job training. It does not take very long to train drivers or typists, for instance, as long as they can read reasonably well.

On the other hand, reading, writing, grammar and speech skills, logical (computer-programming and mathematical) skills are mostly learned in school. These skills, in turn, are essential prerequisites for the accumulation of subject-area (disciplinary) knowledge. It is the possession of communication skills and logical skills that enables people to acquire the substantive knowledge that (along with social skills, mostly learned at home) qualifies people for better paying jobs in today's society. It is the lack of these learning skills, primarily, that holds back displaced factory workers.

The answer, for displaced workers in the long run, is not "retraining" (retraining for what?) but upgraded education. The middle-class currently includes some blue collar workers who would not be as well paid as they now are if it were not for the strength of their unions. These workers, if they lose their present jobs, are likely to slip back in the socio-economic hierarchy. On the other hand, several job categories that have recently been poorly paid, primarily because the jobs in question have been largely held by women—teachers, nurses and secretaries—are likely to improve both in prestige and income. Also, some service jobs will gain in prestige as the public gradually learns to distinguish good from poor service. A noteworthy example of a skilled profession that is grossly undervalued in the U.S. (but not in France) is that of waiter.

To summarize the social and economic impacts of robotics, there will probably be a shift in the composition of the middle class but no major decline in its long-run importance. The role of formal education will probably continue to grow in importance, as the role of on-the-job training declines. Adult education for the under-educated and mid-career educational "sabbaticals" for obsolescent professionals will be much more common.

A final comment: since robots are not on the verge of taking over all jobs, or even a very large fraction, there is no question of humans becoming, in effect, parasites or living off the efforts of hoards of robotic "slaves." This is a purely literary fantasy, not a realistic possibility. Robots will play an important but definitely subsidiary role in the industrial economy of the 21st century, in much the same sense that gasoline engines play an important but subsidiary role today.

3. Recent Developments in Robotics and Flexible Manufacturing Systems

Steven M. Miller[a]

Carnegie-Mellon University

March 13, 1984

This paper was prepared as part of a technology assessment of industrial robots supported by the National Science Foundation, Division of Policy Research and Analysis, grant # PRA-8302137 826. I am especially grateful to Robert Ayres, principal investigator of the technology assessment project, for ideas and for editorial assistance. The contributions of Gerald Agin, for his comments on machine vision research and capabilities, and Thomas Morton, for reviewing the manuscript, are also greatly appreciated.

a: Assistant professor, Engineering and Public Policy and Industrial Administration

31

Abstract

This report reviews some of the important developments in robotics and flexible manufacturing technology since 1979, the year in which the first NSF sponsored technology assessment of robots appeared [Eikonix Corp. 79]. Attention is focused on three broad classes of developments:

1. Extensions of the capabilities of commercially available robotic manipulators

2. Integration of *sensory information processing* with robotic devices, and

3. Integration of robots, machine tools, sensory information processing system, parts handling and transport devices, and computers into manufacturing *systems*.

The first development is covered by discussing the increase in the number of robot models designed for precision placement and in the use of computer controlled robots. In regard to the second development, there are three broad types of applications where the integration of sensory information processing and robotic devices has already proven to be of some commercial importance:

- controlling robot positioning and manipulation in parts acquisition, tool handling, and assembly.

- inspecting objects

- guiding a mobile robot

The discussion of capabilities and limitations of sensor based systems (both commercially available and under development in research labs) focuses on the first of these three application areas.

The third major technological development covered is the integration of machines and robots into flexible manufacturing systems. Three topics within this area are discussed :

- The integration of multiple FMS for machining in one plant

- The use of FMS in manufacturing processes other than machining

- The addition of artificial intelligence and sensor-based-systems to FMS

Definitions of robot technology and a discussion of limitations of commercially available robot manipulators immediately follow. These preliminaries provide a foundation for appreciating the recent developments.

1. Background on Robotic Manipulators

1.1 The Robotic Manipulator

An industrial robot is essentially a device which can move materials, parts, or tools from one point in space to another under programmed control without human intervention. Tracing the current industrial robot's roots back to both the teleoperator and the NC machine tool, Paul (1981) refers to a robot as " a device combining the articulated linkage of the teleoperator with the servo axis of the numerically controlled milling machine." A robot is distinguishable from a teleoperated mechanical arm in that it is programmable and can execute its program automatically. By contrast, the teleoperator serves as an extension of the human controller who is either attached to it physically, or remotely via radio waves.

The robot is able to manipulate parts with an "end effector"--a mechanical gripper-- that can grasp an object, pick it up, reorient it, and reposition it. The end effector is attached to a mechanical arm that can be instructed to move from one position to the next to pick up and move things. Thinking of a robot as a programmable manipulator seems to easily distinguish a robot from a numerically controlled machine tool, but the difference becomes ambiguous on closer inspection. Both the robot and the NC tool have several independent axes of motion. The types of servo mechanisms used to control the motion along each axis are similar in each case. Both can be programmed to move through a specified cycle. Once set-up and programmed, both the robot and the NC tool can complete the designated task without human intervention. While a robot is generally recognized for its ability to pick up an object and reposition it, some NC tools can effectively perform these functions, although not so easily. While an NC tool does not have a gripper, some can still acquire parts from a specially designed pallet and move the pallet onto the machine bed. The machine bed may also have several degrees of freedom, allowing the part to be moved and rotated. A robot must also be multifunctional in that it can manipulate an object in different ways and manipulate different types of objects. Yet a sophisticated machine tool can do the same, since it can also be programmed to move its workhead to carry various types of tools. An NC machine tool is used to alter the physical state of an object by changing its shape or adding or subtracting to its material content. By controlling the position of a tool, a robot can also be programmed to alter the physical state of an object. These comparisons show that a robot basically is a type of numerically controlled machine tool. However, a robot is usually more versatile in grasping, manipulating, and repositioning objects than a conventional (e.g. metal shaping) machine tool because it is designed principally to alter the spatial state of an object.

1.2 Defining Levels of Robot Technology Based on Internal Vs External Sensing

A useful way to delineate levels of robot sophistication is to view a robot as an information processing system. The sophistication of the system depends on two factors: 1) from where it acquires information and 2) how it processes the information it acquires. Any robot can acquire some types of information from its own internal mechanisms, including its mechanical, electrical, or electronic subsystems. Examples are whether the robot is "on" or "off", the temperature of its various subcomponents, and whether its gripper is open or closed. Servo controlled robots also monitor variables such as the position of a particular joint in the arm or wrist and the load on a particular joint. Our first concern is to distinguish between a robot that can only obtain information from its own internal workings and a robot that can also obtain information from its external environment. We refer to the former as an *insensate* robot. and refer to the latter as a *sensate* robot.

Various types of information can be acquired from the external environment. The simplest example is a binary signal (e.g. absent or present, yes or no, on or off, positive or negative). For example, when a part advances to the end of a conveyor, it may push against a switch which sends a "present" signal to the robot. The signal is used to sequence prespecified programs. We use the term Level I to refer to the insensate robot defined above, as well as to the robot that can sense externally generated binary signals and use them to sequence predetermined programs. A definition is given below:

Level I robot: A robots system where the goals *can not* be modified through the result of a feedback and control process that senses the external environment. *All* aspects of the task must be explicitly specified in advance. Predetermined alternatives may be invoked via limit switches, timing devices or simple transducers which transmit a binary inputs or measure predetermined thresholds.

An example of complex sensing and processing of external information would be the use of visual information to identify a part on a conveyor, to determine its location, and finally, to have the robot move to this location and grip the part. We refer to such a robot as a Level II robot. It can sense more than a single binary signal and make decisions more complex than sequencing prespecified alternatives. A definition is provided below:

Level II robot   A robot system where the goals *can* be modified through the result of an adaptive feedback and control process that senses the external environment via some type of transducer. Instruction can be adapted and generated in response to changes in the external environment. Thus, it is possible to perform some kinds of tasks without having to explicitly specify all of its aspects in advance.

Capabilities generally required within each of the major categories of robot applications in

Table 1-1: Major Categories of Robot Applications in Manufacturing
and Capabilities Generally Required Within Each Application

Major Application Area	Examples Within Area	(Capabilities Required To Perform Application)		
		Transport	Manipulation	Sensing
MATERIAL HANDLING	Parts handling	X		
	Palletizing	X		
	Transporting	X		
	Heat Treating	X		
MACHINE LOADING	Die cast machines	X	X	
	automatic presses	X	X	
	NC milling machines	X	X	
	Lathes	X	X	
SPRAYING	Spray painting		X	
	Resin application		X	
WELDING	Spot welding		X	
	Arc welding		X	
MACHINING	Drilling		X	X
	Deburring		X	X
	Grinding		X	X
	Routing		X	X
	Cutting		X	X
	Forming		X	X
ASSEMBLY	Mating parts		X	X
	Acquiring parts		X	X
INSPECTION	Position control			X
	Tolerance			X

Source: Tech Tran Corp. (1983: 63).

manufacturing are shown in Table 1-1. Three types of capability are distinguished: 1) the ability to transport an object from one place to another, 2) the ability to manipulate an object and 3) the ability to use external sensing for feedback control. The table indicates that a robot which can only transport and manipulate objects (Level I robot) is usually adequate to perform material handling, machine loading, spraying, and welding tasks. A robot with external sensing, however, could also perform machining (tool positioning), assembly and inspection tasks. This is only a rough guide, since there are clearly applications in material handling, machine loading, spraying and welding that require (or could benefit from) external sensing and applications in machining, assembly, and inspection that can be engineered to eliminate the need for external sensing. Nonetheless, the table conveys the general notion that the the the use of external sensing would would broaden the range of manufacturing tasks that could be performed by a robot.

1.3 Limitations of Level I Robots

To understand the limitations of Level I robots, we focus on the elementary task of moving the robot's end effector from one position to another. It is surprisingly difficult for an insensate robot to perform this basic function. To begin with, instructing the robot where to move its gripper is a cumbersome process. With "walk-through" programming, an operator takes hold of the end effector and moves it through a specific continuous path while recording the pattern of motion. With "lead-through" programming, an operator commands the robot to move, using a teach pendant or a joy stick (similar to those used in video games), recording a series of discrete points. Both programming procedures require that the robot be taken out of production each time it is taught a new task. Once the program is recorded, however, it can be quickly recalled.[1] Another problem is that after an insensate robot carries out its instructions, it does not know the precise position of its most critical component, its end effector. In short, it is cumbersome to tell a conventional robot where to go, and after it moves, it does not know if it went exactly to the right place.

A schematic of the position control system for an insensate robot (with servo control), shown in Figure 1-1, is useful in illustrating limitations. The position of the tip of the end effector in the real world is given by X_r, the reference position. While programming the robot, the operator thinks in terms of the position of the end effector in (real) world coordinates. The robot system is actually recording the position and angles of each of the separate joints within the arm in its own internal coordinate system. The robot controller contains a model consisting of a set of linked equations that

---

[1]Transmitting new instructions to a robot by inputing coordinates into a computer (even while the robot is in operation) is becoming more widespread and is discussed later.

Figure 1-1: Positioning System for an Insensate Robot with Servo Control

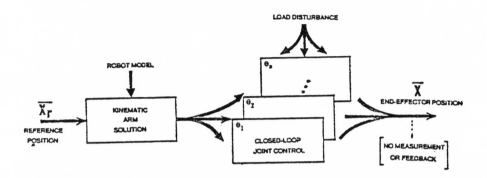

Conventional (insensate) manipulator position control system. An arm solution maps the reference end-effector position into coordinates for each of the individual joints. Closed-loop joint controllers control the positions of each joint independantly. There is no direct closed loop feedback between the final position of the end effector and the reference position.

Source: Sanderson and Weiss (1983: 109)

specifies the position of the end effector as a function of the rotations and positions of all the joints. The joint motions required to achieved a desired end effector position can be calculated by solving these equations. This process is represented by the box labelled *kinematic arm solution*. Once all the reference positions are specified by the operator, the robot must repeat the desired sequence of motions. The box labelled *closed-loop joint control* represents this process. Each of the separate joints is moved to the positions given by the robot model and kinematic arm solution. Using potentiometers, optical encoders, and similar devices, the current position of the joint is sensed and compared to the desired position. Each joint is moved until the difference between the current position and the desired position is reduced to zero.

There are *load disturbances* at each joint, as indicated in the figure. Internal parts, such as gears or linkages, can wear and introduce inaccuracies into the positioning process. Also, physical forces, such as friction and backlash, cause distortions that must be compensated for in the calculations. The effects of these disturbing forces can only be approximated and cannot be completely eliminated. The final end effector position, given by X in the figure, is not independently measured and is therefore not known by the robot itself. It is assumed that by approximately positioning the individual joints in the manipulator arm, the end effector will end up in the desired position. Because of inaccuracies in the robot model (the original specification of the joint movements required to achieve the programmed reference position) and because of many types of disturbances, the end effector is not always exactly in the predicted position. This is known as *open-loop* control. It is characteristic of an insensate robot. More detailed discussions of the basic control and structure of the insensate industrial robot are given in [Paul 81], [Luh 83], and [Coiffet 83a].

Even with their inherent limitations, almost all of the robots in the field today are insensate. In his description of the the earliest machines manufactured by Unimation in the 1960's, Paul (1981: 2) explains how such a robot is typically utilized.

> The industrial robot could be taught to perform any simple job by driving it by hand through the sequence of task positions, which were recorded in digital memory. Task execution consisted in replaying these positions by servoing the individual joint axes of the robot. Task interaction was limited to opening and closing of the tongs or end effector, and to signaling external equipment or waiting for a synchronizing signal. The industrial robot was ideal for pick and place jobs such as unloading a diecasting machine. The part would appear in a precise position, defined with respect to the robot; it would be grasped, moved out of the die, and dropped on a conveyor. The success of the industrial robot, like the numerically controlled (NC) milling machine, relied on precise, repeatable digital servo loops. There was no interaction between the robot and its work. If the diecast machine were moved, the robot could in no way adapt to the new position, any more than an NC milling machine could successfully cut a part if the stock were arbitrarily relocated during cutting. If the diecasting machine were moved, the robot could, however, be retaught. The success of the industrial robot lay in its application to jobs in which task positions were absolutely defined, and in its reliability and positioning repeatability in lieu of adaptation.

In order that task positions be *absolutely* defined with no unplanned variations, the workplace within reach of the robot had to be carefully controlled. The requirement for workplace structuring to use insensate robots was emphasized in an assessment of robot technology report prepared by the Eikonix Corporation in 1979 [Eikonix Corp. 79]:

> Present day industrial robots have achieved their most successful application in such areas as machine loading and unloading, particularly die casting machines, and spot welding in the automobile industry. All of these applications require complete structuring of the environment in order that the appropriate reference of motions and functions can be specified before-hand. While these robots can accommodate changes, these changes have to be completely specified in all their detail...

> Planning for variations requires a complete structuring of the environment in which the robot is to function. The environment need not be totally static, but all aspects of its dynamics must be explicitly known. This structuring of the environment significantly adds to the cost of using a robot. Not only must physical modifications be made, but the production engineers must come to an explicit understanding of the relationship between task and the environment in ways that are totally unnecessary when human workers are used.[2]

Much engineering effort and additional accessory hardware is required to insure that *all* aspects of the task can be prespecified in complete detail. For example, a robot programmed to move to a specific location and pick up a part requires parts feeding and orientation devices to insure that each part arrives at the same position *in the same orientation*. A part slightly out of position or orientation might bring the system to a stop. Grippers and motion sequences for gripping parts have to be designed to compensate for uncertainties in the precise position of the end effector. When a robot is required to position a tool precisely with respect to a workpiece, (as in arc welding), it is often necessary to secure the workpiece to a table that can be precisely positioned, minimizing the requirements for moving the robot arm. The tools themselves often have to be specially designed so that they can be held by the robot.

As a result of these requirements, the cost of an insensate robot is typically only a fraction of the total cost of an application. A recent estimate of the cost of a typical robot application is shown in Table 1-2 for six different types of applications. The basic robot cost includes the cost of the arm, the power supply and the controller. According to this estimate, the basic robot typically comprises only 40 percent of the cost in an assembly application, 55 percent of the cost in a machine loading or welding application, and 70 percent of the cost in a spray painting application. The "accessories cost" includes optional equipment, special tools, grippers, and maintenance and test equipment. Installation includes site preparation, work rearrangement, utility connections, and interfaces. These

---

[2] [Eikonix Corp. 79], pages 26-27.

Table 1-2: Typical Robot System Cost Breakdown in Selected Applications

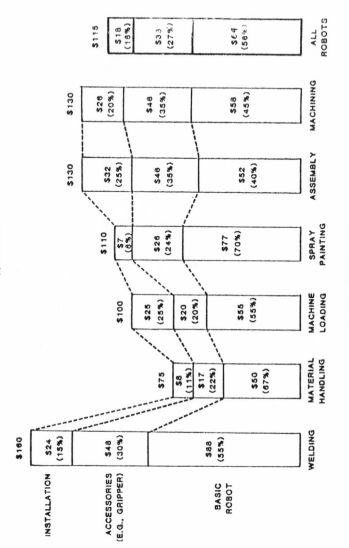

BY APPLICATION ($000)

Source: Tech Tran Corp. (1983: 97).

Table 1-3: Assumptions For Calculating Total Implementation Costs

TYPE OF ROBOT	BASE PRICE	TOTAL IMPLEMENTATION COST ROBOT BASE PRICE
lower cost	$20,000	4
medium cost	$60,000	3
high cost	$100,000	2

Table 1-4: Summary of Cost Assumptions for Retrofitting Level I Robot Systems

Robot Hardware Cost (R)	Development Cost (D)	Total Implementation Cost (I = R + D)	Operators Replaced Per Shift
20,000	80,000	100,000	1
60,000	180,000	240,000	1-2
100,000	200,000	300,000	1-3

Source: Miller (1983:44).

other types of expenses account for the remainder of the total installation cost. The estimates do not include costs required for the study and selection of the application to be robotized (planning) or for selecting the robot to be used (applications engineering). Planning and applications engineering work required before a robot is purchased can add tens of thousands of dollars to the total cost.

Other studies suggest that the the basic robot accounts for even a smaller percentage of total cost than suggested by Table 1-2. In one study of robot use, which focused on applications in the auto industry, Hunt and Hunt (1983) reported that the basic (insensate) robot usually represented less than 40 percent of the total costs in applications areas such as welding, where many robots have already been installed. Miller (1983) estimated that in machine loading and unloading, the basic robot typically comprises only 20 to 30 percent of total cost for the first installation, including the cost of planning and applications engineering.[3] Miller's estimates of the ratio of total implementation cost to robot base price, and of the total cost of implementing a single low, medium and high priced robot in machine loading are shown in Table 1-3. High cost robots generally have more degrees of freedom, a long reach, a large working envelope, and more sophisticated control capabilities. Somewhat less accessory hardware and engineering is required to enable such a robot to acquire, load, and unload parts. Conversely, lower cost robots are more limited in their degrees of freedom, reach, working envelope, and control capabilities. Hence more accessory hardware and engineering is required to compensate for these limitations.[4] Engelberger (1980: 102) has also emphasized this point.

> Generally speaking the higher priced robots are capable of more demanding jobs and their control sophistication assures that they can be adapted to new jobs when original so assignments are completed. So too, the more expensive and more sophisticated robot will ordinarily require less special tooling and lower installation cost. Some of the pick and place robots are no more than adjustable components of automation systems. One popular model, for example, rarely contributes over 20 % of the total system cost.

Since 1979, there are many more models on the market as a result of existing manufacturers (Cincinnati Milacron, Unimation, Prab, etc.) broadening their product lines, and as a result of many new firms who have started to sell and manufacture robots over the last five years (Bendix,GE, GM-Fanuc, IBM, and a number of smaller firms.).[5] One significant aspect of there being more models available is that a user can shop around in the market until he finds a robot with almost exactly the required bundle of attributes (e.g. reach, load, accuracy, type of control etc.) for the application in mind. This means that a user need not pay for more capability than is really necessary.

---

[3]Miller notes that the average cost per application decreases as more robots are installed in a factory, especially when there are multiple installations of a similar application (See [Miller 83a], chapter 2.) If more than one robot were installed, the cost of robots would comprise a smaller proportion of the total installation cost.

[4]There are exceptional cases where the the limited capabilities of the low cost robot are well suited to the task and a minimum of accessory hardware is required.

[5]Engelberger asserts that as of late 1983, there are approximately 230 companies in the world selling robots.

2. Extensions in the Capabilities of Insensate Robots

2.1 Increased Availability of High Precision Robots for Light Manufacturing

One important development has been a sharp increase in the available robot models specifically designed to precisely position lightweight workpieces. These robots are primarily intended for small parts assembly, although they can be used in any application where precise positioning of the end effector is required. Selected characteristics of some of the earlier, as well as more recent high precision robots are shown in Table 2-1. Such machines generally have a repeatability in the range of .001 to .008 inches.[6] Most of these machines are designed to only handle lightweight payloads within a relatively small working envelope. In contrast, robots with longer reaches and with greater payload capacity, such as those used for loading machines and press or for welding, typically have a repeatability in the range of .01 to .08 inches.

The machines made by Seiko Instruments, which have been in use in Japan since 1960, were the first high precision robots developed for small parts assembly. They are extremely precise because they use mechanical stops for positioning instead of servo control. For the same reason, they cannot be reprogrammed without adjusting the hardware. In 1973, ASEA introduced the IRb-6 model which was the first reprogrammable, servo controlled robot with a repeatability approaching that of the Seiko. The repeatability of .002 inches was made possible in part by using an electric, as opposed to a hydraulic power system. In 1978, Unimation introduced the first version of the PUMA, the first electrically powered servo controlled high precision robot that can be controlled by downloading a program from a computer as well as by the more conventional method (teach pendant). Because it can easily be interfaced with a computer, the PUMA quickly spread throughout industrial and academic robotics research labs. In 1981, American Robot Corp. introduced a computer controlled robot that can position a 50 pound payload with a repeatability of .001 inches. This unusual combination of high precision, a relatively large payload and computer control are the result of using a non-servo control system that is powered by electric stepper motors. Several of the other recently introduced models shown in the Table 2-1 are licensed from robot producers in Japan and Europe.

---

[6]Repeatability is a measure of a robot's ability to return to the same position over many repetitions. A repeatability of .005 inches means that the robot will return to within .005 inches of its original position on each successive replication. See [Albertson 83] for a more extensive discussion of repeatability.

Table 2-1: High Precision Robots Designed for Small Parts Assembly

Manufacturer	Model	Payload (Pounds)	Repeatability (Inches)	Approximate Base Price ($ 000)	First System Installation Date
Seiko	100	3	.0004	6	1960
Seiko	400	9	.002	10	1960
ASEA	IRb-6	13	.002	60	1973
Unimation	PUMA 550\650	5	.004	41	1978
Gen. Electric	Allegro	14	.001	125	1980
Unimation	PUMA 260	2	.002	41	1980
Westinghouse	4000	22	.008	65	1980
American Robot	Merlin	50	.001	60	1981
Automatix	AID 600	17	.003	85	1981
Bendix	AA	45	.002	70	1981
Yaskawa	Motoman L3	7	.004	60	1981
C. Milacron	T^3 726	14	.006	56	1982
IBM	7535	13	.002	29	1982
IBM	7565	5	.005	95	1982
Hitachi	A3020	4	.002	28	1983

Sources: Compiled from Hunt (1983) and Tech Tran Corp. (1983).

2.2 Increased Availability of Computer-Controlled Robots

Programmable robots first commercialized in the early 1960's , such as the Unimate and the AMF Versatran (now Prab), had computer-like functions, such as memory, but were made up of special purpose electronic logic components designed to perform a specific set of tasks. Electronic controls were only used to duplicate the functions of other special purpose (hardwired) control functions. Robots controlled by general purpose digital computers were not commercialized until the mid 1970's. The first minicomputer controlled robot was the T^3 (566 and 568) models commercialized in 1974 by Cincinnati Milacron. The first microprocessor controlled robot was developed in 1976 by VICARM, a small company founded by inventor Victor Scheinman, who founded the company to build small electric powered arms primarily for research labs. Unimation acquired the rights to the VICARM robot, further developed the product, and commercially released it as the PUMA 500 robot in 1978.[7] To our knowledge, the T^3 and the PUMA were the only computer controlled robots made by U.S. manufacturers commercially available in 1979. Since that time, many other computer controlled robot models have appeared on the market. For example, all of the high precision robots introduced after 1978 listed in Table 2-1 are computer controlled.

The computer controlled robot is far more flexible than a machine controlled with specialized logic circuits. Because of powerful computational capabilities, it can be programmed in any one of several different coordinate systems. Robots without computer control can only be programmed in one coordinate system. Because instructions can be fed into the controller through a keyboard and stored in computer memory, new robot programs can be developed off-line, even while the robot is engaged in a task. In contrast, "lead-through" and "walk-through" programming, the methods that must be used if the robot is not computer controlled, are on-line methods. The instructions are entered into the controller by physically positioning the robot arm and recording the position. To develop new programs with on-line programming methods requires that the robot be taken out of production.

Perhaps the most significant aspect of robots controlled by general purpose digital computers is that they can more easily and more extensively communicate with other computers, *and therefore, with other machines controlled by computers*. This has made it possible to develop robot programs on a computer in one location (at a centralized engineering facility) and to ship the program through a

---

[7]The 500 model had 5 degrees of freedom. The 600 model had six degrees of freedom. The model numbers were later changed to the PUMA 550 and 650.

network to the robot's computer on the floor of a factory in another location.[8]

The actions of a computer controlled robot can be monitored and supervised by an external host computer which can also monitor and supervise the actions of other machines. This capability has made it possible to more extensively control and coordinate the interactions of robots with other machines in manufacturing cells and so called manufacturing systems (FMS). In addition, computer controlled robots can more extensively process and utilize external information provided from visual, force, and tactile sensors than robots which are not computer controlled.

2.2.1 Experimental Improvements in Robotic Arms

Improvements in the basic mechanical design of rigid robot arms are not highlighted in this report. In fact, such research seems to have been deemphasized lately.[9] Nonetheless, there are noteworthy developments. Virtually every robotic manipulator in use today is a rigid structure that uses transmission mechanisms-- gear trains, lead screws, steel belts, chains and linkages-- to transmit power from motors to the load. Physical forces such as friction and backlash introduce unwanted disturbances which cannot be completely eliminated and performance capabilities are constrained as a result. This limits the speed, accuracy, and versatility of the arm. One significant development in the mechanics of arm design is the the direct drive manipulator developed at Carnegie-Mellon University [Asada and Kanade 81]. Lightweight but high powered direct drive electric motors are directly coupled to each joint, eliminating transmission mechanisms between motors and the load. Backlash is eliminated and frictional forces are minimized. The behavior of this type of arm can be modelled more accurately than an arm with conventional power transmission mechanisms. More accurate modelling provides for improved control, enabling the end effector to be moved at a higher speed and to be more precisely positioned. Eliminating the transmission mechanisms also eliminated mechanical parts, simplified maintenance requirements and potentially improved the reliability of the manipulator. The direct drive arm paves the way for a new generation of light weight, high performance robot arms. At least one manufacturer is experimenting with a direct drive arm for commercial use.

---

[8]For example, in the robotic assembly system reported in [Sanderson and Perry 83], robot programs are developed on a Vax 11/750 and then transmitted to a smaller computer, a PDP 11/23, and then finally transmitted to the controllers of the individual robots. The Vax is also networked to other computers in the research environment.

[9]According to Engelberger, the technology underlying virtually all insensate robots in use today is the servo control technology developed during World War II. He believes there will be no major breakthroughs in the robot arm itself as long as robots are built as stiff structures with the positioning based on classical open loop servo control theory. Given this constraint, he argues that emphasis should be placed on redesigning (or designing) the product for ease of manufacturing, and on integrating robots and other types of machinery into manufacturing systems. A major development, in robot arm technology in his opinion, requires the development of a lightweight flaccid arm that is precisely positioning by sensing the position of the end effector and the target position [Engelberger 83].

3. The Integration of Sensory Information Processing with Robotic Devices

Nitzan (1981: 2) has given a general definition of sensing:

> We define sensing (regardless of whether it is human or robotic, internal or external) in a broad way (i.e., as perception). Sensing is the translation of relevant physical properties of surface and volume elements into the information required for a given application. The physical properties are electric, magnetic, optical (e.g. surface reflectance and transmittance at different wavelengths of the incident radiation), mechanical (e.g. presence/absence, range, position, velocity, acceleration, stress, and pressure), temperature, and so forth.

The previous chapter focused on the use of internal sensing to monitor the internal systems (e.g. joint positions, joint loads; internal temperature, etc.) of a robot. This chapter focuses on the sensing of external information, and on how this extends the capabilities of the Level I robot.

3.0.1 The Importance of External Sensing in Factories

Figure 3-1 provides a framework for understanding how the sensing and processing of external information extends the robot's capabilities. Within the boxed region are elements representing the robot control system (mechanisms for spatio- temporal coordination and effector processing) and the end effector. These elements comprise the insensate robot with open-loop control described earlier. Outside the boxed region are elements representing devices for sensing external information, computer programs for processing and interpreting information (sensory processing and knowledge) and means for communicating the results with the robot manipulator (systems integration). With external sensing, knowledge can be acquired from the external world and communicated back to the robot controller. The position of the end effector can be measured and adjusted using closed-loop control.

External sensing and closed-loop control are important capabilities because they extend the range of manufacturing tasks that can be accomplished by a robot. For some tasks it is not possible to prespecify all motions to the level of exactness required to use an insensate robot. For example, surfaces of many natural materials, such as wood and stone, as well as surfaces of parts shaped in metal forming operations (e.g. sand casting, hammer forging) have irregular and unpredictable variations which often have to be smoothed over by grinding, routing, or deburring. In assembly processes, often there are variations in the tolerances of parts which have to be inserted into one another. Even tiny variations can complicate the insertion process significantly and require the use of sensory feedback. Moreover, many tasks, as currently performed, involve the selection of one or more small parts (washers, nuts, bolts, resistors, etc.) from a bin or pile of randomly oriented parts.

Figure 3-1: Schematic Diagram of a Sensor-Based Robotic System-

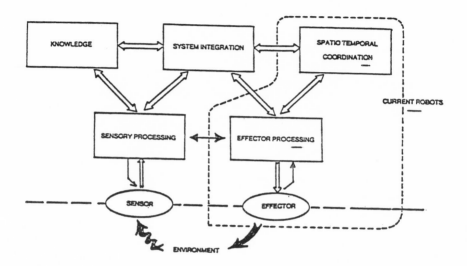

Source: Sanderson (1983: 33).

Finally, an important function in manufacturing is to detect flaws in components about to be assembled, or in completed parts or subassemblies. External sensory information is inherently necessary for all of the above tasks.

More specific reasons for why external sensing would extend the range of robot applications in manufacturing are given by Rossol (1983):

- Vision systems will provide positioning information to robots so that parts, robots and surrounding equipment need not be precisely located.

- Vision systems will perform a range of inspection tasks including incoming part inspection, in-process inspection, and final quality control inspection.

- In a related area, force sensing and control systems will provide fine positioning control, allowing for the detection of unusual conditions such as collisions and make possible high precision assembly.

The use of external sensing might eventually enable precision positioning to be achieved at a lower cost than by existing methods. The total cost of an application would be reduced if the cost of external sensing systems were to fall below the cost of mechanical positioning devices. There is also the possibility of using less precise robot manipulators and achieving precise positioning by measuring and adjusting the final position of the end effector, much the way a "floppy" human arm is positioned.

3.1 The Fundamental Problem of Machine Sensing

"Sensing" is not merely the detection of signals reflecting physical properties of the world. Referring back to Nitzan's definition at the beginning of this chapter, sensing is the *translation* of relevant physical properties of the world into information required for carrying out a task. External data detected through the human sensory organs (eyes, ears, hands) is automatically processed and translated into the information required for a particular problem in the brain. Because processing and interpretation is often unconscious, humans typically do not distinguish between detecting signals from the external world and interpreting what is detected. With machines, however, it is essential to make this distinction. Computers must be explicitly instructed on how to represent and store the physical signals, process them, extract required information, and communicate the information with the robot or other machines. Thus, for robot to "see" requires not only a TV camera, but sophisticated computer programs to process and translate the electronic signals into useful information. Detecting physical signals from the external world is a relatively straightforward engineering task, and there are great variety of devices for doing this, as indicated in Table 3-1. The

Table 3-1: Devices and Methods Available for Robots to
Detect Signals From the External World

VISUAL DETECTION DEVICES

Photo-detectors

Linear arrays

Area arrays

TV cameras

Lasers (triangulation)

Lasers (time of flight)

Optical fibers

FORCE DETECTION DEVICES

Probes

Strain gauges

Piezoelectric materials

Carbon Materials

Discrete arrays

Integrated arrays

ACOUSTIC DETECTION DEVICES

Ultrasonic detectors/emitters

Ultrasonic arrays

Microphones

OTHER DETECTION METHODS

Infra red

Radar

Magnetic proximity

Ionizing radiation

Source: Pugh (1983:6).

primary difficulty with machine sensing, however, is interpreting the signals from the detectors and in translating these signals into useful information.

Another issue is that the human brain is extremely flexible with respect to the types of external information it can process and interpret. It can apply different algorithms and appropriately filter data according to the circumstances. At one instant, a tennis player or a baseball hitter may concentrate totally on the approaching ball, assessing its speed, direction, and spin. At a later instant, the athlete may totally concentrate on the position of his opponent, on the playing surface, or on the fans in the grandstand. An accountant can scan a page of figures looking for the one or two numbers that are not within the expected range. At one moment, an editor can skim a page of print looking only for errors of punctuation and misspelled words. At another moment, the editor reads the page for content. A photographer may look at a landscape and concentrate on the overall balance of light and shadow. A birdwatcher can select out small moving creatures in the air, amongst clouds and trees. The same person may be capable of all of these kinds of visual processing. In addition, the person is also collecting and processing information received through the hands and ears.

A machine, on the other hand, is extremely limited with respect to the types of information that can be processed and interpreted. As yet, there is no such thing as a general purpose machine system that could even remotely approach the full sensing capabilities of a human. At first glance, the competition between the machine and the human is highly unequal because the available sensing systems are extremely crude in comparison to human capabilities. However, many tasks do not require the full power of human sensing capabilities (e.g. stereoscopic color vision). For extracting particular specialized types of information from a scene, a machine sensing system can actually be far more efficient than a human brain. In general, machine senses can compete in effectiveness with human sensing (disregarding cost for the moment) in one of two cases:

Case I   The machine is able to detect signals of a kind *that are not directly accessible to humans*. For instance, the machine may be able to directly sense voltages, currents, magnetic fields, frequencies, amplitudes, or intensities or electromagnetic radiation outside the optical range, pressures, temperatures, sounds, etc. Humans can only detect such signals if they are electronically converted into analog or digital readouts.

Case II   The machine is able to interpret a mass of data more efficiently and faster than a human. For instance, computers can filter and suppress much of the noise in a digitized satellite picture, revealing salient features invisible to the unaided eye. Computers are more efficient at matching fingerprints, because their speed enables them to test quickly for a few key features. Computer vision systems can be designed to efficiently identify the edge of objects in two dimensional patterns and measure the length and width of objects faster and more accurately than humans.

Case I situations are prime candidates for substitution of computer sensing systems for human workers. The reason is that external sensors are required in any event to convert the physical signal into a form that can be comprehended by a human. All that is needed to eliminate the human operator from the control loop is a means of inputing the detected signal into a microprocessor (e.g. an analogue-to-digital converter).[10]   This substitution has already occurred in computer numerical control machines because the basic "state-of-the-system" data for such machines is in a form more accessible to machine sensors than to human eyes, ears, and hands. The vast majority of the use of external sensing systems, especially machine vision systems, fall into Case II. Here, we are talking about using external sensing systems as direct substitutes for human sensing.

3.2 Increases in the Use of Machine Vision

Pugh (1983: 4) comments that "it is perhaps vision more than the senses of touch and hearing which has attracted the greatest research effort." For this reason, our overview of developments in the use of external sensing concentrates on the use of machine vision. Kanade (1983) describes the general process of vision as follows:

> Vision involves visual sensing and interpretation. Visual sensing is a *projection* of a physical environment into a form of representation called images. Projection can vary from the most ordinary picture taken by cameras to active sensing such as by laser rangefinders. Images can also range from a single-point light flux measurement to 3-dimensional range data.   Then vision is defined as a process of understanding the environment through the projected images, or in short *inverse projection*.[11]

Typically, when referring to machine vision, we mean a two-dimensional pattern, characterized by varying degrees of intensity projected from a three dimensional scene. Visual information is often used to inspect critical features of objects, to recognize recognize and identify objects, and to locate the position of objects in space.

3.2.1 Extent of Machine Vision Use Reported by Eikonix in 1979

Potential users of robotics interviewed in the Eikonix report were awaiting the development of machine vision systems. The report states[12]

> It is the sensory capability which has been most anticipated by potential industrial users, and the one which, rightly or wrongly, most users feel will significantly increase the role of robotics in manufacturing.

---

[10]This is discussed in more detail in [Ayres 84].

[11]Page 59.

[12] [Eikonix Corp. 79], page 69

What was lacking at the time were proven applications and commercially available systems. In 1979, there were only a handful of industrial applications of machine vision, which were still development projects, within General Motors, Cincinnati Milacron and Westinghouse. Only one commercially available machine system, the Autoplace Opto-Sense system, was mentioned in [Eikonix Corp. 79]. Research efforts in the field were also reviewed. SRI International reported several test applications using their machine vision system to locate and inspect parts. The University of Rhode Island vision program was working on the acquisition of of parts from a bin or hopper (bin picking), but no significant test applications were mentioned. Basic research efforts on general purpose vision underway at the major university centers of AI research- CMU, MIT, Stanford- and at Cal Tech's Jet Propulsion Lab were described as being truly in the research stages in [Eikonix Corp. 79]. Based on the state of the art in 1979, the report concluded, "despite these examples of industrial vision systems currently in use, robot vision is still in the research and development stage; particularly if we consider general purpose vision (page 70)."

3.2.2 Commercial Use of Machine Vision Systems Since 1979

From several indicators, it is clear that there has been a large increase in the application of machine vision systems in industry since 1979. As of the end of 1982, there were 400 to 500 machine vision systems reportedly installed in U.S. manufacturing operations compared to only a handful of installations reported in 1979 [Sanderson, R. 83]. Whereas at most a few firms supplied vision systems in 1979, 22 manufacturers of machine vision systems and over 20 university or government research institutes with machine vision programs (outside of industrial R & D) in the U.S. alone are listed in [Sanderson, R. 83]. Over 128 manufacturers, developers, and distributors of vision systems, image processing systems and components worldwide are listed in [Zimmerman, Van Boven and Oosterlinck 83].

One important factor contributing to the increased usage in vision systems since 1979 was the commercialization of the system developed at SRI [Agin 80]. The SRI system provided the basic product technology for many manufacturers of vision systems, included Machine Intelligence Corp. and Automatix. Recent applications of industrial applications of vision systems can be found in [Zimmerman and Oosterlinck 83], in the proceedings of the International Conference on Robot Vision and Sensory Controls (1979 through present) and also in the the proceedings of the International Symposium on Industrial Robots, especially the more recent ones (e.g. the 11th, 12th, 13th Symposiums).

3.3 Overview of Machine Vision Technology

Machine vision technology can be described in a variety of ways. Several alternative perspectives are to concentrate on

- the physical hardware and software components

- the types of visual information that can be detected and processed

- the types of industrial tasks that can be performed

Our prime concern is to describe vision technology in terms of the types of industrial tasks that can be performed because of our goal of assessing the impacts of the new technology. We briefly review key concepts associated with the first two aspects in this section, and in the next section describe capabilities of industrial vision systems in a selected application.

The major hardware and software components of a machine vision system are

- the image input system

- the digital processing system

- the signal output system

The image input system consists primarily of one or more cameras and a method of illumination. Cameras are used to convert visual information into electronic signals. The electronic signals are then digitized and organized into a matrix of picture elements (or pixels). typically ranging from 64 x 64 pixel elements to 512 x 512 pixel elements. Vidicons, the standard TV imaging device based on vacuum tube technology, have been widely used as cameras for vision systems because of their widespread availability and reasonable price. Vidicon cameras have many drawbacks. Most important, they are fragile because they contain vidicon tubes. They also suffer from electronic noise, lag, blooming, parabolic distortion, signal drift and frequency distortion and have to be adjusted often.[13] The new generation of cameras used for machine vision systems, based on solid-state electronics technology, consists of an array of photo sensitive electronic elements. They do not contain fragile glass elements. They are smaller, lighter, and more rugged, and have a longer operating life than vidicon cameras. Their performance characteristics are comparable to vidicon cameras, and are steadily being improved. In total, solid state cameras are more suitable for robotic applications in rugged industrial environments [Gonzalez and Safabkhsh 83], [Coiffet 83b].

---

[13]See [Gonzalez and Safabkhsh 83] and [Coiffet 83b], chapter 6.

Ideally, an object is illuminated in a way that enhances the presence of the required information and minimizes the complexity of processing the resulting electronic image. Each specific task that vision systems are used for (e.g. identifying a part, locating a part, measuring a part, inspecting for scratches, etc.) requires different types of visual information. The notion of the complexity of an electronic image is not an absolute one. but depends on the specific types of image processing techniques used to process the electronic image. A wide variety of illumination techniques have been developed, each designed to enhance specific types of visual information while minimizing the complexity of the image processing. For example, backlighting is used when the relevant information can be extracted from the silhouette of an object. When the object of interest has smooth, regular surfaces, it is sometimes possible to detect the relevant information with diffuse (normal overhead) lighting. Detecting a scratch on the surface of an object is sometimes easily accomplished by projecting light onto the object's surface from a known angle (directional lighting). Sometimes the relevant information can be enhanced by spatially modulating the light that illuminates an object. Structured light illumination involves projecting points, strips, or grids of light onto an object and measuring the curvature of the distortions [Gonzalez and Safabkhsh 83].

As mentioned earlier, the processing of physical signals and interpreting them into useful information is the critical part of sensing, and the part that attracts much of the research efforts. The array of digitized pixel elements formed by the image input system is often preprocessed to reformat the information in a way that speeds up subsequent image processing. This might include operations to eliminate electronic noise, geometrical distortions, and other defects introduced by the camera. Preprocessing might also include various types of filtering to reinforce elements that enrich the required information and to suppress other unwanted elements [Coiffet 83b].

Following preprocessing steps, the digitized array of pixel elements is further processed by the digital processing system. Gonzalez and Safabakhsh (1983) describe image processing in terms of the steps of segmentation, description and representation. Segmentation is the process of breaking up the sensed visual scene into its constituent parts or objects. One general approach is to segment the image by analyzing discontinuities in the digitized pixel array. The other general approach is to segment by analyzing similarities within the array. Description is the process of extracting features from the segmented image for the purpose of recognition. Descriptors are based primarily on shape and amplitude (intensity) information. Shape descriptors attempt to capture invariant geometrical properties of an object. Recognition is the process of identifying each segmented object in an image and assigning it a label (e.g. a wrench, seal, bolt). One general approach to recognition is to summarize the features identified in the description stage using a statistical discriminant function. The discriminant function is then used to match the unidentified image against the images of known objects stored in the computer. Another general approach is to summarize the features identified in the description stage as fundamental relationships among patterns in the image. These patterns are then used to match the unidentified image against patterns of known object stored in the computer.

Coiffett (1983) describes image processing in the following way. The first step is the detection of outlines. This is the central problem of image processing since the information contained in an image is essentially contained in the outlines. The second step is to distinguish the different objects in the image. This is done by segmenting the image into a number of lines (the outlines) and surfaces demarcated by these lines. This step is clear if the objects are clear and if they have non-overlapping images. It is more complicated of the images of separate objects in the image overlap. The third step is to represent each segmented object in terms of characteristic parameters. Various types of characteristic parameters can be used, depending on the particular problem at hand. Three classes of parameters used are

- parameters based on geometrical considerations

- parameters based on functions of points in the image

- parameters based on topology

There are a large number of geometrical parameters used to represent an object, which also serve to identify the position, orientation, and identity of an object. For example, geometrical parameters such as the center of the area of the minimal rectangle circumscribing the image, and the center of the area of the solid image are extracted from an image to determine its position. Geometrical parameters such as the direction of the diagonals of the circumscribed, minimal rectangle are used to determine the orientation of an image. Geometrical parameters such as the image area, perimeter, dimensions of the circumscribed, minimal rectangle, number of holes, are used to identify an object. Functions of points of images, such as Fourier transforms, polar moments, and rectangular moments may be used to construct a representation of an image. Positional relationships among different regions of the images, or topological features, may also be used to represent an image. In summary, Coiffet (1983: 146) comments "whilst it cannot be definitely stated which is the best representation system for robotic vision, a combination of available methods, depending on the application, is at our disposal."

The signal output system is the means of communicating the results of the visual information processing to other machines. A robot might use this information to control which program is executed next, or even to control its movements. A computer might use this information to sequence the actions of other machines. The hardware and methods used to interface the vision system computer with other machines depends on the nature of the information being communicated. Sending simple binary messages indicating the absence of presence of a part would require the simplest interface. Sending the position and orientation of an object, and using this information to control the movements of a machine, such as a robot, would require a very sophisticated interface.

A machine vision system can be described in terms of the types of information that can be processed. Three broad types of information are distinguished, following [Zimmerman, Van Boven and Oosterlinck 83]:

- spectral

- spatial

- temporal

More detailed types of information within each of the broad types is shown in Table 3-2. One type of spectral information is frequency, or the range of wavelengths that can be processed. Another type is intensity, or the range of brightness levels (or grey levels) that can be processed. Current machine vision systems can only process one wavelength at at time and therefore cannot handle color data. Most commercially available systems can only process two gray levels in one image. In these so-called binary systems, all gray levels above a specified brightness threshold are converted to "black" and all gray levels below the threshold are converted to "white". There are "pseudo gray scale systems" that process a series of binary images at different brightness thresholds, and infer information about changes in brightness intensity from the differences in the pictures. Some of the commercially available systems which claim to process multiple shades of gray, such as the Automatix Autovision II, are actually "pseudo gray" binary systems. Experimental systems and a few recently commercialized ones can actually process multiple shades of gray in one image [Kinnucan 83], [Sanderson, R. 83], [Sanderson and Perry 83], [Makhlin and Tinsdale 83].

One aspect of spatial information is the shape and position of an object. An object's shape is described in term of its geometry and its topology. Most commercially available systems can only process two dimensional shapes and determine the position and orientation of an object if its distance to the camera is known. Information about the third dimension, such as an object's distance or 3-D profile, comprises another aspect of spatial information. Most commercially available systems cannot process depth and range information. Recently, a few systems have been commercialized which can infer three dimensional information, such as depth or shape, by processing a series of two dimensional images [Levine 83], [Kinnucan 83], [Sanderson, R. 83]. One technique is to construct the 3-D profile of an object by illuminating it with pattern of light (or "structured" light). The pattern is

Table 3-2: Hierarchy of Types of Information
Extractable from the Image of a Single Object

```
VISUAL INFORMATION

        SPECTRAL
                frequency
                        color
                intensity
                        grey tones

        SPATIAL
                shape and position (1-, 2-, and 3 dimensional)
                        geometrical
                                shape
                                dimensions
                        topological
                                holes
                        spatial coordinates
                                position
                                orientation
                depth/range
                        distance
                        3-D profile

        TEMPORAL
                stationary
                        presence/absence
                time-dependent
                        events
                        motions
                        processes
```

Source: Zimmerman, Van Boven, and Oosterlinck (1983).

processed as a two dimensional image. By moving the pattern over the object, and comparing the distortions in the two dimensional pattern from one image to the next, information about the three dimensional shape of the object can be inferred.   Many prototype systems in laboratories are experimenting with detecting and processing depth,range and shape information using structured light, stereo vision, and other techniques [Ballard and Brown 82], [Kanade 83], [Sanderson 83], [Crowley 82], [Jarvis 82].

A graphical representation of the current capabilities of machine vision systems is shown in Figure 3-2. In the figure, systems are described in terms of their ability to resolve color, spatial and temporal information.[14] Capabilities of existing research and commercial systems are designated by the small circles on the axes. The circles representing commercially available systems are enclosed by the dotted lines, and are pointed out in the figure. Three types of color information are distinguished: saturation, hue and intensity. Available systems can process a wide range of intensity (brightness) information. However, available systems cannot process hue (frequency) or saturation information. A distinction is made between spatial information that is effectively one, two and three dimensional. Available systems can process one and two dimensional information, but not information that is effectively three dimensional. Techniques for inferring information about the third dimension from two dimensional information have already been mentioned.

The location of a system on the 1-D and 2-D axis indicates the size of the smallest 1-D or 2-D feature that can be recognized (or resolved). The limits of resolution in a vision system are determined by the number of picture elements (or pixels) in the camera used to detect the image. The standard camera transmits an image of 256 by 256 pixels to the image processing system. If the camera is viewing an area that is one inch square, each side of the square is divided into 256 equal segments. Thus, the size of the smallest feature that can be resolved in this area is 1/256 or .0039 inches. Using a camera which formed an image of 512 by 512 pixels, the size of the smallest feature that could be resolved in a one inch square would be .002 inches. The more pixels, the greater the resolution. The higher the resolution, the finer the details that can be discerned in the field of view. The spatial axes in Figure 3-2 indicates that commercially available systems, which are restricted to processing two dimensional information, can only do so at relatively low resolution. Experimental systems with arrays of 1024 picture elements, can process two-dimensional information at resolutions down to .001 inches.

The time axis designates image processing speed. Because of the time required to process an

---

[14]Sanderson's use of the term "color information" in [Sanderson 83] is nearly the same as the use of the term "spectral information" by Zimmerman, et al in [Zimmerman, Van Boven and Oosterlinck 83].

Figure 3-2: Parameter Space of Computer Vision Systems in Terms of
Ability to Resolve Color, Space and Time

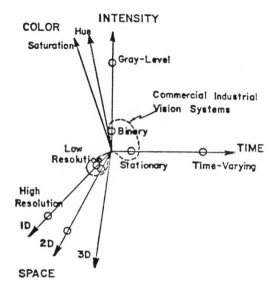

Source: Sanderson (1983).

image, commercially available systems are limited to processing single frames of information and can analyze a stationary event, such as determining whether an object is absent or present in the field view. Commercially available systems cannot operate fast enough to process time dependent information, such as events, motions and processes High speed experimental systems are used to process a series of images which change over time.

According to Sanderson (1983: 34), current commercial industrial vision systems occupy a small volume of this space of imaging capabilities. These systems incorporate thresholded binary imaging of two-dimensional arrays with a processing capability for a single image frame.

Capabilities of vision systems commercially available and in research labs are compared in Table 3-3, based on [Bolles 81]. This comparison is another approach to describing the types of information that can be processed by vision systems. The general problem solving approach of commercially available systems is to simplify the scene to be analyzed, thereby simplifying the information processing requirements. Research systems are designed to operate under less structured conditions and therefore require more sophisticated information processing. A partial list of factors affecting the degree of difficulty of processing visual information are the properties of the scene, the lighting, the camera-object relationship, the range of spectral data processed, the way in which images are represented within the computer, the way in which knowledge is used for problem solving, and the strategy used to make decisions about the image. For each of these factors, the table indicates in the first column the restricted conditions required for most commercial systems to function. The less restricted conditions under which research systems can function are indicated in the second column.

3.4 Machine Vision Applications

In this section we concentrate on the types of industrial tasks that can be performed by machine vision systems. Rosen (1979) divides machine vision applications into two broad areas:

Sensor-controlled manipulation
> Applications in which machine vision is an essential part of a manipulation task involving a robot, or of the control functions in a production process

Inspection   Applications in which machine vision supplants or supports the human in performing quality control inspection.

His designation of major problems within these two broad areas is shown in Table 3-4. Rosen (1979: 4) offers the following rationale for his classification of machine vision applications:
> ...I have attempted to select major classes of problem areas in industry in which successful application of machine vision will have a significant, if not revolutionary, impact

Table 3-3: Comparison of The Capabilities of
Commercially Available Vision Systems and Research Vision Systems

Attribute of System	Current Commercially Available Systems	Research Systems
General problem solving approach	extensive use of task specific engineering to simplify the scene to be analyzed	use of more sophisticated vision systems that can function under less structured conditions
Properties of scenes that can be analyzed	·constrained viewpoint ·isolated parts ·known set of parts ·rigid parts	·unconstrained viewpoint ·partially constrained content ·arbitrary set of parts ·nonrigid parts
Lighting	Very carefully controlled: ·uniform illumination over field of view ·high contrast ·no shadows ·no specular reflectance ·no photometric distortions ·fixed position and intensity of light sources	Less carefully controlled: ·non uniform illumination over field of view ·low contrast ·shadows ·specular reflectance ·distortions ·varying position and intensity of light sources
Camera-Object relationship	constrained viewpoint: ·known range and orientation	unconstrained viewpoint: ·varying range and orientation

Table 3-3, Continued

Attribute of System	Current Commercially Available Systems	Research Systems
Data	·often binary	·high spatial and intensity resolution ·occasionally multifrequency data ·occasionally range data
Representation of image	·usually just one type	·often several redundant types and levels ·use of parameterized object models
Knowledge	·compiled into program	·combination of compiled and explicit knowledge
Strategy	·extract features and make decisions based on feature classification and table look up	·requires search and some reasoning
Computers	·small and inexpensive	·large and expensive

Source: Adapted from Bolles (1981).

Table 3-4: Rosen's Classification of Machine Vision Applications

Sensor-Controlled Manipulation

- Manipulation of separated workpieces on conveyors
  - workpieces lying stably on a belt
  - workpieces hung on hooks partially constrained
- Bin picking
  - completely random spatial organization
  - highly organized and separated
  - partially organized spatially and unseparated
- Manipulation in manufacturing processes
  - finishing, sealing, deburring, cutting, process control, flash removal, liquid gasketing
- Assembly
  - in-process inspection
  - fastening (spot welding, riveting, arc welding, bolting, screwing, nailing, gluing, stabling)
  - fitting (parts presentation, mating of parts)

Table 3-4, Continued.

Inspection

- Highly quantitative mensuration of critical dimensions

    o critical exterior and interior dimensions of key features of workpieces

    o tool wear

- Qualitative and Semiquantitative Mensuration

    o Label reading and registration

    o Sorting

    o Integrity and Completeness

        - All parts and features present, right part handedness

        - Burrs, cracks, warping, defects, approximate size and location of key features

    o Cosmetic and surface finish properties (stains and smears, colors, blemishes, surface discontinuities)

    o Safety and monitoring

Source: Rosen (1979).

on productivity, product quality, and even the mass-production process itself. Although the grouping of tasks in each class is somewhat arbitrary, each class is real, having been identified and described by competent factory personnel during visits to plants and in subsequent discussions between the author and his colleagues over the past six years. Further, there are sufficient instances with comparable requirements in each class, so that general rather than ad-hoc techniques can be applied.

Notably, inspection applications are separated from those which involve the use of vision to manipulate objects. Applications involving an interface between the vision system and a robot manipulator are typically referred to as robot vision. Other classifications of the types of problems machine vision is applied to have been proposed. R. Sanderson (1983) divides applications into three broad areas:

Inspection   Applications in which machine vision is used to examine the characteristics of an object

Part Identification Applications in which machine vision is used to make a positive identification of an object.

Guidance and Control
    Applications in which machine vision is used to measure the relative position of objects in order to provide feedback to the robot.

The distinction between inspection tasks and tasks involving an interface with a robot manipulator is still maintained. The applications in the two areas of part identification and guidance and control are roughly equivalent to Rosen's list of applications in sensor-controlled manipulation, although there are also some differences in the ways applications are defined.[15] Thirty one examples of the use of vision systems in production are shown in Appendix I. The examples are separated into two main groups: sensor controlled manipulation and inspection.

3.5 The Use of Machine Vision in Sensor-Controlled Manipulation

In this section, the intent is to distinguish between the types of part identification and location problems that can be solved with current technology and the types of problems that require improvements in technology.

Table 3-5 lists application areas generally included under sensor-controlled manipulation and describe what the vision system is used to do. Two basic functions are identified:

---

[15] R. Sanderson (1983) separately lists the application of part identification, where Rosen (1979) essentially assumes that part identification is part of the task of acquiring objects from a conveyor or bin. Part sorting falls under part identification in R. Sanderson's classification but is part of inspection in Rosen's scheme.

Table 3-5: Functional Tasks Performed by Vision System
in Applications Involving Sensor-Controlled Manipulation

Application Area	Functional Task Performed by Vision System
Part Sorting	Identifying parts; Locating parts;
Conveyor picking	same
Bin picking	same
Processing/Machining	Measuring relative position and orientation
Weld seam tracking	of tool with respect to workpiece
Assembly fastening	Identifying parts; Locating parts;
and fitting	Measuring relative position and orientation of
	one part with respect to another

- Identifying, locating (and sometimes counting) objects

- Measuring the position and orientation of one object relative to another

In the first group of applications in Table 3-5 (part sorting, conveyor picking, bin picking) the robot has to pick up a part whose location is not precisely known and move it to another location. This is commonly referred to as parts acquisition. Often, the identity of the part must be checked. Sometimes, counts of part types are required. If it is the desired type of part, its position and orientation must be ascertained and communicated to the robot so that it can move to that position and pick up the part. The vision system is used to identify and count parts, to determine position and orientation, and to communicate this spatial information to the robot controller.

In the second group of applications in the table (routing, gluing, grinding, deburring, weld seam tracking, etc.) the robot has to move a tool over the surface of a workpiece. We assume here that, to some degree, the workpiece surface or seam is irregular and variable and its precise shape is difficult to prespecify--otherwise external sensing would not be required. The vision system is used to guide the path of the robot so that the tool is always properly positioned and oriented as it moves across the surface of the workpiece.

The third group of applications, fitting and fastening in assembly, may require both functions of the vision system. First the robot has to acquire a part. (In some cases, the part acquisition step can be engineered so as not to require a vision system. However, if a vision system is needed for other reasons, it is likely to be worthwhile to use it for parts acquisition.) Then, one part has to be mated with another in order to complete the assembly or fastening process. This requires positioning one part relative to another. It might also involve measuring and adapting to forces.

Machine Vision Technology Requirements for Identification and Location of Parts

Levels of machine vision technology are described in a way that is especially relevant for the problem of identifying and locating parts. Two key aspects of image processing are distinguished; 1) the way in which an object is represented within the machine, and 2) the way in which the represented image is analyzed. Taken together, the representation and analysis constitute an approach toward image processing.

Two standard methods of representing objects as images are as follows:

Binary image (BI) representation
            An object is represented by its silhouette.

Edge detection (ED) representation

An object is represented by a line drawing which includes internal as well as external edges. This representation is often referred to as a "gray-scale" image.

While there are many specific algorithms and procedures for analyzing images, one can identify three very general approaches. The first is global feature analysis (GFA). This is the image processing approach used by the basic SRI system described in [Agin 80]. In this approach, designated geometrical and topological features in the object (e.g. number of holes, area of holes, areas of "blobs", centroid, perimeter, length of bounding rectangle, width of bounding rectangle, etc.) are measured. "Global" in this sense means that there is no attempt to make inferences about spatial relationships among the features. GFA does not specify how a particular feature, such as a hole, is spatially oriented with respect to another hole, or how two holes are spatially oriented with respect to a corner. Objects are recognized by comparing the global feature vector for each observed object with stored models. This is pure *numerical* pattern recognition, since matching is based solely on a statistical measure of how closely the observed global feature vector matches feature vectors stored in memory. In general, this type of analysis can only be used to identify isolated parts. In special situations, though, image processing "tricks" can be used that will enable parts that are not isolated to be identified and located by means of global features. For instance, it may be possible to identify some special structure such as a hole or a projection that always occurs in a given situation and to use that information to help perform necessary image analysis. BI/GFA can not, in general, count parts correctly if they are touching. But if we are only counting nuts, we can do so by counting holes, even if the bolts are touching, as long as they do not overlap in a way that obstructs the view of the hole. The GFA approach to image analysis is applicable to both BI and ED image representations.

The second approach is local feature analysis (LFA);it has been described in [Bolles and Cain 83]. In this approach, an object is defined as a collection of local features with specified *spatial relationships* between various combinations thereof. This approach is restricted to considering 2-D spatial relationships (e.g. the big hole is always to the right of the small hole). Objects are identified by a comparison of local features and spatial relationships in the image and in stored prototypes. The key functional capability of this approach is that an object can be identified by considering only some of its local features. Thus, under some conditions, this approach can be used to identify parts that are not fully visible, such as overlapping parts. The LFA approach to image analysis is also applicable to both BI and ED image representations.

The third general approach is three dimensional analysis (3-D); This approach requires a complete geometric model of objects stored in a computer, and software to recognize the model from any arbitrary orientation and from any arbitrary perspective. Descriptions of general purpose systems for model based vision are given in [Kanade 83] and [Kanade and Reddy 83]. A system for a very particular type of three dimensional analysis (bin picking of cylinders) has been developed [Kelly and Birk, etal 82]. Three dimensional analysis extends notion of spatial relationship among local features to the third dimension. The analysis of stereo views, shading patters, shadow patterns, surface texture, occluded views, and views of moving objects are used to infer three dimensional information from two dimensional images [Crowley 84]. A goal of 3-D vision systems is to recognize an object from any arbitrary perspective, and to distinguish it from the background.

The various combinations of representation and analysis methods are summarized in Table 3-6.[16] The BI/GFA systems are widely available. The first generation of ED/GFA systems have been commercialized during 1983. Systems using the LFA image analysis approach are still in the development stage. General purpose 3-D systems are still in laboratories, although there is at least one commercially available system that can analyze three dimensional images, using ED representation, under highly constrained situations.

A partial taxonomy of part identification/location problems is shown in Table 3-7. For each problem situation, we designate the minimum level of technology that would be required to accomplish the task. A distinction is made between simply counting parts and identifying and locating parts. In some cases, a less sophisticated level of technology is required if the problem is restricted to counting parts. Clearly, we are most interested in the more general capability of locating parts since this is what is required to pick it up with a robot. The full set of tables for the part identification/location taxonomy are shown in Appendix II. A few general comments are worth noting here.

Binary Imaging with Global Feature Analysis (BI/GFA) is almost always limited to working with separated parts, of a few distinctively differentiated species, lying flat, in a high contrast environment. According to Brian Carlisle of Adept Technologies, this set of conditions holds for only about 1 percent of the potential applications in the area of object recognition and location.[17]   Other restrictions of these type systems were already mentioned in Table 3-3.

Edge detection with GFA would make it possible to identify and count separated parts under low contrast lighting, and in some cases, parts which are touching or overlapping. But as long as the

---

[16]An overview of the different generations of vision systems, based on binary image representation, edge detection, and 3-D analysis, is given in [Crowley 84].

[17] [Carlisle 83].

Table 3-6: Availability of Types of Vision Systems

BI/GFA   First appeared commercially in 1978. Now widely available commercially . MIC, Automatix, and many other companies supply systems of varying degrees of sophistication.

These types of systems have already been through extensive field-testing and software development. They are engineered to be user friendly and industrially hardened.

Most systems fast enough for medium cycle time applications. Not fast enough to keep up with high speed, mass production lines (e.g. canning lines, or food processing lines).

ED/GFA   First appeared commercially in 1982-3. Offered by only a small number of companies (e.g. Pixie system of AIS).

Just beginning to be field tested. Software just beginning to be developed. Not yet engineered to be "user friendly" and industrially hardened.

Systems based on cellular logic very fast. Might be applicable to high speed applications.

BI/LFA   Not available commercially

ED/LFA   Not available commercially

Special purpose 3D
                Commercially available in 1982-83 (Object Recognition Systems "I-bot"). Can only work with specially constrained problems. For example, locating a part in a bin of identical cylindrical parts.

General purpose 3D
                Not commercially available.

Table 3-7: Taxonomy of Problem Classes for Part Identification and Location

Conditions	Tasks:	
	Counting Number of Parts	Locating Position of Parts
1] IDENTICAL PARTS		
a] Rigid parts.		
Each part is in the same predetermined stable position.		
i] separated		
high contrast	BI/GFA	BI/GFA
low contrast	ED/GFA	ED/GFA
ii] touching (without overlapping or tilting)		
high contrast	BI/GFA, Trick	BI/LFA
low contrast	ED/GFA,Trick	ED/LFA
iii] overlapping and/or only slightly tilting (< 20 degrees)		
high contrast	BI/GFA, Trick maybe	BI/LFA
low contrast	ED/GFA, Trick maybe	ED/LFA
iv] Tilting		
high contrast	Part cannot be in known position	
low contrast	" "	
b] Rigid Parts.		
Each part is in any one of n distinct stable positions.		
(e.g. in any one of two, three, or four stable positions)		
Each stable position has a unique silhouette.		
i] separated		
high contrast	BI/GFA	BI/GFA
low contrast	ED/GFA	ED/GFA
ii] touching		
high contrast	BI/LFA	BI/LFA
low contrast	ED/LFA	ED/LFA
iii] overlapping and/or slightly tilting		
high contrast	BI/LFA	BI/LFA
low contrast	ED/LFA	ED/LFA
iv] tilting		
high contrast	BI/3D	BI/3D
low contrast	ED/3D	ED/3D

image analysis is restricted to GFA, the capability provided by "gray scale" edge detection systems is not significantly larger than that available with BI. The impact of ED/GFA systems on the market for machine vision applications is not likely to be great.

As long as image analysis algorithms are restricted to GFA, machine vision will only be used for part identification and location in highly structured workplaces where great efforts (and large sums of money) have gone into engineering the domain to insure that such problems are solvable. This appears to limit applications of GFA to the following situations:

- In existing high volume operations where the system is required to make only one type (or at most a few) decisions per part on large numbers of parts. For example, on lines for auto components, separating small parts from large parts , or separating parts with a hole from those those without a hole, or checking the length and width of each part.

- In existing low volume operations where there is a need to make many very precise measurements per part. For example, counting and measuring the hundreds of holes in a jet engine oil filter.

- New applications where it is possible to engineer the environment so that parts always appear separated and in high contrast lighting.

In general, to locate parts which are touching or parts which are overlapping, local feature analysis and the matching of spatial relationships is needed. The appearance of LFA on the machine vision market *would* have a large impact on broadening the applicability of machine vision systems, especially if combined with edge detection.   According to Carlisle, a system that could locate touching parts in low contrast lighting would solve approximately 50 percent of the potential applications in the area of object recognition and location.[18]

Systems with ED/LFA would make it possible to identify parts which are randomly placed or "thrown" onto a conveyor line, or which are stacked in bins in a semi-orderly arrangement (partially organized spatially and unseparated). This would widen the applicability of machine vision in retrofit situations since it would reduce the need to reengineer the domain.

General purpose 3D systems are needed in order to solve the general bin picking problem. Such a system would make it possible to integrate robots into many existing factories without having to make major modifications to how materials flow through the factory. However, by the time general purpose 3D systems are commercially available at an affordable price (which some experts predict is almost 10 years away), many users of machine vision systems and robots might already have built new applications and engineered them to work with 2-D types of analysis, such as LFA.

---

[18] [Carlisle 83].

Rossol (1983) has predicted the year in which vision-based robot systems could become commercially available for acquiring parts from floor conveyors, pallets and trays, overhead conveyors and bins. He believes that any of these capabilities could be available in the year indicated in Table 3-8 given appropriate demand and R&D resources. This does not mean that all of these features will necessarily be available in the predicted times.

A correspondence can be made between the most of the part feeding applications listed in Table 3-8 and the taxonomy of problem classes for part identification and location in Table 3-7. Parts on a floor conveyor would typically be separated or touching. Parts in pallets and trays would typically be touching or overlapping. Parts in bins would be tilting (at an arbitrary angle). The one case for which there is no direct correspondence between the two tables is the case of parts on an overhead conveyor. Here, parts are typically separated but sway back and forth as the overhead line moves. (Imagine auto tires supported by hooks on an overhead line). We use these correspondences to infer when systems will be commercially available for the selected classes of parts feeding problems shown in Table 3-7.

Commercially available systems already exist for acquiring separated parts on a floor conveyor under high contrast lighting. Rossol's predictions suggest that systems could be available in 1984 that can pick separated parts off of floor conveyors under low contrast lighting, and even when parts are touching. Also, that systems could be available to acquire touching and overlapping parts in 1986 and tilting parts (out of bins) in 1988.

Jim Crowley, a research scientist at the Carnegie-Mellon Robotics Institute, predicts that in 1986-87, a second generation of vision machines will be available. These will be grey scale (edge detection) systems. Although they will still be essentially restricted to recognize 2-dimensional images, they will be able to recognize a larger set of objects. He predicts that in 1988-90, crude 3-D vision systems will start becoming available. They will have simple motion tracking capabilities and recognize a still larger set of objects [Wright 83].

Table 3-8: Rossol's Prediction of the Year of First Commercial
Available for Vision-Based Robot Systems for Parts Feeding

PART FEEDING APPLICATION	PREDICTED AVAILABILITY (YEAR)
Floor conveyors	1984
Pallets and trays	1986
Overhead conveyors	1987
Bins	1988

Source: Rossol (1983).

4. Manufacturing Systems

4.1 An Overview of Manufacturing Systems

We broadly interpret the term "manufacturing system" to mean the following:

- a set of tools for processing materials,

- a means for moving materials from one tool to another

- a means for controlling and monitoring the action of the tools and the movement of materials

Various types of manufacturing systems are distinguished in Table 4-1, based on general attributes of the processing tools, material handling system, and means of supervisory control and monitoring. Of the automated systems in the table (the first four types), the one with the longest history of use is the specialized transfer line. Such systems may be very large, including up to several hundred interconnected process tools. Each tool in the system is single purpose, highly specialized to perform one type of task. Material flow from one process tool to another is automated, but the sequence of the flow is fixed, predetermined in advance, and cannot be altered without stopping production and reconfiguring the system. The loading and unloading of material into and out of each process tool is also automated. Control of the movement of materials from one process tool to another, as well as the control of the parameters of each tool, is implemented via dedicated logic controllers built in (or hardwired) into the machine hardware.

A semi-flexible transfer line differs from a specialized one in that some or all of the process tools in the interconnected line may be multipurpose, capable of performing a range of tasks. Such systems may include many process tools, but typically not as many as used in the specialized transfer lines. Although the sequence of the flow of materials from one tool to another is still fixed, the sequence of operations at each tool is not since several operations can be carried out at each multipurpose tool. This added flexibility at each tool makes it possible for a semi-flexible transfer line to process a wider range of parts than a specialized transfer line. A limited degree of programmability is required to control the timing of the material flow and the parameters of each tool. Typically, a semi-flexible transfer line is controlled by programmable logic controllers (PLC).

A flexible manufacturing system (FMS) differs from a semi-flexible transfer line in that neither sequencing of the flow of materials from one tool to another nor the sequence of operations at each tool is fixed. The material handling system has "random bypassing capability", i.e. a part can be

Table 4-1: An Overview of Manufacturing Systems

Attributes	Specialized Transfer Line	Semi-Flexible Transfer Line	Flexible System	Cell	Stand-Alone Tool
Process Tools:					
-number of interconnected tools	up to several hundred	3 to 100	3 to 50	1 to 10	one
-degree of specialization of each tool	single-purpose	multi-purpose	multi-purpose	multi-purpose	multi-purpose
Handling of Material Flows:					
-across tools in network	automated, fixed sequence	automated, fixed sequence	automated, variable sequence	automated, variable sequence	none
-into and out of each tool	automated	automated	automated	automated	none
Supervisory Control:					
-of material flows	dedicated logic control	programmable logic control	hierarchical computer control	hierarchical computer control	manual
-of tool parameters	same	same	same	same	manual or computer

moved from any one tool in the interconnected system to another because the transport system can bypass any tool along its path, on demand. Thus, each part can traverse a variable route through the system. This flexibility in material handling, in combination with the use of multipurpose tools, makes it possible for a flexible manufacturing system to process a greater diversity of parts than a semi-flexible transfer line. Because of the larger number of possibilities, control of automated material flows and of tool parameters in this type of system is considerably more complex than in a semi-flexible transfer line. Typically, a flexible manufacturing system requires a hierarchical control system where a computer supervises and coordinates the actions of the control mechanisms in the material handling system and in the process tools. Such systems typically have fewer process tools than transfer lines, partly because the requirements for extensive information processing and control limits the number of elements that can be supervised and coordinated.

An outline of a typical hierarchical control scheme used in a flexible manufacturing system for machining is shown in Figure 4-1. The type of parts to be produced, volume requirements, and due dates are inputed into the FMS supervisory computer by the system manager, which is typically a human but which may be another computer. Information on when and how each part is to be processed flows from the supervisory computer at the top level of the hierarchy to the material handling system and machine tools at the lower level. The control mechanisms within the material handling system move the palletized parts around in such a way that the higher level commands of the supervisory computer are satisfied. The material handling system "reports" back information on where parts are, and whether they have been machined or not, to the supervisory computer. The supervisory computer also controls and monitors some actions of the machine tools. The supervisory program determines which NC programs should be residing at each machine tool, and transfers the appropriate program to the appropriate tool when necessary.

The difference between a manufacturing cell and a flexible manufacturing system is more one of degree than of kind. A system with multipurpose tools, variable sequencing material handling, and a computer supervising the control of the material handling system and tool parameters would be called a *cell* if there were only three or four different tools in the system. If there were a larger of number tools, say five or ten or more, it would be be called a flexible manufacturing system or FMS. There is no clear cut boundary between large cells and small flexible systems. The control scheme used in a manufacturing cell would be essentially the same as that shown in Figure 4-1. However, the material handling "system" may now only consist of one or several robots. A cell could even be considered as a single "stand-alone" work station. Within the cell, the flow of materials is automated, but the flow of materials into and out of the cell is handled manually. A flexible manufacturing system could be considered as a network of manufacturing cells connected via a material handling system. The flow of materials is automated within each cell, as well as across all cells within the network.

Figure 4-1: Overview of Hierarchical Control System in
A Flexible Manufacturing System for Machining

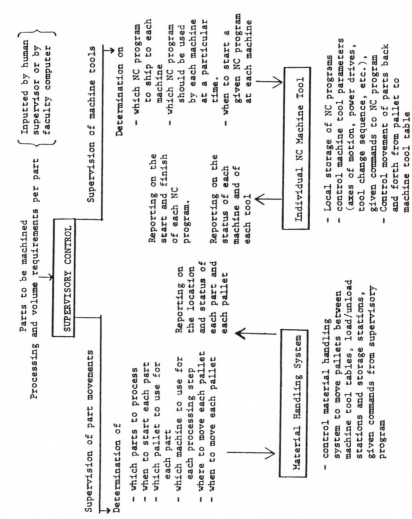

Processing and volume requirements per part

Parts to be machined

Inputed by human
supervisor or by
faculty computer

SUPERVISORY CONTROL

Supervision of part movements

Determination of

- which parts to process
- when to start each part
- which pallet to use for each part
- which machine to use for each processing step
- where to move each pallet
- when to move each pallet

Reporting on the location and status of each part and each pallet

Reporting on the start and finish of each NC program.

Reporting on the status of each machine and of each tool

Supervision of machine tools

Determination on

- which NC program to ship to each machine
- which NC program should be used by each machine at a particular time.
- when to start a given NC program at each machine

Material Handling System

- control material handling system to move pallets between machine tool tables, load/unload stations and storage stations, given commands from supervisory program

Individual NC Machine Tool

- Local storage of NC programs
- control machine tool parameters (axes of motion, power drives, tool change sequence, etc.), given commands to NC program
- Control movement of parts back and forth from pallet to machine tool table

The simplest example of a manufacturing system in this framework would be a single stand-alone process tool. Typically, the machine would be multipurpose. There is no automated flow of materials from machine to machine, since each machine is not interconnected with other any except by manual transfer. The loading and unloading of material into and out of the machine is manual. If the loading and unloading were automated, the stand-alone tool would in effect become a cell. Automatic control is not required to supervise a manual material flow system. While there is usually an operator, some aspects of the machine tool parameters may be taken over by a computer. This is the least sophisticated manufacturing system from the viewpoint of the use of automation, but is the most adaptable. Engineering changes are easily incorporated and the time required to change tools is minimal.

Kearney and Trecker, Corp., a pioneer in the manufacture of FMS, has developed a volume/variety chart (Figure 4-2) to illustrate where it is appropriate to use the various types of manufacturing systems for machining applications. The term volume refers to the number of parts per year manufactured with the system. Variety refers to the number of distinct part designs run through the system.

The specialized transfer line is appropriate when production requirements lie within the the top left quadrant of Figure 4-2, the region of highest volumes per part and only a few different part types. There is almost no limit on the maximum number of identical parts per year a transfer line may be built to produce. There are also special cases in which a line may also be used to produce as few as 2,000 pieces of one part number per year. Stand-alone tools are most appropriate when production requirements fall within the lower right quadrant of the figure, the region of lowest volume per part and a large number of different part types. In machining applications, this typically corresponds to 300 or more part different types per year, with most parts produced in batches of 50 or less. Prior to the development of intermediate forms of manufacturing systems in the mid-1960's, the only alternative choices were the two extremes of the spectrum.

The three more recent variants of automated manufacturing systems, semi-flexible transfer lines (labelled as special systems in Figure 4-2), flexible manufacturing systems and manufacturing cells, are designed for mid-volume and mid-variety production requirements. In machining applications, this typically includes yearly production volumes in the range of 15 to 15,000 units per part combined with varieties ranging from 2 to 800 different part types. Within this range, semi-flexible transfer lines are designed for large batch production. In machining applications, this roughly corresponds to volume requirements ranging from 1,000 to 15,000 units per part, and part varieties ranging from 2 to 10 different part types. Flexible manufacturing systems are designed for medium sized batch

Figure 4-2: Kearney and Trecker Volume/Variety Chart

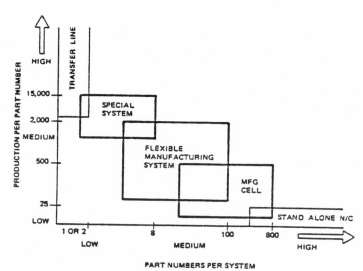

Source: Kearney and Trecker, 1982.

production, typically including volume requirements ranging from 50 to 2000 units per part and part varieties ranging from 4 to 100 part types. Manufacturing cells are designed for small batch production, typically including volumes ranging from 10 to 500 units per part and varieties in the range of 10 to 800 different part types.

4.2 Flexible Manufacturing Systems for Machining

Flexible manufacturing systems for the machining of parts have been in commercial use for over 15 years in the United States. The first systems were built by White-Sundstrand and Kearney & Trecker. A partial list of FMS currently in use in the U.S. is given in Table 4-2. More detailed descriptions of FMS for machining are given in [Jablonowski 80], [Kearney & Trecker 82] [Dupont-Gatelmand 83], [Draper Lab 83], and [Manufacturing Engineering 83]. Klahorst (1983), of Kearney & Trecker Corp., reports that most FMS systems will produce between 5,000 and 75,000 parts per year. He reports that attempts to justify an FMS for fewer than 5,000 workpieces will usually fail because of the resultant high cost of automation per workpiece produced. Justification of an FMS for more than 75,000 parts will usually fail because the system will be a more costly alternative than more specialized forms of automation (transfer lines).[19] Total production volumes of the systems listed in Table 4-2 range between 5,000 and 25,000 parts. Total production volumes of the systems built by Kearney and Trecker range between 5,000 and 60,000 parts [Klahorst 83a].

---

[19] [Klahorst 83a], page 53.

Table 4-2: Partial List of FMS for Machining in
Use in the United States

User	Vendor	Date Installed	Volume:[a] Number of Parts:[b]	Types of Parts
*Sundstrand Aviation	White-Sundstrand	1967	na na	aluminum pump parts
*Ingersoll-Rand	White-Sundstrand	1970	$20,000^a$ 14^b	hoist and motor cases
*Rockwell	K&T	1972	$25,000^a$ 45^b	automotive axle carriers
*Allis-Chalmers	K&T	1970-1973	$23,000^a$ 8^b	agricultural equipment
*Caterpillar	White-Sundstrand	1973	$8,600^a$ 6^b	crank case housings, covers
*AVCO-Williamsport	K&T	1975-1978	$24,000^a$ 9^b	aircraft engines
*AVCO-Lycoming	K&T	1979	$15,000^a$ 10^b	turbine engines
*Caterpillar	Giddings and Lewis	1980	na na	construction equipment
*John Deere	K&T	1981	$5,000^a$ 8^b	farm equipment
*International Harvester	Giddings and Lewis	1981	na na na	
*General Electric	Giddings and Lewis	1981	$5,600^a$ 7^b	motor housings
*AVCO	Giddings and Lewis	1982	na na	na
*Detroit Diesel Allison	White-Sundstrand	1983	na 40^b	transmission housings

a: total number of parts per year machined on system

b: number of different part types machined on system

na: information not available

Source: Draper Labs, 1983.

Available information on investment in FMS hardware and reported savings in operating costs are shown in Tables 4-3, 4-4, 4-5, and 4-6. Investment in system hardware runs into the multimillion dollar range. Users report large savings in operating costs, including reductions in labor cost, reductions in cost associated with in-process inventories, and reductions in cost associated with floor space requirements.

The hardware costs of the Kearney and Trecker systems listed in Table 4-3 ranges between $5 and $18 million. This sum includes the cost of the machine tools, material handling equipment, control systems, tools and fixtures, and testing, training, manuals, and other support services. This sum does not include the costs of planning, engineering, and installation. Their are no available data on these "support" costs. Kearney & Trecker has found that it typically takes three years of planning between the time a customer decides to buy an FMS and the time a system is installed by a vendor.[20] Considering that almost every functional area of manufacturing is affected by the use of the FMS and must be involved to some extent[21], it seems that the costs of planning and engineering are considerable. The FMS users listed in Table 4-3 report a variety of savings in operating costs including substantial reductions in the amount of labor, tooling and floor space required to produce a given level of output. In two cases in the table, the price quote for the FMS was equal to or less than the price quote for a special purpose transfer line. Thus, in some production situations, FMS can actually reduce the investment required to automate, in addition to providing the advantages of flexibility.

A comparison of machining locomotive parts (motor frame housings) with a conventional system of stand alone machines and with a new FMS at General Electric's Erie facility is shown in Table 4-4. The FMS has one third as many machine tools as the system it replaced. Twenty-nine manually operated machines were replaced by nine automated machining centers. As a result, floor space requirements were reduced by 25 percent and the typical number of times a part had to be loaded onto a separate machine was cut in half. The total number of people required to support the machining activity over two shifts (material handlers, operators, maintenance workers schedulers, and supervisors), was reduced from 86 to 16. The capacity of the FMS is 38 percent greater than the capacity of the manually operated stand-alone tools. The new system is designed to produce 5600 parts per year, whereas the old system produced about 4100 parts per year. GE management claims that capacity of the FMS could be increased by an additional 60 percent, to 9000 parts per year, with only minor modifications of the current system. The average in-process time was reduced from 16

---

[20] [Kearney & Trecker 82], page 14.8.

[21] [Kearney & Trecker 82], page 14.8.

Table 4-3: Hardware Investment and Reported Savings in Operating Costs
for Selected Kearney & Trecker Machining FMS

User	Cost (1982 $)	Product Volume and Part Variety	Reported Comparisons with Old System
Rockwell, (truck axles)	$5.6 million	24,000[a] 45[b]	1/4 floor space; set up costs virtually eliminated
AVCO- Williamsport (aircraft engines)	$8.4 million	24,000[a] 9[b]	1/3 floor space; 1/4 labor; 1/2 number of part holding devices;
John Deere (tractor components)	$18 million	50,000[a] 8-12[b]	cost estimates of FMS: $18 million; dedicated transfer line: $28 million;
Mack Truck (truck components	$5 million	65,000[a] 5[b]	cost estimate of FMS about the same as estimate for dedicated transfer line, with comparable cycle time, less flexibility;
Caterpillar[c] (construction equip.)	$5 million)	8,000[a] 8[b]	Total transit time through system: Old system: 8.5 hours New system: .3 hours

a: total number of parts per year machined on system

b: number of different part types machined on system

c: This system is a semi-flexible transfer line. All other systems listed here are flexible manufacturing systems. See Table 4-1 for definitions.

Source: [Klahorst 83b]

Table 4-4: Hardware Investment and Reported Savings in Operating Costs
for Machining FMS at General Electric, Erie

Comparison	Old Machining System	New Machining System
Number of Machines	29	9
Total production worker requirements for two shifts (operators, supervisors and maintenance)	86	16
Typical number of machine loadings required to complete one part	10 to 11	4 to 5
Maximum annual output for family of 7 parts	4100	5600
Average in-process time for a part	16 days	16 hours
Cost	base	$16 million
Productivity Change (total factor)	base	+ 240 %

Source: [Miller 83b]

Table 4-5: Hardware Investment and Reported Savings in Operating Costs
for Machining FMS System at
Yamazaki Machinery Works

Plant	Yamazaki Machinery Works, Ltd., Nagoya, Japan.
Process	Machining
Products	Parts for numerically controlled machine tools and robots.
Total costs	$20 million
Total employment	13
Comparisons	A conventional machining system with similar production volume would require 215 workers and nearly 4 times as many machines. Average in-process time per part was reduced from three months to three days. Over a five year period, the plant is expected to be 15 times more profitable than a conventional plant its size.

Table 4-6: Hardware Investment and Reported Savings in Operating Costs
for Machining FMS a Fanuc Ltd.

Plant	Fanuc Ltd., Fuji complex, Japan.
Process	Machining
Products	Parts for machine tools and robots.
Total cost	$32 million
Comments on hardware	30 machining centers, consisting of computer controlled machine tools loaded and unloaded by robots, along with material handling robots, monitors and programmable controllers.
Employment	About 100.
Comparison	The plant is about five times more productive as its conventional counterpart. It would probably take 10 times the capital investment and 10 times the labor force to produce the same output with a conventional plant.

Source for Tables 4-5 and 4-6: Bylinski (1983).

days to 16 hours. The hardware cost of the system was about $16 million. GE management reports a 240 percent increase in total factor productivity as a result of installing the new system, although they do not specify over what time period this increase has been (or will be) realized.[22]

Two large Japanese flexible manufacturing systems are described in [Bylinski 83]. An overview of the FMS at the Yamazaki Machinery Works, in Nagoya, Japan, is given in Table 4-5. Yamazaki management claims that a conventional system with manually operated stand-alone machine tools would require nearly 15 times as many workers and four times as many machines. They calculate that over time, the new FMS will be 15 times more profitable than a conventional plant its size. The company estimates that over five years of operation, the plant will produce after-tax profits of $12 million compared with $800,000 for a conventional plant that size. The hardware cost of the system is estimated to be $20 million.[23] An overview of the FMS at Fanuc, Ltd., in Fuji, Japan, is given in Table 4-6. Fanuc management reports that a conventional system would require ten times the number of workers and capital investment to produce the same output as the new FMS. They claim that the plant is about five times more productive than its conventional counterpart would be. The hardware cost are estimated at $32 million.[24]   Yoshikawa, Rathmill and Hatvany (1981: 11-12) provide a more detailed description of the Fanuc plant:

> Recently, there has been a remarkable trend toward having systems unmanned during the night shifts. Fanuc has constructed a new factory in Fuji that does this. This plant has 29 cell-like work stations. Seven of them are equipped with robots; 22 are equipped with automatic pallet changers with pallet pools. These stations are connected by unmanned vehicles guided by electromagnetic or optical methods. The plant has two automatic warehouses, one for materials and another for finished parts and subassemblies. The vehicles transport the materials from the warehouse to the unmanned machining stations. Robots or automatic pallet changers load materials onto the stations from the vehicles. Finished parts are transferred again automatically by the vehicles to the second warehouse.
>
> The plant has an assembly floor, where workers work during the day. Transportation between the parts warehouse and this floor is also by unmanned vehicles. In the daytime, 19 workers are working around machining stations, mainly for palletizing, and 63 workers are in the assembly section. Thus, 82 workers are in the plant during the day, but at night

---

[22]Performance comparisons were obtained directly from GE personnel during a visit to the facility [Miller 83b]. These comparisons are corroborated in [Manufacturing Engineering 83], pages 66-67, and in [Bylinski 83].

[23]Tom Klotz, sales manager of Mazak (the U.S. affiliate of Yamazaki) claims that the FMS also reduced floor space requirements by over half, from 70,000 square feet to 30,000 square feet. Klotz also reports the following benefits in another Yamazaki plant (The Seiko Works). Number of machines: reduced from 54 to 24. Number of workers (three shift total) : reduced from 105 to 14. Floor space: reduced by one third. In process time: reduced from 13 weeks to 3 days [Klotz 84].

[24]While visiting the United States in the summer of 1983, Professor Yukio Hasegawa, an experienced Japanese robotics researcher from Wasada University in Japan, corroborated the information on the two Japanese systems reported in [Bylinski 83].

there is only one. The assembly floor is closed, and the machining floor is operated without any workers. Every station is equipped with a monitoring device with a TV camera, and one person sits in the control room to monitor all the working stations. He observes the working status of stations through the camera, without touring the factory. The monitoring device also records the spindle motor current, calculating the cutting force and time to judge cutting conditions.

Insensate (Level I) robotic manipulators are sometimes used in the large FMS to load and unload parts, as indicated above. One robot is used in the GE system to move cutting tools into and out of a machine tool. In general, robotic manipulators do not play a major role in these types of systems. The primary reason is that the basic transformation task, the machining, is already automated by the machine tool. Most of the other tasks are related to material handling and there are often alternative forms of automation to move parts and tools. For example, automated guided vehicles are often used to transport parts from the loading station to the machine tool. There are also many ways of moving a tool from a tool rack to the tool storage compartment in the machine tool. Robots are used more extensively in manufacturing cells, which are smaller than the large scale FMS. If the machines in the cell are properly arranged, it is typically less expensive to use one or several robots to move materials from one machine to another than to use other types of material handling equipment.

An analogy can be made between the first generation of flexible manufacturing systems used for machining and the Level I industrial robot. Like the Level I robot, the flexible manufacturing systems described above are essentially insensate.[25] As a result, much of the cost of the complete system is a result of having to engineer the domain to eliminate all possible sources of unexpected variation out of the system.

Both the Yamazaki and Fanuc plant make parts for machine tools and robots. Given the examples of cost savings cited earlier, one would expect a reduction in the cost of machine tool and robot components manufactured on flexible manufacturing systems. A reduction in the cost of components would have a substantial impact on total robot cost, given that for widely used models, the cost of components comprises three quarters of the total manufacturing cost [Engelberger 83]. Thus, while robots do not play a major role in the large FMS for machining, these systems will eventually have a substantial impact on the cost (and therefore on the use) of robots.

---

[25]These systems may use bar code technology to identify which part has been placed on a pallet. These systems do not yet use machine vision or taction or other forms of external sensing to detect the position and orientation of parts, or to inspect parts.

4.3 Developments in Flexible Manufacturing Systems

Three developments in flexible manufacturing systems are discussed in this section:

- The integration of multiple FMS for machining in one plant

- The use of FMS in manufacturing processes other than machining

- The addition of artificial intelligence and sensor-based-systems to FMS

In the flexible manufacturing systems described in Tables 4-3 through 4-6, one FMS was installed in the factory. The largest of the systems includes about 30 machine tools, while many of the systems include fewer than 20. Using available technology, it is possible to build systems of flexible manufacturing systems with larger numbers of workstations. Mazak Corp., the American affiliate of Yamazaki, has recently built a factory in Florence, Kentucky with five flexible manufacturing systems for machining. Each system produces a different family of parts. In total, there are 60 computer numerically controlled machines in the five systems, and a central supervisory system can communicate with all 60 machines. The hardware cost of the plant, which includes assembly as well as machining, is estimated at over $ 200 million. This does not include planning and engineering costs [Klotz 84].[26]

Flexible manufacturing systems are now being tested and marketed for applications other than machining. Unimation, a subsidiary of Westinghouse, is marketing flexible manufacturing systems for sheet metal fabrication and for wire harness assembly. Researchers at the Carnegie-Mellon Robotics Institute have developed a FMS for open-die forging [Wright, Bourne, et al 82] and for electronics assembly [Sanderson and Perry 83]. Both systems have been transferred to industry. Numerous examples of flexible manufacturing systems for assembly are found in the literature. Westinghouse developed an FMS to assemble small electric motors, but the system was only used for experimental purposes [Stauffer 83]. IBM has developed an FMS to assemble display writers [Stauffer 82]. Other flexible manufacturing systems for assembly are discussed in [Nevins and Whitney 79], [Warnecke and Walther 83], [Kanesaka 83], [Freeman 83]. A hybrid flexible assembly system, using both robots and humans, is discussed in [Lien 83]. All of these systems use robotic manipulators for parts transfer. All of the assembly systems also use robots to perform the basic assembly tasks.

Within the next several years, FMS will be used in a wide range of manufacturing processes. The next logical step would be to integrate the flow of material from one FMS to another. Early attempts to

---

[26]Klotz estimates the design required hundreds of thousands of engineering hours.

integrate material flow across processes in flexibly automated systems have already been made. In the Fanuc plant, automated guided vehicles move material from an automated warehouse to the machining FMS. In the Maze plant, the automatic carts used to transport parts through the machining FMS are also used to move parts to and from manual assembly areas.

The flexible manufacturing systems for open-die forging and electronics assembly developed at the CMU Robotics Institute make extensive use of new technologies: computer programming techniques that emerged from the discipline of artificial intelligence (AI) and external sensing. The open-die forging system is controlled by a new rule based computer language [Bourne 82] that is based on the AI programming technique of production systems. One advantage of this rule based language is that it simplifies the process of designing a control system that is robust and resilient to unanticipated errors in the system. Conventional programming languages (Fortran or Pascal or APT) are procedural. There is a sequential flow of control from one statement in the program to the next. At any random statement in the program, the proposed task to be performed is critically dependent on the previous task performed. If one task were not carried out because of a machine or program error, a subsequent task may be adversely affected or not enacted. If the control system is written in a procedural language, it is difficult to prevent a break in any link of the chain from breaking the whole control system. With a rule based system, which is nonprocedural, the flow of control is determined by detecting which conditions hold and then by executing rules which specify what to do given the particular set of conditions. This is in contrast to following a predetermined sequence. If a machine or part of the program fails, the control program will detect that a new set of conditions now hold, and then execute the rules that govern what to do under those conditions.

A second advantage of rule base systems is that they can easily accommodate changes in the physical system. Additions of new equipment do not require major revisions of the control program as they would with control systems written with conventional programming languages. A third advantage is that the rule based structure makes it possible to control unwanted interactions, such as two robots moving to the same place at the same time. Once programs written in conventional procedural languages get very large, it is nearly impossible to anticipate all possible system interactions. Events such as collisions could happen under a freak set of circumstances. The rule based language makes it possible to add rules that preclude certain interactions from ever happening, even if the reasons giving rise to these interactions are unanticipated. This AI approach to controlling a system, which is robust to failure, easily modified and extended, and protective against unanticipated harmful interactions among system components, is one reason why an FMS could be applied to this new application. The FMS described in [Wright, Bourne, et al 82] also incorporates a machine vision system which is used to inspect the shape of the part that emerges from the open-die forge.

The electronics assembly cell described in [Sanderson and Perry 83] makes extensive use of external sensing systems to detect visual and force information. In addition, the system makes extensive use of techniques that have emerged from AI to process and interpret the sensory information. Vision is used to guide a robotic manipulator in acquiring an unoriented electronic component. Both visual and force sensing are used to guide the manipulator in inserting electronic components into a printed wire board. These technologies make it possible to build an FMS for the assembly of small electronic components. Also, the sensor based technologies make it possible to combine processes, such as assembly and inspection, into one system.

Applications of sensor-based systems and AI are also being developed to improve the capabilities of the flexible manufacturing systems for machining. Vendors are experimenting with machine vision systems to verify if the correct part is being processed on a particular machine, to inspect parts after they have been machined, and to inspect tools to see if they need replacement. Also , vision is being tested to eliminate the need for precisely positioning each part and each tool in the machining system.[27]

These examples indicate that the range of processing tasks that can be integrated into a flexible manufacturing system will expand as the technologies of AI and external sensing improve and become part of FMS technology. There is also indication that these technologies will improve the performance of flexible manufacturing systems in areas where they are already being used.

4.4 The Economic Importance of Automated Manufacturing for Mid-Volume/Mid-Variety Production

A knowledge of the relative importance of batch and mass production in manufacturing provides a backdrop for appreciating the potential significance of automation designed for mid-volume/mid variety production. Miller's (1983) estimate of the percent of value added in the metalworking industries (SIC 34-37) that is generated from small batch, medium batch, large batch and mass production is given in Table 4-7. According to these estimates, nearly 75 percent of the value added in these industries is generated from small, medium and large batch production, with only 25 percent generated by mass production. Nearly 50 percent of the value added is generated from medium and large batch production.

The Kearney & Trecker volume/variety chart (Figure 4-2) indicates that the flexible manufacturing

---

[27] These experimental applications of machine vision systems in machining FMS were provided by representatives from Cincinnati Milacron and Mazak.

systems are economically justifiable in the mid-volume/mid-variety region. Suppose we roughly figure that this region corresponds to Miller's designation of medium and large batch production, which, according to Table 4-7, accounts for over 50 percent of the value added in metalworking. This suggests that nearly 50 percent of the processing requirements in SIC 34-37 could eventually be carried out on some type of flexible manufacturing system. Given the available information on cost and labor savings cited earlier, it appears that the widespread use of FMS would have a great impact on the cost of manufactured goods as well as on the number of people required for production.

Table 4-7: Distribution of Value Added by Mode of Production, SIC 34-37

Region and Mode of Production	Major SIC Groups				Total, 34-37
	34	35	36	37	
PERCENT OF VALUE ADDED FOR INDUSTRIES IN SAMPLE					
Custom and Small Batch	1.1	41.0	58.5	30.6	31.5
Mid-Batch	28.3	42.5	30.2	6.9	26.4
Large Batch	45.5	16.5	3.1	2.9	16.4
Mass	25.1	0.0	8.2	59.6	25.7
Sample coverage of value added in all industries	94.2	95.8	51.2	97.4	86.0

SIC: standard industrial classification
SIC code   Major Industry Group Name

34: Fabricated metal products
35: Machinery, except electrical machinery
36: Electrical equipment and machinery
37: Transportation equipment

Source: Miller (1983: 176)

I. Examples of the Use of Vision Systems in Industry

User	Sensor-Controlled Manipulation Applications	Vendor
Hitachi	Robot-vision system which detects holes for assembly. <u>Includes</u>: solid state optical sensors, CCD-type TV camera mounted on robot arm.	?
Western Electric Atlantic Plant	Color-sorting of telephone receiver caps into bins. (6500/hr). Uses photo diodes and color filters. 99.9% accuracy.	?
G-M Warren, Mich.	Stacks random mix of pre-taught parts, Uses light stripe, PUMA robot system, 3 DEC LSI 11's, video camera and VAL programming language.	

User	Inspection Applications	Vendor
Unknown	Automatic inspection of welded automobile wheel hubs. Checks for integrity of structure.	MIC
Unknown	Off-line Floppy-disk jacket inspection, manually operated. Checks dimensions.	MIC
Unknown	Automatic identification of various models of electrical circuit breakers on a conveyor belt. Checks product type.	MIC
Unknown	Automatic inspection of ceramic supports for cathode ray tubes Checks for dimensions.	MIC
Unknown	Automatic inspection of cathode ray tube displays. Checks for integrity of features.	MIC
Unknown	Automatic inspection of spark plugs on a moving conveyor belt. Checks dimensions.	MIC

User	Inspection Applications	Vendor
Unknown	Automatic Fluoroscopic inspection of cut and welded parts for stress cracks. Checks integrity of internal structure and dye is used to make flaws flouresce	MIC
Unknown	Automated inspection of glass CRT Necks, uses a UV light source to image internal defects. Checks integrity of internal structure.	MIC
Unknown	Automatic inspection of plastic sutures. Checks integrity and dimensions.	MIC
Unknown	Automatic inspection of automotive wheel hubs for conformance to forged dimensions prior to subsequent machining operations. Checks integrity.	MIC
Unknown	Inspection of valve bodies for automatic transmission. Vision is interfaced with robot. Software mask examines internal details. Exact positioning is required. Checks dimensions of a single type of product.	MIC
Unknown	Automatic inspection systems for precision components. Vision is interfaced with a robot. Checks dimensions.	MIC
Unknown	Gray-scale imaging system for paper-cup packaging. Checks for number of cup lips.	Octek, Inc.
Cummins	Inspection of engine blocks. Uses light striping.	RVS
Hitachi Japan	Automatic Reticle System (ARI) which uses semiconductor photomask inspection for products	?

User	Sensor-Controlled Manipulation Applications	Vendor
Westinghouse Winston-Salem, NC	Robot - vision system to pick & place and inspect turbine blades.	In-house with C-MU
G-M	Consight I Vision-Robot System Picks randomly placed parts off of moving conveyor.	In-house
General Motors Janesville, Wis.	Light-stripe sensor or Robot wrist (Robo-Sensor) for welding of J-cars.	RVS
Lockheed - Georgia	Robot-based assembly of cargo air-craft using the Robo-Sensor. Includes: light projector, wrist-mounted camera, computer, software. Hardware cost: $35 - $70,000	RVS
Lockheed - Georgia	Assembly of internal part for C-130 Hercules Cargo aircraft.	RVS
Lockheed - Georgia	Riveting cell for aircraft assembly.	MIC
Kawasaki	Laser-based vision system used for path correction in arc welding of motorcycle parts.	?
Matushita Electric Co., Japan	Robot-vision system for vacuum.	?
Texas Instruments Lubbock, Texas	Calculator assembly lines with robots.	?
United Technologies, Sikorsky Aircraft	Drilling and Riveting for aircraft assembly. Includes: ASEA 1Rb-60 robot mounted on track, DEC LSI 11/23 as system controller, various contact and vision sensors.	?

User	Inspection Applications	Vendor
Delco Electronics Kokomo, Indiana	Determines chip position and orientation, inspects chip structurally, allows for proper alignment of test probes with chip contacts.	In-house
Honeywell	Robot vision station for solder joint inspection of printed circuit boards. Uses TV camera for 2-D image, PUMA 560 robot, Autovision II, Apple II, plus micro-computers.	In-house

Combined Sensor-Controlled Manipulation and Inspection Application

User	Application	Vendor
Automatix Corp. Billerice, Ma.	Robot-vision system for assembly and inspection of keyboard arrays. Uses the Cybervision Assembly Station and the Autovision II processor, with the AID 600 robot and AI 32 controller.	Automatix

Key to Vendor Abbreviations

CMU: Carnegie-Mellon University
G-M: General Motors
MIC: Machine Intelligence Corp.
RVS: Robot Vision System
? :   Vendor of system not specified in literature.

II. Full Taxonomy of Problem Classes for Part Identification and Location

Part of the taxonomy of problem classes for part identification and location was shown in Table 3-7 of the main report. The purpose of this appendix is to include the full taxonomy. Important abbreviations and terms used in the table are defined below. More detailed discussions of these terms are found in section 3.5.

BI   Binary image representation of an image

ED   Edge detection (gray scale) representation of an image

GFA   Global feature image analysis

LFA   Local feature image analysis

Trick   A technique for solving a problem under a particular set of conditions by exploiting some special structure. If the set of conditions were changed, the technique would not work. For example, BI/GFA can not. in general be used to distinguish parts that are touching. However, if we know we are counting screw nuts, and that each nut only has one hole. we could distinguish separate parts by the "trick" of counting the number of holes, even if the nuts are touching.

(Image Representation/Image Analysis)

Conditions	Tasks:	
	Counting Number of Parts	Locating Position of Parts
1] IDENTICAL PARTS		
a] Rigid parts.		
Each part is in the same predetermined stable position.		
i] separated		
high contrast	BI/GFA	BI/GFA
low contrast	ED/GFA	ED/GFA
ii] touching (without overlapping or tilting)		
high contrast	BI/GFA, Trick	BI/LFA
low contrast	ED/GFA,Trick	ED/LFA
iii] overlapping and/or only slightly tilting (< 20 degrees)		
high contrast	BI/GFA, Trick maybe	BI/LFA
low contrast	ED/GFA, Trick maybe	ED/LFA
iv] Tilting		
high contrast	Part cannot be in known position	
low contrast	" "	

(Image Representation/Image Analysis)

Conditions	Tasks:	
	Counting Number of Parts	Locating Position of Parts

b] Rigid Parts.
Each part is in any one of n distinct stable positions.
(e.g. in any one of two, three, or four stable positions)
Each stable position has a unique silhouette.

	Counting	Locating
i] separated		
high contrast	BI/GFA	BI/GFA
low contrast	ED/GFA	ED/GFA
ii] touching		
high contrast	BI/LFA	BI/LFA
low contrast	ED/LFA	ED/LFA
iii] overlapping and/or slightly tilting		
high contrast	BI/LFA	BI/LFA
low contrast	ED/LFA	ED/LFA
iv] tilting		
high contrast	BI/3D	BI/3D
low contrast	ED/3D	ED/3D

c] Rigid Parts.
In a given orientation,
each part may appear in a continuum of stable positions
(e.g. a scissors partly open)

	Counting	Locating
i] separated		
high contrast	BI/GFA	BI/LFA
low contrast	ED/LFA	ED/LFA
ii] touching		
high contrast	BI/LFA	BI/LFA
low contrast	ED/LFA	ED/LFA
iii] overlapping and/or slightly tilting		
high contrast	BI/LFA	BI/LFA
low contrast	ED/LFA	ED/LFA
iv] tilting		
high contrast	BI/3D	BI/3D
low contrast	ED/3D	ED/3D

(Image Representation/Image Analysis)

Conditions	Tasks:	
	Counting Number of Parts	Locating Position of Parts

d] Non rigid string-like parts (perimeter is invariant)
  (e.g. rubber bands, hoses, necklaces)

i] separated		
high contrast	BI/GFA	ED/LFA
low contrast	ED/GFA	ED/LFA
ii] touching		
high contrast	BI/GFA,Trick	BI/LFA
low contrast	ED,GFA, Trick	ED/LFA
iii] overlapping and/or slightly tilting		
high contrast	ED/LFA	ED/LFA
low contrast	ED/LFA	ED/LFA
iv] tilting		
high contrast	BI/3D	BI/3D
low contrast	ED/3D	ED/3D

e] Nonrigid non-string-like parts (no invariant shape features)
  (e.g. loose-link chain, crumbled baggied, crumpled shirt or towel)

i] separated		
high contrast	BI/GFA	LFA, maybe
low contrast	ED/GFA	ED/LFA, maybe
ii] touching		
high contrast	No Solution	No Solution
low contrast	No Solution	No Solution
iii] overlapping and/or slightly tilting		
high contrast	No Solution	No Solution
low contrast	No Solution	No Solution
iv] tilting		
high contrast	No Solution	No Solution
low contrast	No Solution	No Solution

(Image Representation/Image Analysis)

Conditions	Tasks:	
	Counting Number of Parts	Locating Position of Parts

2] 1 CATEGORY OF PARTS, EACH PART WITH UNIQUE SET OF SILHOUETTES
(Parts identical (very smiliar) in shape, varying in size. e.g. nuts of different sizes)

a] Rigid Parts.
Each part is the same predetermined stable position

	Same as case 1 b][1]	Same as case 1 b][1]

b] Rigid Parts.
Each part is in any one of n distinct stable positions

	Same as case 1 b]	Same as case 1 b]
	Longer matching time[2]	Longer matching time[2]

c] Rigid Parts.
Each part may appear in a continuum in stable positions

	Same as 1 c]	Same as 1 c]
	Longer matching time[2]	Longer matching time[2]

d] Non rigid string-like parts

	Same as case 1 d]	Same as case 1 d]
	Longer matching time[2]	Longer matching time[2]

e] Non rigid non-string-like parts

	Same as case 1 e]	Same as case 1 e]
	Longer matching time[2]	Longer matching time[2]

3] N CATEGORIES OF PARTS,
EACH PART TYPE WITH UNIQUE SET OF SILHOUETTES

Same as case 2. conditions a], b], c], d], and e].
Matching time will take even longer.[3]

---

1) In case 1 b], there is only one type of part but it can appear in any of n distinct stable positions. From an image analysis standpoint, recognizing the part in each of its n distinct stable positions is the same as recognizing n different types of parts. Thus, the case of recognizing different parts is already covered in case 1 b].

2) Matching times are longer here than in the comparable condition in case 1 because there are more part models to compare the image against.

3) As the number of different parts to compare grows large. matching time may become too slow for a particular commercial use.

(Image Representation/Image Analysis)

Conditions	Tasks:	
	Counting Number of Parts	Locating Position of Parts

4] MIX OF PARTS WITH THE SAME SILHOUETTE

Rigid Parts:
a] Each part is in the same predetermined stable position

i] separated		
high constast	ED/GFA	ED/GFA
low contrast	ED/GFA	ED/GFA
ii] touching		
high contrast	ED/LFA	ED/LFA
low contrast	ED/LFA	ED/LFA
iii] overlapping or slightly tilting		
high contrast	ED/LFA	ED/LFA
low contrast	ED/LFA	ED/LFA
iv] tilting		
high contrast	ED/3D	ED/3D
low contrast	ED/3D	ED/3D

Summary of Results of Taxonomy of Problem Classes for Part Identification and Location

The implications of the table are summarized in section 3.5. The key point is that as of today, there is only level of vision technology with proven industrial experience, BI/GFA. The table indicates that this technology is suitable for only a very small proportion of the problems of object identification and location. Edge detection (gray scale) systems are just becoming commercially available. While this added capability will increase the range of identification and location problems that can be solved, it will not make a dramatic increase. The table indicates that what is really needed is local feature analysis, or in general. a means of analyzing spatial relationships among parts of an image.   The combination of edge detection and local feature analysis would solve all problems in the table with the exception of those requiring general purpose three dimensional analysis.

References

[Agin 80]   Agin, Gerald J.
                   Computer Vision Systems for Industrial Inspection and Assembly.
                   Computer (IEEE Computer Society) 13:11-20, May, 1980.

[Albertson 83]   Albertson, Paul.
                   Verifying Robot Performance.
                   Robotics Today 5(5):33-36, October, 1983.

[Asada and Kanade 81]
                   Asada, Haruhiko and Takeo Kanade.
                   Design of Direct-Drive Mechanical Arms.
                   Technical Report CMU-RI-TR-81-1, Carnegie-Mellon University Robotics Institute,
                       April, 1981.

[Ayres 84]   Ayres, Robert U.
                   *The Sensory Feedback Dependence of Assembly Tasks as a Measure of
                       Comparative Advantage of Human Workers vis-a-vis Sensor-Based Robots.*
                   Technical Report, Engineering and Public Policy Department, Carnegie-Mellon
                       Univesity, February, 1984.

[Ballard and Brown 82]
                   Ballard, Dana H. and Christopher M. Brown.
                   Computer Vision.
                   Prentice-Hall, Inc., 1982.

[Bolles 81]   Bolles, Robert C.
                   An Overview of Applications of Image Understanding to Industrial Automation.
                   Technical Note 242, SRI International, May, 1981.

[Bolles and Cain 83]
                   Bolles, R.C. and R. A. Cain.
                   Recognising and Locating Partially Visible Objects: The Local-feature-Focus
                       Method.
                   In Pugh, Alan (editor), *International Trends in Manufacturing Technology.* Volume
                       1: *Robot Vision*, pages 43-82. IFS Publications, London, England, 1983.

[Bourne 82]   Bourne, David A.
                   A Numberless, Tensed Language for Action Oriented Tasks.
                   Technical Report CMU-RI-TR-82-12, Carnegie-Mellon University Robotics Institute,
                       October, 1982.

[Bylinski 83]   Bylinski, Gene.
                   The Race to the Automatic Factory.
                   Fortune 107(4):52-64, February 21, 1983.

[Carlisle 83]   Carlisle, Brian.
                   Private Communiction.
                   Phone conversation with Steven Miller, Pittsburgh, PA, August, 1983.
                   1983.

[Coiffet 83a]   Coiffet, Phillipe.
                 Robot Technology. Volume 1: *Modelling and Control*.
                 Prentice-Hall, Inc., 1983.

[Coiffet 83b]   Coiffet, Phillipe.
                 Robot Technology. Volume 2: *Interaction with the Environment*.
                 Prentice-Hall, Inc., 1983.

[Crowley 82]   Crowley, James L.
                 A Representation for Visual Information.
                 Technical Report CMU-RI-TR-82-7, Carnegie-Mellon University Robotics Institute,
                     November, 1982.

[Crowley 84]   Crowley, James L.
                 Machine Vision: Three Generations of Commercial Systems.
                 Technical Report CMU-RI-TR-84-1, Carnegie-Mellon University Robotics Institute,
                     January, 1984.

[Draper Lab 83] Draper Lab Staff.
                 *Flexible Manufacturing System Handbook, Volume II: Description of the
                     Technology*.
                 Technical Report 12703, Charles Stark Draper Lab, 1983.
                 Prepared for the US Army Tank-Automotive Command, Warren, MI.

[Dupont-Gatelmand 83]
                 Dupont-Gatelmand, Catherine.
                 A Survey of Flexible Manufacturing Systems.
                 Journal of Manufacturing Systems 1(1):1-16, 1983.

[Eikonix Corp. 79] Eikonix Corp.
                 Technology Assessment: The Impact of Robots (Final Report).
                 Technical Report EC/2405801-FR-1, Eikonix Corp., September 30, 1979.
                 NSF grant number ERS-76-00637.

[Engelberger 80] Engelberger, Joseph F.
                 Robotics in Practice.
                 American Management Association, New York, 1980.

[Engelberger 83] Engelberger, Joseph F.
                 Private Communiction.
                 Interview with Steve Miller, Danbury, Conn, November, 1983.
                 1983.

[Freeman 83]   Freeman, G.T.
                 P.C.B. Automation in the 1980's.
                 In *15th CIRP Seminar on Manufacturing Systems :Assembly Automation*, pages
                     329-334. International Institute for Production Engineering Research (CIRP),
                     June, 1983.

[Gonzalez and Safabkhsh 83]
    Gonzalez, R.C. and R Safabakhsh.
    Computer Vision Techniques for Industrial Inspection and Robot Control: An
        Overview.
    In Lee, C.S.G., R.C. Gonzalez and K.S. Fu (editor), *Tutorial on Robotics*, pages
        300-324. IEEE Computer Society Press, 1983.

[Hunt 83]   Hunt, V. Daniel.
                Industrial Robotics Handbook.
                Industrial Press Inc., New York, 1983.

[Hunt and Hunt 83]
    Hunt, H. Allan and Timothly L. Hunt.
    Human Resource Implications of Robotics.
    W.E. Upjohn Institute for Employment Research, Kalamazoo, MI, 1983.

[Jablonowski 80] Jablonqwski, Joseph.
                Aiming for Flexibility in Manufacturing Systems.
                American Machinist 124(3):167-182, March, 1980.

[Jarvis 82]   Jarvis, John F.
                Research Directions in Industrial Machine Vision: A Workshop Summary.
                Computer (IEEE Computer Society) 15(12):30-37, December, 1982.

[Kanade 83]   Kanade, Takeo.
                Visual Sensing and Understanding: The Image Understanding Point of View.
                Computers in Mechanical Engineering 1(4):59-70, April, 1983.

[Kanade and Reddy 83]
    Kanade, Takeo and Raj Reddy.
    Computer vision: the Challange of Imperfect Inputs.
    IEEE Spectrum 20(11):88-91, November, 1983.

[Kanesaka 83]   Kanesaka, S.
                Flexible Micro Coil Assembly System CAS-20.
                In *15th CIRP Seminar on Manufacturing Systems :Assembly Automation*, pages
                    308-328. International Institute for Production Engineering Research (CIRP),
                    June, 1983.

[Kearney & Trecker 82]
    Kearney & Trecker Corp., Special Products Division.
    Manufacturing Systems Applications Workbook.
    Kearney & Trecker Corp., Milwaukee, WI, 1982.

[Kelly and Birk, etal 82]
    Kelley, Robert B., John R. Birk, A. S. Martins and Richard Tella.
    A Robot System Which Acquires Cylindrical Workpieces from Bins.
    IEEE Transactions on Systems, Man, and Cybernetics SMC-12(2):204-213,
        March/April, 1982.

[Kinnucan 83]   Kinnucan, Paul.
                Machines That See.
                High Technology 3(4):30-37, April, 1983.

[Klahorst 83a]   Klahorst, H. Thomas.
                  How to Plan Your FMS.
                  Manufacturing Engineering 91(3):52-54, September, 1983.

[Klahorst 83b]   Klahorst, Thomas.
                  Interview with Steve Miller.
                  Private research notes on visit to Kearney & Trecker, Corp., October, 1983.
                  1983.

[Klotz 84]   Klotz, Thomas.
                  The Flexible Manufacturing Factory.
                  Presentation made at the Flexible Manufacturing Systems Seminar sponsored by
                      the Robot Institute of America, January 26, 1984, Pittsburgh, Pa.
                  1984.

[Levine 83]   Levine, S.S.
                  Application of RVSI's 3-d vision systems into industrial robotic manufacturing and
                      inspection operations.
                  In Zimmerman, N.J, and A. Oosterlinck (editor), *Industrial Applications of Image
                      Analysis*, pages 321-331. D.E.B. Publishers, Pijnacker, The Netherlands, 1983.

[Lien 83]   Lien, T.K.
                  Flexible Control of an Integrated Automatic-Manual Assembly Line.
                  Manufacturing Systems: Proceedings of the CIRP Seminars 12(3):218-226, 1983.

[Luh 83]   Luh, J.Y.S.
                  An Anatomy of Industrial Robots and Their Controls.
                  In Lee, C. S. G., R. C. Gonzalez and K.S. Fu (editor), *Tutorial on Robotics*, pages
                      5-25. IEEE Computer Society Press, 1983.

[Makhlin and Tinsdale 83]
                  Makhlin, A.G. and G.E. Tinsdale.
                  Westinghouse grey scale vision system for real-time control and inspection.
                  In Pugh, Alan (editor), *International Trends in Manufacturing Technology*. Volume
                      1: *Robot Vision*, pages 345-354. IFS Ltd., London, England, 1983.

[Manufacturing Engineering 83]
                  Manufacturing Engineering Magazine.
                  Special Issue on Flexible Manufacturing Systems.
                  Manufacturing Engineering 91(3), September, 1983.

[Miller 83a]   Miller, Steven M.
                  *Potential Impacts of Robotics on Manufacturing Costs in the Metalworking
                      Industries.*
                  PhD thesis, Carnegie-Mellon University, May, 1983.

[Miller 83b]   Miller, Steven.
                  The GE Erie Facility.
                  Private research notes on visit to the GE Erie facility, October, 1983.
                  1983.

[Nevins and Whitney 79]
        Nevins, J.L. and D.E. Whitney.
        Robot Assembly Research and Its Future Applications.
        In Dodd, Geoorge D. and Lothar Rossol (editor), *Computer Vision and Sensor-Based Robots*, pages 275-321. Plenum Press, New York, 1979.

[Nitzan 81]   Nitzan, Divad.
        Assessment of Robotics Sensors.
        In Pugh, Alan (editor), *Proceedings of the 1st International Conference on Robot Vision and Sensory Control*, pages 1-11. IFS Publications, London, England, 1981.

[Paul 81]   Paul, Richard P.
        The MIT Press Series in Artificial Intelligence: Robot Manipulators: Mathematics, Programming and Control.
        MIT Press, Cambridge MA, 1981.

[Pugh 83]   Pugh, Alan.
        Second Generation Robotics.
        In Pugh, Alan (editor), *International Trends in Manufacturing Technology*. Volume 1: *Robot Vision*, pages 3-10. IFS Publications, London, England, 1983.

[Rosen 79]   Rosen, C.A.
        Machine Vision and Robotics: Industrial Requirements.
        In Dodd, George G. and Lothar Rossol (editor), *Computer Vision and Sensor-Based Robots*, pages 3-18. Plenum Press, New York, 1979.

[Rossol 83]   Rossol, Lothar.
        Computer Vision in Industry.
        In Pugh, Alan (editor), *International Trends in Manufacturing Technology*. Volume 1: *Robot Vision*, pages 11-18. IFS Publications, London, England, 1983.

[Sanderson 83]   Sanderson, Arthur C.
        Robot Vision and Industrial Automation.
        In Zimmerman, N. and A. Oosterlinck (editor), *Industrial Applications of Image Analysis*. D.E.B. Publications, Pijnacker, The Netherlands, 1983.

[Sanderson and Perry 83]
        Sanderson Arthur C and George Perry.
        Sensor-Based Robotic Assembly Systems: Research and Applications in Electronic Manufacuring.
        Proceedings of the IEEE 71(7):856-871, July, 1983.

[Sanderson and Weiss 83]
        Sanderson, Arthur C. and Lee E. Weiss.
        Adaptive Visual Servo Control of Robots.
        In Pugh, Alan (editor), *International Trends in Manufacturing Technology*. Volume 1: *Robot Vision*, pages 107-116. IFS Publications, London, England, 1983.

[Sanderson, R. 83]
        Sanderson, Ronald J.
        Machine Vision Systems: A Summary and Forecast.
        Tech Tran Corporation, Naperville, Illinois, 1983.

[Stauffer 82]   Stauffer, Robert N.
                  IBM Advances Robotic Assembly in Building a Word Processor.
                  Robotics Today 4(5):19-23, October, 982.

[Stauffer 83]   Stauffer, Robert N.
                  Westinghouse Advances the Art of Assembly.
                  Robotics Today 5(1):33-36, February, 1983.

[Tech Tran Corp. 83]
                  Tech Tran Corp.
                  Industrial Robots: A Summary and Forecast.
                  Tech Tran Corp., Naperville, IL, 1983.

[Warnecke and Walther 83]
                  Warnecke H.J. and Walther, J.
                  Programmable Assembly Station for the Automatic Assembly of Car Aggregates.
                  In *15th CIRP Seminar on Manufacturing Systems :Assembly Automation*, pages
                      290-299. International Institute for Production Engineering Research (CIRP),
                      June, 1983.

[Wright 83]   Wright, J.M.
                  Vision of the Future.
                  1983.
                  Term paper prepared for a class on robot technology.

[Wright, Bourne, et al 82]
                  Wright, Paul K., David A. Bourne, J.P. Colyer, J.A.E. Isasi, and G.S. Schatz.
                  A Flexible Manufacturing Cell for Swaging.
                  Mechanical Engineering Magazine :76-83, October, 1982.

[Yoshikawa, Rathmill, and Hatvany 81]
                  Yoshikowa, Hiroyuhi Keith Rothmill, and Josef Hatvany.
                  Computer-Aided Manufacturing: An International Comparison.
                  Technical Report, National Academy Press, 1981.
                  Sponsored by the National Research Council.

[Zimmerman and Oosterlinck 83]
                  Zimmerman, N. and A. Oosterlinck (editors).
                  Industrial Applications of Image Analysis.
                  D.E.B. Publishers, Pijnacker, The Netherlands, 1983.

[Zimmerman, Van Boven and Oosterlinck 83]
                  Zimmerman, N.J., G. J. R. Van Boven and A. Oosterlinck.
                  Overview of Industrial Vision Systems.
                  In Zimmerman, N. and A. Oosterlinck (editor), *Industrial Applicatins of Image
                      Analysis*, chapter 10, pages 203-229. D.E.B. Publications, Pijnacker, The
                      Netherlands, 1983.

4. Impacts of Robotic and Flexible Manufacturing Technologies on Manufacturing Costs and Employment

Steven M. Miller[a]

Graduate School of Industrial Administration
Carnegie-Mellon University
Pittsburgh, PA 15217

March 13, 1984

This is a condensed version of the Ph.D thesis, *Potential Impacts of Robotics on Manufacturing Costs in the Metalworking Industries,* submitted to the Engineering and Public Policy Department, Carnegie-Mellon University, in May, 1983. The original research was supported by the Carnegie-Mellon Robotics Institute, the Engineering and Public Policy Department, and the Carnegie-Mellon Program on the Social Impacts of Information and Robotic Technologies. I am especially grateful to Robert Ayres, committee chairman, and to Larry Westphal for serving as advisors to the research which lead to the Ph.D. thesis. The present condensed version was supported by the National Science Foundation, Division of Policy Research and Analysis, grant # PRA-8302137 826.

a: Assistant professor, Engineering and Public Policy and Industrial Administration.

Abstract

The issues analyzed in this paper are the extent to which unit costs and production labor requirements might be reduced in manufacturing industries if there is more widespread use of industrial robots and flexible manufacturing systems. The analysis is reported in detail in [Miller 83]. These issues are analyzed from two different perspectives. The technological focus of the first perspective is narrowly confined to the use of robotic manipulators. It is assumed that robotic manipulators will be "retrofitted" into existing production facilities without making major changes in the organization of production within the factory, other than modifying individual work stations so that robots can replace one (or perhaps several) operators. The critical variable in this perspective is an estimate of the percent of the production worker jobs that will be replaced by robots. Reductions in unit cost are calculated by assuming that a given percentage of labor costs is reduced. Cases are also considered where robot use results in a moderate increase in output as well as a decrease in production labor requirements. The question of whether decreases in production labor requirements could be offset by a increase in demand stimulated by a reduction in price is also analyzed.

The technological focus of the second perspective is much broader than the first, and is concerned with the impacts of integrating robots with other types of computer assisted manufacturing (CAM) technologies into flexible manufacturing systems. It is assumed that a factory using general purpose machines to produce specialized products in batches can be reorganized and integrated so that machines are fully utilized and used more efficiently. One critical variable in this perspective is an estimate of the potential increase in output that could be realized if all of the time in a year available for production were utilized. The other critical variable is an estimate of the unit cost and of the labor requirements in a fully utilized batch production plant. Based on an analysis of a large cross section of metalworking industries, a relationship is specified between the level of output and the level of unit cost. Reductions in unit cost for a given increase in output are derived from this relationship. Reductions in unit labor requirements are calculated in a similar manner.

1. Metalworking Industries

To date, 80 to 90 percent of the robots used in the United States and in Japan, as well as in the rest of the world, have been installed within a subset of manufacturing industries referred to as the metalworking sector. For this reason, the analysis of the potential impacts of robot use on unit cost and on production labor requirements focuses on the industries included in the metalworking sector.

What is a *metalworking* industry? This simply means that all or most of the establishments classified within the industry are involved to some degree in the shaping, finishing, and assembling of metal products.[1] Which industries are metalworking industries ? One way to answer this question is to identify those industries which use the "metalworking equipment" - metal cutting machines, metal forming machines, joining equipment, and other types of inspecting, and finishing equipment. Every five years since 1925, the American Machinist Magazine has conducted a census of metal shaping, metal forming and related metalworking equipment. Industries within the following major SIC groups were included in the American Machinist Inventory conducted between 1976-1978:

SIC CODE	Major Group Name
25	Furniture and fixtures
33	Primary metals
34	Fabricated metal products
35	Machinery, except electrical machinery
36	Electrical equipment and machinery
37	Transportation Equipment
38	Precision Instruments
39	Miscellaneous Manufacturing

Only the industries in major groups SIC 34-37 are included in most of the subsequent analysis. These four "core" groups include over 85 percent of the units of metalworking machinery counted in the American Machinist Inventory, and nearly 85 percent of the total employment. One major group, SIC 33, primary metals, can be distinguished from the other major groups because the major activity of most of its industries is the conversion of unprocessed metal ores into standard shapes (bar stock, sheets, tubes, pipes, plates, etc.). By contrast, the primary activity of all of the industries in major groups SIC 34-37 includes either the fabrication, finishing or assembly of products from standard metal shapes, and from other purchased parts and subassemblies. Industries in major SIC 33 are omitted from most of the subsequent discussion since the metal refining process is very different from the processes of fabrication, finishing and assembling. Only some of the industries in major groups SIC 25, 38 and 39 are classified as being in metalworking. Since these industries account for a relatively small percentage of the machines used and of the people employed, they too are excluded from the definition of metalworking used here.

In 1980, almost 40 percent of the 20 million people employed in manufacturing and almost 40 percent of the value added in manufacturing were concentrated in the the four major groups of metalworking industries, SIC 34-37. About 50 percent of manufacturing employment and of value added are concentrated in SIC 33-38.

Vietorisz (1969) has described the metalworking sector as "the bellwether of economic growth" for an industrial society because all of the tools and capital equipment used by all manufacturing industries (including itself), and by all other sectors of the economy are produced within it. It is the place within the industrial system where new knowledge is embodied into a physical form, enabling it to be utilized throughout the entire economic system. Since all new products and processes require these capital goods, it is not farfetched to claim that much of the knowledge that becomes part of the economic system enters through the metalworking sector. To the extent that one believes that capital goods, and the role they play in the creating of new products and processes, are essential to economic survival and growth, one can argue that the importance of this sector goes beyond the number of people directly employed within it.

2. The Impacts of Robotic Manipulators

Surveys of the percentage of workers within selected occupations that could be replaced by robots have been collected from 22 manufacturing establishments where robots were either being used or where being seriously considered for use (Table 1). These survey estimates are used to estimate the percent of production workers in metalworking industries that could be replaced by Level I (insensate) and by Level II (sensor-based) robots.[2] The survey results are used as the basis for estimating the percent of jobs in all production worker occupations that could be performed by robots (Table 2).

It is estimated that about 10 percent of the jobs of manufacturing production workers could be performed by Level I (insensate) robots and about 30 percent by Level II (sensor-based) robots. The Bureau of Labor Statistics estimates there were 5.1 million production workers in

Table 1: Summary of Survey Responses of the Percent of Jobs
That Could be Robotized by Occupation and by Level of Robot Technology

Job	Level	Number of responses	Min response	Max response	Aver, simple	Aver, weighted by distribution of employees	Aver, weighted by batch size distribution

ORDERED BY AVERAGE (SIMPLE) RESPONSE FOR LEVEL I

Job	Level	Number of responses	Min response	Max response	Aver, simple	Aver, weighted by distribution of employees	Aver, weighted by batch size distribution
Dip plater	I	6	20	100	48.3	55.7	43.7
	II	6	50	100	78.3	79.7	81.5
Punch press op.	I	5	10	100	45.0	44.3	39.0
	II	5	60	100	76.0	75.0	67.8
Painter	I	16	0	100	40.0	43.5	37.7
	II	15	0	100	62.3	66.8	60.5
Rivetor	I	3	5	100	38.3	40.5	25.2
	II	3	10	100	50.0	51.8	35.9
Shotblaster/	I	6	10	100	35.8	35.6	31.9
sandblaster	II	6	10	100	35.8	35.6	31.9
Drill press op.	I	5	25	50	33.0	32.5	30.1
	II	5	60	75	67.0	67.0	64.8
Etcher-Engraver	I	5	0	100	27.0	29.9	24.3
	II	5	0	100	53.0	59.2	40.3
Welder	I	17	0	60	23.8	25.5	22.0
	II	17	10	90	45.6	45.7	47.8
Coil Winder	I	7	0	40	23.6	24.5	24.8
	II	7	15	50	38.6	40.2	39.7
Heat treater	I	3	5	50	21.7	22.7	16.8
	II	3	40	90	60.0	61.1	52.5
Machine tool-NC	I	20	0	90	19.8	21.7	18.4
	II	19	0	100	44.7	46.5	41.2
Grinding/abrading	I	5	10	20	18.0	18.2	19.3
machine op	II	5	30	100	58.0	57.5	53.5
Lathe/turning	I	5	10	20	18.0	18.2	19.3
machine op.	II	5	25	65	50.0	50.4	50.0

Table 1, Continued

Conveyor operator	I	14	0	50	17.5	14.9	18.7
	II	14	15	65	33.2	41.9	33.2
Electroplater	I	6	5	40	17.5	18.1	15.2
	II	5	15	60	43.0	42.9	44.5
Milling/planning	I	5	10	20	16.0	16.1	16.9
machine op.	II	5	40	60	52.0	52.1	50.7
Filer/grinder/buffer	I	13	0	35	12.1	9.8	11.6
	II	13	5	75	27.7	27.6	26.2
Packager	I	15	0	40	11.8	10.8	8.7
	II	15	0	70	27.1	26.5	20.5
Pourer	I	3	5	20	11.7	10.9	13.1
	II	3	10	30	20.0	21.4	20.0
Assembler	I	19	0	40	10.3	8.9	9.5
	II	19	15	60	31.1	28.8	29.4
Composites and	I	1	10	10	10.0	10.0	10.0
bonded structures	II	1	40	40	40.0	40.0	40.0
Sheet metal op.	I	1	10	10	10.0	10.0	10.0
	II	1	40	40	40.0	40.0	40.0
Inspector	I	19	0	25	8.2	7.5	7.9
	II	19	5	80	29.2	30.4	28.3
Caster	I	4	5	15	7.5	8.6	7.2
	II	3	10	20	15.0	15.2	15.9
Electronic wirer	I	3	0	10	6.7	7.0	7.3
	II	3	10	50	30.0	27.6	32.0
Order filler	I	9	0	20	6.7	8.9	5.5
	II	9	0	80	29.4	31.7	25.3
Tester	I	17	0	10	5.8	4.8	5.1
	II	17	0	30	11.4	11.3	10.8
Mixer	I	3	0	10	5.0	6.0	5.2
	II	3	10	10	10.0	10.0	10.0
Tender	I	2	0	10	5.0	6.4	5.5
	II	2	20	20	20.0	20.0	20.0
Millwright	I	4	0	15	3.7	4.4	3.0
	II	4	0	15	3.7	4.4	3.0

Table 1, Continued

Kiln-furnace op.	I	3	0	10	3.3	2.9	2.0
	II	3	5	20	13.3	14.7	10.5
Tool and	I	8	0	5	1.5	1.3	0.9
die maker	II	8	0	60	16.6	15.3	9.2
Oiler	I	3	0	0	0.0	0.0	0.0
	II	3	0	0	0.0	0.0	0.0
Rigger	I	2	0	0	0.0	0.0	0.0
	II	2	0	0	0.0	0.0	0.0
Trader/helper	I	1	0	0	0.0	0.0	0.0
	II	1	50	50	50.0	50.0	50.0

Table 2: Summary of Survey Estimates of Potential
Displacement of Production Workers: All Occupations

Occupation	Percent Displacement	
	Level I	Level II
Tool handlers	27.2	46.7
Metalcutting Machine Operators	15.5	42.6
Metalforming Machine Operators	26.2	55.0
Other Machine Operators	13.2	26.2
Assemblers	8.9	28.8
Laborers	3.8	27.7
Miscellaneous Craft Workers	2.8	13.2
Maintenance and Transport Workers	0.0	0.0
Totals	10.6	28.6

The average percentage displacement within each group of occupations is based on an analysis of the occupational employment within one particular industry group, SIC 351. The percentages would vary somewhat if they were based on the occupational employment of all industries with SIC 34-37.

Table 3: Number of Jobs Displaced and Robot Population
Implied by Ayres/Miller Estimates of Potential Displacement

NUMBER OF JOBS DISPLACED

Industries	Employment (1980)	Potential Level I	Displacment by Level II
Metalworking (SIC 34-37)	5,091,800	539,731	1,456,255
All Manufacturing (SIC 20-39)	14,190,289	1,504,171	4,058,423

ROBOT POPULATION IN METALWORKING AND IN ALL MANUFACTURING

Industries	Number of robots assuming 1 robot replaces		
	2 Workers	3 Workers	4 Workers
Metalworking, SIC 34-37			
Level I .54 million workers displaced)	270,000	180,000	135,000
Level II (1.5 million workers displaced)	728,100	485,400	364,000
All Manufacturing, SIC 20-39			
Level I (1.5 million workers displaced)	750,000	500,000	375,000
Level II (4.0 million workers displaced)	2,000,000	1,333,333	1,000,000

SIC 34-37 in 1980. Based on the estimates of the percent of jobs that could be performed by robots and of the number of production workers, it is estimated that Level I robots could potentially perform the jobs of 540,000 workers in SIC 34-37 and that Level II robots, if available, could potentially perform the jobs of 1.45 million workers in these same industries (Table 3). Extrapolating the job displacement data within metalworking to the 14.2 million production workers in all of manufacturing, it appears that level I robots could theoretically replace about 1.5 million jobs and that level II robots could theoretically replace about 4 million jobs. Assuming 1 robot replaced two production workers, these estimates of the potential for robot use, based on an analysis of robot capabilities and job requirements, imply that there is a potential use for over 700,000 Level I robots or for over 2 million Level II robots throughout all manufacturing.

Most market forecasts place the cumulative robot population for 1990 within the range of 50,000 to 150,000 units (Table 4). Assuming each robot is used to displace two workers, on average, this implies that that only 100,000 to 300,000 workers will be lost, displacing only 0.7 to 2 percent of manufacturing production workers. Considering only Level I robots, the estimate of the potential number of applications is 5 to 15 times larger than the market forecasts of the Level I robot population for the year 1990.

An attempt is made to explain this large difference between the estimate of the technical potential for robot use and the estimates of actual robot sales. The cost of installing robots for loading and unloading machine tools is analyzed for the purpose of identifying the conditions under which there would be a strong economic incentive to use robots given that there is a technical potential for doing so. The analysis considers the purchase price of the robot as well as the additional implementation costs that are typically required. The assumptions made in calculating total implementation cost, for low cost, medium cost and high cost robots, shown in Table 5, are based on the observations that

- the ratio of total implementation cost\robot base price ranges from a factor of 3 to 5 in retrofit situations,

- application costs are a larger multiple of the robot base price for lower cost robots then for the higher cost robots,

and on the estimates of 1982 robot base prices. The summaries of the total implementation cost for retrofitting low cost, medium cost and high cost Level I robot systems into a factory to load and unload machine tools is given in Table 6.

Table 4: Forecasts of the Population of Robots in the the U.S. in 1990

Source of Estimate	Cumulative Population
Hunt and Hunt (Upjohn Institute)	50-100,000
Conigliaro (Bache, Halsey and Shields)	122,000
Aron (Dawia Securities)	94-95,000
University of Michigan/ Society of Manufacturing Engineers Delphi Survey	150,000
Engelberger (Unimation, Inc.)	150,000
Robot Institute of America	75-100,000

Source: Hunt and Hunt (1982: 25).

Table 5: Assumptions For Calculating Total Implementation Costs

TYPE OF ROBOT	BASE PRICE	TOTAL IMPLEMENTATION COST ROBOT BASE PRICE
lower cost	$20,000	4
medium cost	$60,000	3
high cost	$100,000	2

Table 6: Summary of Cost Assumptions for Retrofitting Level I Robot Systems

Robot Hardware Cost (R)	Development Cost (D)	Total Implementation Cost (I = R + D)	Operators Replaced Per Shift
20,000	80,000	100,000	1
60,000	180,000	240,000	1-2
100,000	200,000	300,000	1-3

2.1. Calculation of Payback Periods Based on Direct Labor Savings

A cost-benefit framework is adopted where the benefits are *narrowly* defined as labor savings, and where costs are the total costs of implementing the robot system. While other benefits are sometimes realized when robots are used, such as more consistent and higher quality processing, increased throughput, and improved conditions for wokers moved out of unpleasant jobs, labor savings are widely regarded as the primary (and often the only) variable to consider. Engelberger (1980: 103) makes the point quite clearly:

> The prime issue in justifying a robot is labor displacement. Industrials are mildly interested in shielding workers from hazardous working conditions, but the key motivator is the saving of labor cost by supplanting a human worker by a robot. So very much the better if a single robot can operate for more than one shift and thereby multiply the labor saving potential.

The respondents to the CMU robotics survey [Carnegie-Mellon 81] and all other available evidence [Whitney et al 81], [Industrial Robot 81], [Ciborra, Migliarese, and Romano 80] strongly supports this view. Respondents to these survey overwhelmingly ranked efforts to reduced labor costs as their main motivation for installing robots. Payback periods are calculated under various assumptions regarding the total annual cost of a worker and the number of workers replaced per robot (Table 7). Based on comments in the literature on the economic justification of robots [Smith and Wilson 82], it is assumed that a robot would only be installed if the projected payback periods were three years or less.

First the cost of installing one robot is considered. This also includes the case of multiple installations if the cost of installing *n* robots is *n* times the cost of installing one robot. If the three year payback period were really a hard and fast rule (which it is not, of course), one would conclude from this simplified analysis that given the technical feasibility of using Level I robots, the only users would be

- those plants with enough demand to operate on a three shift bases.

- those plants where it was possibly to eliminate two or more workers per shift on two shifts with one robot.

- and those plants which could use the low cost (low capability) robots to eliminate one worker per shift for two shifts.

Taking a conservative outlook, suppose it were the case that one robot only eliminated one worker per shift, that paybacks were calculated on a two shift basis, and that the "heavy duty" robots were required for most machine loading applications, especially in "heavy" manufacturing. Payback periods would range between 4 and 5 years, depending on total

Table 7: Simple Payback Periods For Level I Robots
Based On Labor Savings

25 K PER WORKER PER YEAR

Scenario:	Number of shifts per day		
Replacement rate per shift	1 shift	2 shifts	3 shifts
1 robot:1 worker			
25 K robot	4.0	2.0	1.3
60 K robot	9.6	4.8	3.2
100 K robot	12.0	6.0	4.0
1 robot:2 workers			
60 K robot	4.8	2.4	1.6
100 K robot	6.0	3.0	1.9
1 robot: 3 workers			
100 K robot	4.0	2.0	1.3

30 K PER WORKER PER YEAR

Scenario:	Number of shifts per day		
Replacement rate per shift	1 shift	2 shifts	3 shifts
1 robot:1 worker			
25 K robot	3.3	1.7	1.1
60 K robot	8.0	4.0	2.7
100 K robot	10.0	5.0	3.3
1 robot:2 workers			
60 K robot	4.0	2.0	1.3
100 K robot	5.0	2.5	1.7
1 robot: 3 workers			
100 K robot	3.3	1.7	1.1

worker costs. These longer than desirable payback periods would probably discourage many financial analysts from giving the "go ahead" on robot application. The conclusion here is that if one takes the most conservative view of the the economics of robot use (e.g. robots are only viewed as labor savers and must pay for themselves in a very short time period), than it appears that too long of a payback period (or correspondingly, too low of a return on investment) will restrain the growth of robot use over the next several years. Given the assumptions of this cost-benefit model, the conclusion is that substantially fewer robots would be installed than could be used. This would mean that the number of jobs displaced would be closer to the levels implied by the current market forecasts than by the survey based estimates of the potential for robot use.

A key assumption in the first cost-benefit analysis is that the cost of installing n robots is n times the cost of installing one unit. According to applications engineers and consultants, this is not the case if additional installations are similar to one another. The development cost (planning, tooling, design of accessory hardware) for the second and subsequent applications are lower than for the first one. If one large establishment were to install many robots, or correspondingly, if one large firm were to install many robots across several plants, the average cost per robot would be less than if only one or several units were installed. In a second series of cost-benefit calculations carried out, the cost of installing n robots in an establishment (or across a company) is adjusted so that the development cost component of total costs decreases by 10 percent for each subsequent installation (Table 8).

Given this revised cost model, payback periods are calculated by size of establishment for one industry. Within establishments of a given size class, the number of workers displaced per establishment is derived from the total number of workers potentially displaced, the distribution of employment by size of establishment, and from the number of establishments within the size class. The result of this analysis is that payback periods are substantially shorter in the largest sized establishments (1000 and more production workers) than in the other size classes (Table 9). The reason being that the average cost per robot decreases as the number of robots installed increases, and it is assumed that more workers would be displaced in the establishments with more production workers.

An important conclusion of this analysis is that, given the cost assumptions, only the largest establishments could justify the use of robots under the conservative assumptions that one robot replaces a total of two workers, that "heavy duty", higher cost, robots are required, and

Table 8: Total and Average Cost For Multiple Installations
of Similar Applications

Number of Robots	60 K ROBOT*		100 K ROBOT*	
	Total Cost	Avg. Cost per robot	Total Cost	Avg. Cost per robot
	(x 1000)	(x 1000)	(x 1000)	(x 1000)
1	240	240	300	300
9	1642	182.5	2125	236.1
46	4429.6	96.3	6455	140.3

*) Base price

Total cost of installing n robots is approximated by

$$I_n = n \cdot R + D \cdot \sum_{0}^{n-1} (.9)^i$$

Assume: The development cost for each successive application decreases by
10 percent for <u>similar</u> applications.

I_n = total cost of installing n robots.
R = robot base price
D = development cost for first installation

Table 9: Payback Periods Based on Production Labor
Savings by Size of Establishment

25 K PER WORKER PER YEAR

60 K ROBOT (Base Price)

Size of Establishment	1R:1W:2S	1R:1W:3S	1R:2W:2S	1R:2W:3S
1-19	9.6	9.6	9.6	9.6
20-99	4.8	4.8	4.8	4.8
99-249	4.5	3.1	3.1	1.6
250-499	4.4	3.1	2.4	1.7
500-999	3.1	2.4	2.0	1.5
1000 and >	1.9	1.5	1.3	1.0

100 K ROBOT (Base Price)

Size of Establishment	1R:1W:2S	1R:1W:3S	1R:2W:2S	1R:2W:3S
1-19	12.0	12.0	12.0	12.0
20-99	6.0	6.0	6.0	6.0
99-249	5.6	3.9	3.9	2.0
250-499	4.4	4.0	3.1	2.1
500-999	4.2	3.1	2.6	1.6
1000 and >	2.8	2.1	1.7	1.3

1R:1W:2S reads as follows:

One robot replaces one worker per shift for 2 shifts.

that labor savings are the only quantifiable economic benefit. Smaller size establishments, with only one or two applications, would not be able to realize the "scale economies" realized when multiple units are installed. Payback periods would be too high to obtain the financial approval for robot use.

Survey responses from 52 members of the Robot Institute of America indicate that as of 1981, robot use was heavily concentrated in establishments with 1000 and more production workers. The survey also showed that these large establishments typically used many robots. This lends support to the hypothesis that the economic incentives for robot use are much stronger in the largest sized establishments than in the smaller ones. It is also noted that the automobile industry, which has the largest proportion of production workers in large establishments, is also the largest user of robots. (The auto industry also has the highest wage rates of any industry in SIC 34-37).

Suppose it were assumed that robot use will continue to be heavily concentrated in the largest establishment and also in the metalworking industries (SIC 34-37) until the end of the decade. Almost 40 percent of the 5.1 million workers employed in these industries as of 1980 are located in establishments with 1000 or more production workers. To displace 10 percent of these workers would require almost 100,000 robots, assuming one robot displaces two people. Clearly, if some robots were used in smaller sized establishments, as well as outside of the metalworking industries, somewhat more than 100,000 robots would be required. (If robots were used to displace 10 percent of production workers in all manufacturing industries, it would imply a robot population of 180,000 units.) Most market forecasts predict there will be a total of 75,000 to 150,000 Level I robots in use throughout industry by the year 1990. It is plausible that these forecasts are predicated on the assumption that robots will mostly be used within the largest establishments in the metalworking industries, and that roughly 10 percent of the jobs in these establishments could be robotized.

This example shows that there is not necessarily an inconsistency between the estimate that 10 percent of production worker jobs could potentially be performed by Level I robots (implying a potential market of 750,000 robots) and that there are only expected to be 50,000 to 150,000 Level I robots in use throughout industry by 1990. It appears that the key to explaining the difference between the survey based estimates of technical potential for robot use and the market forecasts of the number of robots actually sold is an understanding of how large a segment of the potential market will have a strong enough economic incentive to install robots given that there is a technical potential for doing so.

Will robot use continue to be heavily concentrated in large establishments in metalworking as it has been, and as many market forecasters apparently expect that it will be ? it is important to know whether future robot use will follow the same market patterns as past use to understand the extent of potential labor impacts-- whether 10 percent of a small segment of the workforce will be displace or whether 10 percent of the total manufacturing workforce will be displaced. An understanding of the likely patterns of robot diffusion over the next several years would also help to understand whether or not initiatives might be required to promote robot uses in places where it might otherwise be indefinitely deferred, such as in smaller size establishments.

Is the use of Level II (sensor-based) robot systems going to alter the extent of robot diffusion and make it necessary to reevaluate the potential impacts on job displacement ? At this point, the answer appears to be no. Currently, sensor based systems with enough capability to acquire randomly oriented parts are substantially more expensive than Level I systems so the payback periods are much longer. Sometime within the next several years, the cost of sensor-based systems will drop substantially, and the answer may be yes.[3] Suppose, as a result of future technological improvements, that the cost of installing a Level II system was the same as the cost of installing a Level I system. There are perhaps three times as many applications for Level II systems as for Level I systems. There would be more potential applications per establishment, and even the medium size and smaller establishments would have use for several (or more) robots. Then installing three times as many robots means that the average cost per installation would be less than for Level I robots, assuming, as before, that the total cost of installing a robot decreases as the number of robots installed increases. Payback periods would decrease, especially for the medium and smaller size establishments. If this were to happen, there would be good reasons for reconsidering the market forecasts that predict that their will be *at most* 150,000 robots by the end of 1990. Given the plausibility of this scenario, future studies of robotic impacts should consider the rate at which the cost and capabilities of Level II robot systems are changing.

2.2. The Price Elasticity Argument

An analysis has been made of what would happen to production worker employment in an industry if robots were used to replace workers and if a decrease in price resulting from higher levels of productivity stimulated demand for the industry's output. The question of interest is whether or not price induced increases in demand could be expected to increase total labor requirements by a sufficient amount to offset job displacement in an industry using robots. In the first scenario, it is assumed that robots decrease cost only as a result of reducing labor requirements and that that throughput is held constant. In the second scenario, it is assumed that robot use results in a 20 percent increase in throughput as well as a reduction in total labor requirements. In this scenario, unit labor requirements and production costs decrease by substantially more than in the case where throughput is held constant.

Given assumptions on the decrease in production labor requirements. and on the increase in the throughput of the factory, the amount by which the demand for output would have to increase in order to reabsorb all workers whose jobs are displaced is calculated. It is assumed that the demand for output increases as its price decreases. For every Δp percent decrease in price, demand for output is assumed to increase by ν percent, where the parameter ν is referred to as the price elasticity of demand. Given the calculated price change, the magnitude of the price elasticity of demand which would be required to induce enough of an increase in output so that employment levels would be maintained is calculated. The magnitude of this "break even" price elasticity is of particular interest. Clearly, if the magnitude of the price elasticity of demand were large enough, and if there were no limitations on how large the demand for output could increase, any decrease in labor requirements could be offset by price induced increases in the demand for output. The concern relevant to public policy is whether the calculated values for the price elasticities of demand required to maintain employment levels are near the levels of price elasticities normally observed in the "real world" marketplace.

For both scenarios, the increase in demand required to reabsorb all displaced workers and the value of the "break even" price elasticity is calculated with and without assuming that there are job turnovers as a result of attrition. In the case with "attrition", it is assumed that 15 percent of the workers in the industry leave the workforce as a result of death, retirement, sickness, disability, etc. during the period in which workers are replaced by robots. When attrition is considered, a smaller increase in demand and a smaller magnitude of the breakeven price elasticity is required to reabsorb workers displaced by robots since there are job openings created by job turnover.

The conclusion on whether or not jobs displaced by robots could be reabsorbed within the same industry is not conveniently summarized, since it depends on several variables, including

1. the percent of jobs that are displaced.

2. whether or not robot use increases throughput as well as decrease labor requirements.

3. whether or not the there is job turnover due to attrition.

First consider the case of no attrition in the workforce to focus on whether price induced increases in demand, by itself, could be expected to offset job displacement. If throughput is held constant and price decreases are due only to decreasing a fraction of production labor cost (Scenario I, no attrition), it is concluded that very few of the displaced workers would be reabsorbed. If throughput were also to increase, thereby causing a larger decrease in price (Scenario 2, no attrition), it seems that a 10 percent displacement of jobs could be offset by price induced increases in output. However, demand would have to be relatively price elastic. Without considering attrition, the conclusion as to whether or not the potential job displacement of Level I robots could be offset depends on the extent of the economic benefits of robot use. Price induced increases in output would not fully offset the potential job displacement of Level II robots.

Now consider the case where the size of the workforce decreases by 15 percent as a result of attrition over a 3 to 5 year period. If throughput is held constant (Scenario 1, attrition), and if 10 percent or fewer workers were displaced, job openings from turnovers would outnumber jobs displaced by robots, even without considering the effects of price decreases. If the use of robots were to also increase throughput (Scenario 2, attrition), it appears that a potential displacement of up to 20 percent could conceivably be offset through the combined effects of job turnover and price induced increases in output. Even with attrition, though, it is unlikely that a potential displacement of 30 percent could be offset. The conclusion here is that Level I robots could be fully utilized in an industry and displace 10 percent of the workers over a several year period without resulting in any unemployment. However, there would still be a significant numbers of jobs lost if Level II robots were fully utilized and 30 percent of the workers were displaced.

This analysis suggests that a more thorough and precise understanding is required of how robotics will alter labor requirements in order to further analyze employment issues in industries using robots. It is important to know if the economic benefits of robot use are restricted to savings in labor cost, or whether they might also increase throughput. It is also very important to know about the rate of job turnover, since the attrition is often cited as the way of offsetting displacement effects of robots. More detailed information on rates of attrition are needed to confirm the conclusion made here that up to 10 percent job displacement could be fully offset by attrition if there is no increase in throughput. This analysis also suggest there is a need to take a more detailed look at the rate of development of sensor-based robot systems.

Some of the assumptions underlying these conclusions need to be clarified in order to make the limitations of this analysis more transparent to the reader. First, issues relating to changes in skill requirements for a given occupation or changes in the overall occupation profile are not considered here. In this simplified framework, it is assumed that if a production worker is displaced by a robot and if there is a need for an additional production worker either as a result of an increase in demand or job turnover, then the displaced worker can be reabsorbed by the firm. Thus, it is assumed that the skills required by production workers after the implementation of robots do not pose a barrier to reabsorbing the displaced workers.

Second, all aspects of cost changes are not carefully considered. Increases in capital cost required to install robots is ignored in this analysis so the calculated price decreases can only be viewed as upper bounds. If capital cost was included and the decrease in price was smaller, then the magnitude of the price elasticity required to maintain employment levels would be larger. If the price elasticities were larger, it is possible that the percentage of job displacement that could be offset is smaller than indicated.

Third, only one industry is considered here and interindustry transactions are ignored. An important characteristic of most metalworking industries is that they sell most of their output to other industries (especially to other metalworking industries) to be used as capital or material inputs. Suppose all industries were to reduce their cost by a given amount, say 2 percent, in one period as a result of reducing labor cost, and bought and sold materials and equipment from one another. In the next period, all purchased materials and capital equipment would be 2 percent cheaper, so each industry would realize an additional one to two percent cost reduction. If these interaction affects result in larger price decreases than are are considered here, then the "breakeven" value of the price elasticity required to maintain employment levels would be smaller than is indicated. If the price elasticity were smaller, then it is possible that a larger percentage job displacement could be offset as a result of price induced increases in demand.

Fourth, relationships between the cost of manufactured goods and the level of economic activity in other sectors of the economy which use these goods are not considered. The possibility that employment losses in manufacturing might be offset by employment gains in other sectors of the economy which expand as a result of decreases in the cost of capital and consumer goods is not explored. More definitive conclusions require that these four factors be considered.

3. The Impacts of the Fully Utilized, Flexibly Automated Factory for Batch Production

The relationship between the level of unit cost and the level of output produced is examined across 101 different metalworking industries. An estimate of the pounds of basic metals and of processed metal inputs purchased is used as a surrogate measure of the level of output of each industry.[4] Value added per unit and units of output are computed for each industry using pounds of metal processed as the standardized unit of output (Figure 1). Regression relations between unit cost and unit cost components are summarized in Table 10. The basic structural relationships that underlie the shape of a "neoclassical" long run unit cost curve for a particular product are also apparent in the comparison of unit cost versus output *across* industries. These basic relationships are:

1. Capital costs for equipment and machinery per unit of output decrease across industries as the units of output produced increases.

2. Production labor costs per unit of output decrease across industries as the units of output produced increases.

3. Value added per unit of output decreases across industries as the units of output produced increases.

4. Machine utilization increases across industries as units of output produced increases.[5]

The implication is that the tradeoffs which most strongly affect the organization of production within a particular plant--either organizing to make small volumes of specialized products at· a high cost or organizing to make large volumes of standardized products at a low cost-- are also affecting the organization of production across industries.

Because it appears that the *custom-batch-mass* paradigm characterizes the organization of production across industries (as well as within specific plants), it is argued that that the dominant mode of technology used within an industry can be inferred from the industry's measures of pounds of metal processed and unit cost. Industries with the highest levels of

Figure 1: Value Added Per Pound of Metal Vs Pounds of Metal/Establishment for Metalworking Industries, 1977

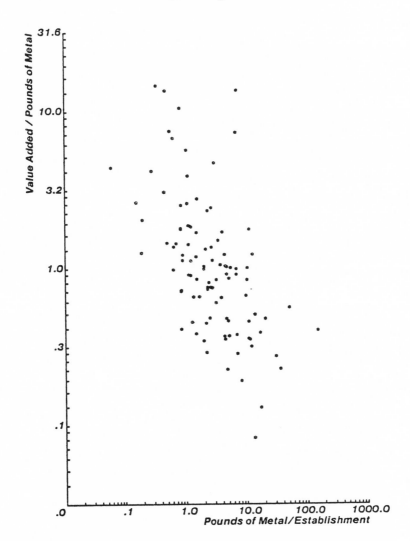

Table 10: Summary of Regression Results of Unit Cost Versus
Pounds of Output/Number of Establishments Across Metalworking Industries

Dependent Variable	Constant b_0	Output elasticity b_1	Significance level for output elasticity	Goodness-of fit measure: R^2 (percent)
Pooled four digit data set, SIC 34-37: 101 four digit industries				
output = pounds of metal/number of establishments				
k/m	-0.7243 (-6.75)	-0.371 (-5.49)	0.99	22.5
l/m	-0.835 (-8.95)	-0.440 (-7.49)	0.99	35.5
va/m	0.401 (3.96)	-0.436 (-6.83)	0.99	31.4
k/l	-0.056 (-1.41)	0.051 (2.06)	0.95	3.1
k/employee	2.160 (44.17)	0.156 (5.08)	0.99	19.9
s/l	-0.473 (-9.34)	-0.114 (-3.57)	0.99	10.5

( ) = t ratio for estimate.

Unit Cost Components (Dependent Variables)
k/m = gross value of equipment and machinery / pounds of metal
l/m = all included production worker costs / pounds of metal processed
va / m = value added / pounds of metal
k/l = gross value of equipment and machinery / all included production worker costs.
k/e = gross value of equipment and machinery /total employees
s/l = salaries / hourly production worker wages

Output Measure (Independent Variable)
m/e = pounds of metal processed/number of establishments

Output elasticity for each cost component is the estimate of b_1 in
$E[\ln(\text{unit cost component})] = b_0 + b_1 \ln(m/e)$

production cost per pound of output and with the fewest pounds of output are classified as being comprised of custom and small batch producers. Industries with the lowest levels of production cost per pound of output and with the most pounds of output are classified as being comprised of mass producers. The remaining industries, those with mid range levels of production cost per pound of output and with mid range levels of pounds of output, are classified as being comprised of batch producers. With these assumptions, the proportion of value added and of output accounted for by metalworking products which are custom, batch and mass produced is estimated. The result of this analysis is that the industries with the highest levels of output and with the lowest levels of unit cost (which are assumed to be the mass producers) account for *at most* 25 percent of the value added and for less than 35 percent of the total output of the 101 industries in the sample[6] (Tables 11 and 12). This analysis corroborates the widely cited claim that most of the value added in the metalworking sector is accounted for by products which are batch produced.

The significance of the claim that most industries in metalworking produce batches of specialized products can only be appreciated by considering the difference in unit cost between batch and mass production. One finding is that a previously published estimate claiming by Cook (1975) that for the case of a typically machined product, the unit cost using the most efficient mass production techniques would be 100-500 times lower than if the produced were produced in a "one-of-a-kind" mode, and 10-30 times lower than if it were batch produced seems reasonable. Considering that 1) most of the value added in metalworking is accounted for by batch production, and 2) products which are batch produced are much more expensive than products mass produced in large volumes, it is typically argued that much of the value added within the metalworking sector can be viewed as a type of penalty cost that has been unavoidable because of the inherent inefficiencies of custom and batch production relative to mass production. This is the foundation for many of the arguments citing the need to accelerate the development and use of "robotic" and other types of flexible production technologies which are applicable to batch production.

Table 11: Distribution of Value Added by Mode of Production, SIC 34-37

Region and Mode of Production	Major SIC Groups				
	34	35	36	37	Total, 34-37
PERCENT OF VALUE ADDED FOR INDUSTRIES IN SAMPLE					
Custom and Small Batch	1.1	41.0	58.5	30.6	31.5
Mid-Batch	28.3	42.5	30.2	6.9	26.4
Large Batch	45.5	16.5	3.1	2.9	16.4
Mass	25.1	0.0	8.2	59.6	25.7
Sample coverage of value added in all industries	94.2	95.8	51.2	97.4	86.0

Table 12: Distribution of Output by Mode of Production, SIC 34-37

PERCENT OF TOTAL OUTPUT FOR INDUSTRIES IN SAMPLE

	34	35	36	37	Total, 34-37
Custom and Small Batch	1.0	36.3	55.1	20.1	24.8
Mid-Batch	25.3	44.3	31.4	4.9	23.3
Large Batch	45.1	19.4	3.8	2.8	19.9
Mass	28.6	0.0	9.7	72.2	35.7
Sample coverage of output in all industries	95.1	96.1	51.0	98.0	88.0

An analysis is made of the decrease in unit cost and in labor requirements that would result if the use of robots, in conjunction with other types of automation, made it possible to substantially increase the capacity of a factory which uses general purpose types of machines to produce specialized products in batches. Previously published estimates of machine utilization in conventionally organized factories producing low and mid-volumes of specialized products [Mayer and Lee 80] are used to estimate the potential for increasing output. The conclusion is that output in batch production facilities could theoretically be increased by 150 to 550 percent if all of the productive time available in a year were fully utilized and if the plant were organized to work more efficiently (Table 13).

Published information on flexible manufacturing systems indicate that with the most advanced types of flexible automation currently available, parts of the manufacturing process can be *fully* automated even when making specialized products in batches . In these systems, the output is several times that of its conventional counterpart, which is consistent with the range of increase derived from the analysis of theoretical capacity. Robotic manipulators, per say, are only a very small part of the total automation used in these plants. This suggests that when analyzing the case of a fully utilized plant running around the clock, one should more appropriately address the potential impacts of flexible automation *systems* on cost and employment, as opposed to the impacts of robotic manipulators. Also, these examples indicate that very large capital investments are required to design and install such systems.

To date, it has not been possible to make a detailed comparison of the capabilities and economics of flexibly automated plants against those of conventionally organized ones because the published information on the handful of flexibly automated plants throughout the world is too sparse. Given that the unit cost in the proposed high volume batch production plant can not be directly observed, it must be inferred or approximated through some indirect means. The framework used here for estimating the potential reduction in unit cost is to assume that a flexibly automated plant producing specialized products in small and medium sized batches would have some of the characteristics of conventional plants producing more standardized products in larger volumes. Thus, the unit cost observed in industries dominated by plants using conventional (e.g. specialized) types of automation to make more standardized types of products in larger volumes is used to infer the level of unit cost in a fully utilized flexibly automated batch production.

A regression relationship between unit cost and units of output produced across

Table 13: Summary of Potential Increases In Output.

Type of Plant		Potential Capacity Increases	
	Base Case	Robots Only	Robots with CAM
High volume			
Available hour index	1.00	1.31	1.31
Throughput index	1.00	1.11	1.39
Output index	1.00	1.45	1.82
Increase in output (%)		45	82
Low-volume: double shift			
Available hour index	1.00	2.17	2.17
Throughput index	1.00	1.16	1.52
Output index	1.00	2.52	3.30
Increase in output (%)		152	230
Mid-volume			
Available hour index	1.00	2.98	2.98
Throughput index	1.00	1.14	1.55
Output index	1.00	3.40	4.62
Increase in output (%)		240	362
Low-volume: single shift			
Available hour index	1.00	4.35	4.35
Throughput index	1.00	1.16	1.52
Output index	1.00	5.05	6.61
Increase in output (%)		405	561

Available hour index: The relative amount by which the time available for production could be increased. This includes the effects of recouping the days per year that the plant is not scheduled for production, as well as recouping the shifts per day that that are idle during those days that the plant is scheduled for production. One hour per day is allotted for preventive maintenance.

Throughput index: The relative amount by which the time available for production could be increased during those times that the plant is operating. This includes the effects of reducing set-up time, loading/unloading time, tool change time, and idle time.

metalworking industries, estimated in Table 10, is used as a starting point for this analysis. The elasticity of unit cost with respect to output is used to derive the percent reduction in unit cost that would result from increasing output. Similarly, the elasticity of production labor cost with respect to output is used to derive the percent reduction in production labor requirements that would be realized with an increase in output.

A more detailed analysis of the variation in unit cost across industries is also carried out. The explanatory variables used in the expanded multiple regression model are summarized in Table 14. The proposed effect on unit cost is also shown by indicating whether the sign of the estimated elasticity of unit cost with respect to each variable should be positive or negative. A key feature of the more detailed analysis is that two surrogate measures of processing complexity are constructed which are believed to indicate important differences in the nature of the processing requirements across industries. One complexity measure is the average unit cost of the basic metals purchased by an industry, called the basic metal cost index (bmci). An increase in the index means that more expensive metals are used, which is taken as an indication that the difficulty of the shaping operations increases. Since more difficult operations require more capital and/or labor inputs to accomplish, unit cost is assumed to increase and the proposed sign of this elasticity is positive. The second complexity measure is the ratio of processed metal input cost to basic metal input cost. It is argued that this variable is an indicator of the relative proportions of assembly to metal shaping. It is included to account for the difference between industries which are primarily involved in shaping and forming versus those primarily involved in assembly. Two reasons for believing that a higher ratio of processed metal costs to basic metal costs indicates a more "complex" process are as follows. First, the higher the ratio of processed metal inputs to basic metal inputs, the greater the diversity of material inputs used in an industry. There is some tendency for the ratio of salary costs/production worker cost to increase across industries as the ratio of processed metal inputs/basic metal inputs grows larger. This provides some evidence that it takes more organizational control and supervision to coordinate production when there is a a larger proportion of processed metal inputs. Second, the inspection of assembled products requires more than just the verification of dimensions. Since subcomponents must be properly integrated with one another, testing is required to verify that final product performs its designated functions properly. With electronics equipment, and computers, this can be a fairly extensive and complicated processes. Also, several other variables which introduce noise into a cross sectional comparison, such as differences in wage rates and in the coverage of material inputs used to construct the output measure, are introduced into the multiple linear regression model.

Table 14: Summary of Explanatory Variables in Multiple Regression Model

FACTOR	VARIABLE	NOTATION	EFFECT ON UNIT COST (SIGN OF COEFFICIENT)
Level of output	pounds of metal processed / number of establishments	m\e	$\frac{d \ln(u)}{d \ln(m\e)} < 0$
Complexity of metal shaping activities	dollars of basic metal / pounds of basic metal = basic metal cost index	bmci	$\frac{d \ln(u)}{d \ln(bmci)} > 0$
Degree of assembly	dollars of purchased metal / dollars of basic metal	pmbm	$\frac{d \ln(u)}{d \ln(pmbm)} > 0$
"All included" hourly wage	production worker wages + benefits / production worker hours	w	$\frac{d \ln(u)}{d \ln(w)} > 0$
Material coverage	dollars of metals / dollars of total materials	c	$\frac{d \ln(u)}{d \ln(c)} < 0$

```
THE REGRESSION EQUATION IS
ln(u) = - 1.17 - 0.295 ln(m/e) + 0.983 ln(bmci) + 0.488 ln(pmbm)
      + 0.948 ln(w)  - 0.765 ln(c)
                                    ST. DEV.   T-RATIO =
VARIABLE           COEFFICIENT      OF COEF.   COEF/S.D.
Constant            -1.1707          0.4625     -2.53
ln(m/e)             -0.2947          0.0387     -7.61
ln(bmci)             0.9827          0.0909     10.81
ln(pmbm)             0.4881          0.0490      9.96
ln(w)                0.9806          0.1911      5.13
ln(c)               -0.7644          0.2051     -3.73

THE ST. DEV. OF Y ABOUT REGRESSION LINE IS
S = 0.3717
WITH ( 101- 7) =  95 DEGREES OF FREEDOM

R-SQUARED = 87.0 PERCENT
R-SQUARED = 86.3 PERCENT, ADJUSTED FOR D.F.
```

It is observed that industries with low levels of output tend to use more expensive basic metal inputs than high levels of output. This suggests that "scaling up" means more than just increasing the volume of production. There also tends to be a change in the mix of material inputs as well. It is believed that more standard material inputs are suggestive of simplified and standardized product designs. If this were the case, it would indicate that processing requirements themselves are simplified and standardized as the volume of output increases. For this reason, it is argued that the unit cost elasticity derived from the simple regression of unit cost against output without including the complexity parameters incorporates the effects of both increasing the average batch size and of standardizing the material inputs. When the basic metal cost index and the other explanatory variables are included in the regression analysis, the magnitude of the unit cost elasticity is lessened. The reason for this is that the effects of increasing the average batch size are separated from the effects of standardizing the material inputs when the complexity parameters (principally the basic metal cost index) are included in the multiple regression model.

If it were the case that each product is optimally designed for ease of manufacture in the flexibly automated batch production factory, as is typically the case with standardized products made in a conventional mass production factory, one would want the effects of both increasing batch size and of altering material inputs (and product design) to be included in the unit cost elasticity. If, however, each product were not designed to minimize the complexity of processing requirements, as is typically the case with making specialized products in a conventional batch production factory, then one would want to separate the effects of standardizing the material inputs from increasing the average batch size.

The unit cost elasticities estimated from the regression equations without and with the complexity parameters are shown in Table 15. The elasticities of total value added per unit of output are used to derive high and low estimates of the percent reduction in unit cost that would result from an increasing output from 50 to 1000 percent in a batch production plant (Table 16). The result is that severalfold increases in output would lead to a very substantial decrease in unit production cost. For example, if output were to increase by 100 percent, the estimated decrease in unit cost ranges from 18 to 26 percent. If output were to increase by 1000 percent, the estimated decrease in unit cost ranges from 50 to 65 percent.

It is emphasized that the analysis of the economics of the flexibly automated factory is more speculative than the analysis of the economics of robotic manipulators. Since the relationship

Table 15: The Elasticity of Unit Cost Components with Respect to Output

Unit Cost Component	Average Share of Unit Cost[a] (%)	Estimated Elasticity Unit Cost Component With Respect to Output:	
		With Complexity Variables[b]	Without Complexity Variables[c]
Total value added	100.0	-0.295	-0.436
1] Labor value added	55.9	-0.345	-0.461
1 a] Production worker costs	35.4	-0.306	-0.440
1 b] Salary costs	20.4	-0.406	-0.536
2] Nonlabor value added	44.1	-0.232	-0.408

value added = labor value added + nonlabor value added

labor value added = production worker costs + salary costs

a) Average share of total cost for 101 metalworking industries included in sample.

b) Output elasticity for each cost component is the estimate of b_1 in
$E[\ln(\text{unit cost component})] = b_0 + b_1 \ln(m/e) + b_2 \ln(bmci) + b_3 \ln(1 + pmbm) + b_4 \ln(w) + b_4 \ln(c)$

c) Output elasticity for each unit cost component is the estimate of b_1 in
$E[\ln(\text{unit cost component})] = b_0 + b_1 \ln(m/e)$

Table 16: Percent Decrease In Unit Cost Derived from Estimate of Output Elasticity

Percent Increase in Output	Percent Decrease in Unit Cost Assuming Elasticity Equals:	
	-0.295[a]	-0.436[b]
50	11.3	16.2
100	18.5	26.1
200	27.7	38.1
300	33.6	45.4
400	37.8	50.4
500	41.0	54.2
1000	50.7	64.8

a) Output elasticity derived from estimate of b_1 in

$$E[\ln(va/m)] = b_0 + b_1 \ln(m/e) + b_2 \ln(bmci) + b_3 \ln(1+pmbm) + b_4 \ln(w) + b_5 \ln(c)$$

b) Output elasticity derived from estimate of b_1 in

$$E[\ln(va/m)] = b_0 + b_1 \ln(m/e)$$

Δ unit cost $= (1 + \Delta output)^{-b_1} - 1$

b_1 = elasticity of unit cost with respect to output

between unit cost and the level of output is derived from an analysis of industries using the *current* generation of production technology, this is only an indirect analysis of unit production costs in a factory making use of the new generation of flexible production technologies. Hence, the analysis is, at best, suggestive of the economics of production in a newly designed, flexibly automated plant. It is not known whether a more direct and detailed analysis would yield the same conclusions. Nonetheless, if the inference of this analysis is correct, and it is the case that a flexibly automated batch production plant would have a substantial cost advantage over a conventionally organized facility, one would expect that these new types of plants would rapidly diffuse throughout manufacturing industries.

If unit cost in a fully utilized flexibly automated factory is so much less than in a conventional factory, one wonders why so few have been built. Is it too difficult and too expensive to build such a plant, or is it the result of other less tangible factors? No formal analysis has been carried out to address this question. However, a few informal interviews with major manufacturing companies revealed that several companies have plans on the drawing boards to build such plants. This suggests that within the next few years, more attempts will be made to construct flexible manufacturing systems in the U.S. Some executives commented that organizational barriers have stopped plans for building such plants. One interesting comment is that there are situations where such a plant would have more capacity than could be utilized by one division of a company. To be fully utilized, it would have to be shared across divisions. It has been suggested that this generates organizational resistance because the plant is no longer "captive" to one manager.

If a plant could be built that has several times the capacity of a conventional batch production plant, there is the possibility that several old plants could be closed down and their production consolidated into the new facility which has the flexibility to produce a mix of different products. This seems to be a likely scenario if the flexibly automated plant were built in a mature industry where the potential for market growth was limited. One example worked out, using the results presented here, shows that if three plants were closed down and their output consolidated into one high volume, flexibly automated plant, total labor requirements would decrease by 30 to 40 percent. This decrease in labor requirements is based on the elasticity of unit production labor cost estimated from the regression analysis (which is based on the use of conventional types of technologies across low, medium and high volume industries). The available information on the existing flexible manufacturing systems suggests that the one flexibly automated plant might have substantially fewer workers than even one of the smaller plants it replaces. If this were the case, the percentage decrease in total labor requirements would be much larger.

Since the flexible factory scenario holds the largest promise for reducing unit cost, and potentially poses the largest threat to employment in an industry, it warrants more serious analysis. Further research should focus on a more refined and direct analysis on the economics of production in flexibly automated factories, and on forecasts of their use throughout specific industries.

References

[American Machinist 78]
American Machinist.
The 12th American Machinist Inventory of Metalworking Equipment, 1976-1978.
American Machinist 122(12):133-148, December, 1978.

[Ayres and Miller 83]
Ayres, Robert U. and Steven M. Miller.
Robotics: Applications and Social Implications.
Ballinger Publishing Co., Cambridge, Mass., 1983.

[Bureau of Labor Statistics 82]
Bureau of Labor Statistics, U.S. Dept. of Labor.
The National OES Based Industry-Occupation Matrix for 1980.
Government Printing Office, Washington, D.C., 1982.

[Bylinski 83]   Bylinski, Gene.
The Race to the Automatic Factory.
Fortune 107(4):52-64, February, 1983.

[Carnegie-Mellon 81]
Carnegie-Mellon University.
The Impacts of Robotics on the Workforce and Workplace.
Department of Engineering and Public Policy, Carnegie-Mellon University, Pittsburgh, Pa., 1981.
A student project cosponsored by the Department of Engineering and Public Policy, the School of Urban and Public Affairs, and the College of Humanities and Social Sciences.

[Census of Manufacturers 81a]
Bureau of the Census, Industry Division, U.S. Dept. of Commerce.
1977 Census of Manufacturers, MC77-SR-1: General Summary.
Government Printing Office, 1981.

[Census of Manufacturers 81b]
Bureau of the Census, Industry Division, U.S. Department of Commerce.
Industry Statistics, Part 3, SIC Major Groups 35-39. Volume II: *1977 Census of Manufacturers.*
Government Printing Office, Washington, D.C., 1981.

[Census of Manufacturers 81c]
Bureau of the Census, Industry Division, U.S. Department of Commerce.
Industry Statistics, Part 2, SIC Major Groups 27-34. Volume II: *1977 Census of Manufacturers.*
Government Printing Office, Washington, D.C., 1981.

[Ciborra, Migliarese, and Romano 80]
Ciborra, Claudio, Piero Migliarese, and Paulo Romano.
Industrial robots in Europe.
Industrial Robot 7(3):164-167, September, 1980.

[Cook 75]   Cook, Nathan H.
                 Computer-Managed Parts Manufacturing.
                 Scientific American 232(2):22-29, 1975.

[Engelberger 80] Engelberger, Jospeh F.
                 Robotics in Practice.
                 American Management Association, New York, 1980.

[Industrial Robot 81]

                 Robotics in the UK.
                 Industrial Robot 8(1):32-38, March, 1981.

[Mayer and Lee 80]
                 Mayer, John E. and David Lee.
                 Estimated Requirements For Machine Tools During the 1980-1990 Period.
                 In Arthur R. Thompson, Working Group Chairman (editor), *Machine Tool
                     Systems Management and Utilization*, pages 31-41. Lawrence
                     Livermore National Laboratory, October, 1980.
                 Volumn 2 of the Machine Tool Task Force report on the Technology of
                     Machine Tools.

[Miller 83]   Miller, Steven M.
                 *Potential Impacts of Robotics on Manufacturing Cost Within Metalworking
                     Industries.*
                 PhD thesis, Carnegie-Mellon University, 1983.

[Smith and Wilson 82]
                 Smith, Donald N. and Richard C. Wilson.
                 Industrial Robots: A Delphi Forecast of Markets and Technology.
                 Society of Manufacturing Engineers, Dearborn, Michigan, 1982.

[Vietorisz 69]   Vietorisz, Thomas.
                 *UNIDO Monographs on Industrial Development: Volume 4: Engineering
                     Industry.*
                 United Nations Industrial Development Agency, Vienna, Austria, 1969.

[Whitney et al 81] Whitney, Daniel E., et al.
                 Design and Control of Adaptable-Programmable Assembly Systems.
                 Technical Report R-1406, Charles Stark Draper Laboratory, Inc., 1981.
                 Prepared for the National Science Foundation, Grant No. DAR77-23712.

[Yoshikawa, Rathmill, and Hatvany 81]
                 Yoshikowa, Hiroyuhi Keith Rothmill, and Josef Hatvany.
                 Computer-Aided Manufacturing: An International Comparison.
                 Technical Report, National Academy Press, 1981.
                 Sponsored by the National Research Council.

[1]Throughout this paper, the Standard Industrial Classification System (SIC), use by the Bureau of the Census, is used to define industries and products.

[2]For one type of processes, metalcutting machine tool operations, an estimate is made of

the percent of tools that could be operated by Level I and Level II robots in order to check the validity of the survey estimates. The two estimates of the potential for robot use in metalcutting machine operations, derived independently of one another, are in close agreement. It appears that the survey based estimates are good indicators of the potential for using robots to operate metalcutting machine tools, and there is no strong reason to disbelieve the survey based estimates of potential robot use in the other application areas either.

[3]In a state-of-the-art Level II application for machine loading developed at the CMU Robotics Institute, much of the added expense was the result engineering effort required to to improve the communication between the commercially available robot and vision system and the control of the overall system. The actual Level II hardware, the vision system, only accounted for a small part of the cost difference. It appears that if the vendors made minor modifications to their commercially available systems, it would be possible to achieve the degree of communication and control required for sophisticated applications without extensive engineering efforts. This would substantially reduce the cost of a Level II installation.

[4]The term "basic metals" refers to inputs of "raw" metal stock-- steel, brass and aluminum in the form of bars, billets, sheets, strips, plates, pipe, tubes, etc., as well as casting and forgings made of the three basic metals. The term "processed metals" refers to inputs which are themselves the products of the industries in major groups SIC 34-38. In general, these products are basic metals which have been further processed within the metalworking industry.

Pounds of metal processed is divided by the number of establishments within the industry to adjust for differences in the number of establishments across industries.

There are an additional 31 industries in SIC 34-47 which are excluded because of inadequate data on their material inputs.

[5]This regression result is estimated with data aggregated at the three digit SIC level, and is not shown in Table , which gives results for data aggregated at the four digit SIC level.

[6]Industries in the sample account for almost 90 percent of the total value added in the "universe" of metalworking industries that are considered.

5. Custom, Batch and Mass Production in the Metalworking Industries

Steven M. Miller[a]

Carnegie-Mellon University

March 14, 1984

This paper is adapted from the author's Ph.D thesis, *Potential Impacts of Robotics on Manufacturing Costs in the Metalworking Industries,* submitted to the Engineering and Public Policy Department, Carnegie-Mellon University, in May, 1983. This research was supported by the Carnegie-Mellon Robotics Institute, the Engineering and Public Policy Department, and the Carnegie-Mellon Program on the Social Impacts of Information and Robotic Technologies. I am especially grateful to Robert Ayres, committee chairman, and to Larry Westphal for serving as advisors to the research which lead to the Ph.D. thesis. The present condensed version was supported by the National Science Foundation, Division of Policy Research and Analysis, grant # PRA-8302137 826.

a: Assistant professor, Engineering and Public Policy and Industrial Administration.

Abstract

The relationship between the level of unit cost and the level of output produced is examined across 101 four digit industries in SIC 34-47. Using data from the 1977 Census of Manufacturers, an estimate is made of the pounds of basic metals and of processed metal inputs purchased by each industry. This measure is used as a surrogate measure of the level of output of each industry. Value added per unit and units of output are computed for each industry using pounds of metal processed as the standardized unit of output. A comparison of cost versus the scale of output across the cross section . of metalworking industries exhibits the same key structural patterns that underlie the shape of the long run average cost curve for a particular product. Since there is a correspondence between the the level of output, the level of unit cost and the mode of technology used in the long-run unit cost curve for a particular product, it is assumed that there should also be similar correspondences in a comparison of cost versus output *across* industries. Thus, the dominant mode of technology used within an industry is inferred according to its position on a graph of average cost versus output. Each of the 101 industries is explicitly classifed as either being dominated by custom and small batch production, mid-batch production, large batch production, or mass production.   Given the assignments of industries to a mode of production, the distribution of value added, output, and production worker employment by mode of production is calculated. This analysis corroborates the claim that most of the value added in the metalworking sector is accounted for by products which are batch produced.

1 Constructing Measures of Output and Unit Cost

The relationship between the level of unit cost and the level of output produced is examined across 101 different metalworking industries.[1] An estimate of the pounds of basic metals and of processed metal inputs purchased is used as a surrogate measure of the level of output of each industry.[2] Average unit cost is approximated by the quantity value added per unit of ouput. Value added per unit and units of output are computed for each industry using pounds of metal processed as the standardized unit of output.

---

[1] Industries from the following groups are included in this analysis (followed by their standard industrial classification code): fabricated metal products (SIC 34), Machinery, except electrical machinery (SIC 35), electrical and electronic equipment, machinery and supplies (SIC 36), and transportation Equipment (SIC 37).

[2] The term "basic metals" refers to inputs of "raw" metal stock-- steel, brass and aluminum in the form of bars, billets, sheets, strips, plates, pipe, tubes, etc., as well as casting and forgings made of the three basic metals. The term "processed metals" refers to inputs which are themselves the products of the industries in major groups SIC 34-38. In general, these products are basic metals which have been further processed within the metalworking industry.

The pounds of output produced by each industry is not directly measured, and must be estimated. Because of the limitations of available industry wide data, the pounds of output must be approximated from the available information on material *inputs* purchased by each industry. This presents a problem because metal removal is one of the primary means of production in the metalworking industries. If material is removed form metal bars, plates and sheets to shape an industry's product, then the weight of the output will be less than the weight of the purchased material inputs. (In fact, the difference between the weight of the purchased materials and the weight of the final product is a good indicator of the complexity of the production process.) Given available data sources, there is no way to obtain a measure of the weight of the final output, so one has to settle for a measure of the pounds of metal inputs. Even with this compromise, there are still problems with obtaining a reliable measure of the pounds of material inputs used. First, while the Census of Manufacturers lists the total value of materials purchased by each industry, only part of the total is "specified in kind", where a subtotal for each specific type of material purchased is shown.[3] Not all of the materials can be used in the estimate of pounds of input purchased. Second, the percentage of total materials purchased that are specified in kind varies across industries, so the material coverage for each industry is not the same. Third, and most important, for some categories of specified materials, the amount purchased is given only in dollar terms, making it necessary to estimate the pounds of these materials received. For the other categories of materials specified in kind, the pounds of material received by all establishments within the industry, along with the dollar value of the purchase is given by the Census of Manufacturers.

For most metalworking industries, metal inputs account for most of the value of materials which are specified in kind. Only metal inputs, including both basic and processed metals, are used to estimate the pounds of materials used by each industry. This simplifies the laborious task of estimating the pounds of materials used by each industry by a considerable amount, and in most cases, excludes only minor quantities of materials. If purchases of basic metals and of processed metals account for less than 50 percent of an industry's total material purchases, in dollar terms, the industry is excluded from the analysis. This rule excludes 6 industries from SIC 34, 3 industries from SIC 35, 18 industries from SIC 36 (half of all 4 digit industries), and 4 industries from SIC 37 because of low material coverage. The remaining industries are appropriately designated as metalworking industries, since specified metal inputs account for at least half of total material cost.

---

[3]Throughout this report, the term "material purchases" or "material cost" refers to the direct charges paid (or payable) for all raw materials, as well as for semifinished goods, parts, components, containers, scrap and supplies put into production or used as operating supplies during the year, including freight charges and other direct charges incurred by the establishment in acquiring these materials. The following items are not included in whenever the item material cost is discussed: 1) Electric energy purchased, 2) fuels consumed for heat, power or electricity, 3) materials bought and subcontracted out to be worked on by other manufacturing establishments (contract work), and 4) materials bought and resold in the same condition (resales). For the most part, the raw materials and semifinished goods included here are physically embodied in the product itself.

Material inputs purchased from other metalworking industries ("processed" metal inputs) are only given in dollar terms, making it necessary to estimate the pounds of basic metal "embodied" in a dollars worth of purchase of a processed metal input. The estimation procedure is summarized here. Basically, the estimate of basic metals purchased by each industry is used to approximate the metal content of processed metal inputs. The total pounds of metal processed by industry i can be estimated by

$$m_i = b_i + \sum_j a_{i,j}b_j + \sum_k \sum_j a_{j,k}a_{i,j}b_j + \sum_l \sum_k \sum_j a_{k,l}a_{j,k}a_{i,j}b_j + ... \qquad (1)$$

with

m_i = pounds of metal processed by industry i
b_i = pounds of basic metals processed by industry i

$$a_{i,j} = \frac{\$ \text{ sales from industry j to industry i}}{\$ \text{ output of industry j}}$$

= proportion of industry j's output sold to industry i

These equations can be summarized in matrix form:

$$M = [I + A + A^2 + A^3 + ...]B$$

where the solution to above equation is given by the well known result in input-output theory,

$$M = [I - A]_1 B \qquad (2)$$

with

M = column matrix of all m_i values. Dimension (n x 1)
I = n x n identity matrix
A = matrix of output coefficients, $a_{i,j}$. Dimension (n x n).
B = column matrix of all b_i values. Dimension (n x 1).

Equation (1) is explained with the assistance of Figure 1, which illustrates how basic metal inputs are "embodied" in purchases from other metalworking industries. The pounds of basic metal directly purchased by industry 1 is given by b_1 in the figure. The proportion of industry 2's output sold to industry 1 is given by $a_{1,2}$. The pounds basic metal purchased by industry 2 is given by b_2, and the other output coefficients, $a_{2,3}$, and $a_{2,4}$ are derived from the sales of industries 3 and 4 to industry 2. Similarly, the proportion of industry 3's output sold to industry 1 is shown ($a_{1,3}$), along with the basic and processed metal inputs used by industry 3. The first summation in equation (1),

$$\sum_j a_{i,j}b_j, \qquad (3)$$

which equals

$$a_{1,2}b_2 + a_{1,3}b_3$$

gives an estimate of the basic metal directly embodied in the processed metal inputs purchased by industry 1. The second summation in equation (1) would add the amount of basic metal inputs embodied in the processed metal inputs of the processed metal inputs. For example, in Figure 1, industry 1 buys inputs from industry 2, and industry 2 buys outputs from industry 3. The basic metal embodied in a dollers worth of purchase from industry 2 depends on b_2, but also in the basic metal embodied in the processed metal inputs purchased by industry 2 which is given by $a_{1,2}a_{2,3}b_3$. In matrix form, equation (2) would count all of the indirect contributions of basic metal embodied in the processed metal inputs.

The basic metal directly embodied in the processed metals can easily be calculated from the first summation (equation (3) without explicitly using the full (150 by 150) matrix of output coefficients, A. The calculation of all of the indirect contributions of basic metals embodied in the processed metal inputs purchased by each industry requires the matrix A to be inverted. In this analysis, equation (2) is not solved due to the problems associated with forming and inverting a very large matrix. Only the basic metals directly embodied in the processed metal inputs are included by calculating equation (3) for each industry in the analysis. If industry i's inputs were comprised mostly of processed metals, as opposed to basic metals, and if the processed metals themselves were comprised mostly of processed metals, the estimate of the pounds of metal processed would be substantially understated because the indirect chains are not included here.

Differences in the number of establishments within each industry introduce noise into a comparison of average cost versus output when industries are used as the unit of analysis. Individual manufacturing establishments are grouped together, based on their primary product, to define a industry, and there are differences in the number of establishments assigned to each industry. Since an industry's output is the aggregated total of the output of these individual establishments, it depends on the output per establishment, and on the total number of establishments within the industry. An industry with a small number of establishments, each producing a large volume of output, could have the same, or even less total output as another industry with a large number of establishments, each producing a small volume of output. If there were two such industries, the first industry (few establishments, large output per establishment) would have a lower unit cost than the second one (many establishments, small output per establishment) even though its total output was less than that of the second industry. To adjust for this problem, the measure of output is redefined from pounds of metal to pounds of metal/number of establishments.

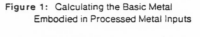

Figure 1: Calculating the Basic Metal
Embodied in Processed Metal Inputs

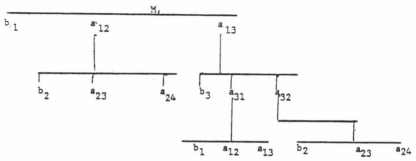

M_1 = total pounds of metal processed by industry 1.
b_1 = pounds of basic metal processed by industry 1.

$$a_{i,j} = \frac{\text{\$ of sales from industry } j \text{ to industry } i}{\text{\$ output of industry } j}$$

2 Identification of Custom, Batch and Mass Production Industries

For a particular product, unit cost decreases in a regular fashion as output increases over a wide range of volumes if one considers the lower envelope of the "long-run" average cost curve, where the optimal (i.e. cost minimizing) technology is used for each level of output.   The underlying determinants of the shape of long run average cost curve are as follows:

- Capital cost per unit of output decreases as the level of output increases.

- Direct labor cost per unit of output decreases as the level of output increases.

- Value added per unit of output decreases as the level of output increases.

- The level of machine utilization increases as the level of output increases.

A comparison of cost versus the scale of output across a cross section of industries exhibits the same key structural patterns that underlie the shape of the long run unit cost curve for a particular product:

- Capital cost for equipment and machinery per pound of metal/establishment decrease across industries as the pounds of metal/establishment processed increases.

- Production labor cost per pound of metal/establishment decreases across industries as the pounds of metal/establishment increases.

- Value added per pound of metal/establishment decreases across industries as the pounds of metal/establishment increases.

- Machine utilization increases across industries as the pounds of metal/establishment increases.

A comparison of cost versus the scale of output across the cross section of metalworking industries exhibits the same key structural patterns that underlie the shape of the long run average cost curve for a particular product. Since there is a correspondence between the the level of output, the level of unit cost and the mode of technology used in the long-run unit cost curve for a particular product, we assume that there should also be similar correspondences in a comparison of cost versus output across industries. This is the basic rational for proposing that the dominant mode of technology used within an industry can be inferred from the the measures of the level of output and of average cost per unit. We assume

- industries with the highest levels of unit cost and with the lowest levels of output are comprised mostly of custom and small batch producers,

- industries with the lowest levels of unit cost and with the highest levels of output are comprised mostly of mass producers, and

- industries with mid-range levels of unit cost and with mid-range levels of output are comprised mostly of batch producers.

3 Rules for Classifying Industries

Within each major group of industries, the variables cost per pound and pounds of metal/establishment are partitioned into low, medium and high levels, as shown in Table 1. There are nine (3X3) cells if all combinations of levels for both variables are considered. Pounds of metal/establishment increases from cell to cell moving across the table from right to left. Unit cost decreases from cell to cell moving down the table from top to bottom. The top right corner cell for observations with the highest levels of unit cost and with highest levels of output is always empty, as is the bottom left corner cell for observations with lowest levels of unit cost and lowest levels of output. Referring back to the idealized long run cost curve for a particular product, unit cost decreases as the scale of output increases. Considering this, these two cells should be empty, since in the theoretical unit cost curve for a particular product, the highest levels of unit cost occur with the lowest level of output, and the lowest level of unit cost occur with the highest level of output. These two cells *are* empty, with one exception. There is one industry which has a high level of unit cost and a high level of output. It is believed that that the large unit weight of the products of this one industry (engines and turbines) results in a distortion of the output measure. This raised the question of whether differences in unit weight of products across industries exerts a larger influence on the measure of output than the number of units produced. There are enough examples where industries with products with small unit weights have greater measures of output than industries with products with large unit weights to suggest that differences in the number of units produced, and not in the size of the units produced, is mostly responsible for the relative levels of output across industries.

The remaining seven cells are labelled region 1 through region 7. Each region is associated with a mode of production, as is indicated in the Table 1. Starting with the top left corner, the three cases along the diagonal are designated as custom production (region 1), medium batch production (region 3) and mass production (region 7). These three cells represent the ideal correspondences between the level of cost, the level of output, and the mode of technology which have been discussed previously. If there were not large differences in the nature of the processing requirements across industries, in the unit weights of products, or errors in the estimation of pounds of output, one might expect all observations to fall within the three diagonal cells. However, a fair number of observations fall in region 2 and region 3, which both adjoin region 1, as well as into region 5 and region 6, which

Table 1: Clustering of Industries by Mode of Production

POUNDS OF METAL / NUMBER OF ESTABLISHMENTS

(millions)

	LOW	MIDDLE			HIGH	
	(m_1)	(m_2)	(m_3)		(m_4)	(m_5)	(m_6)

HIGH (u_1)	Region R1		Region R2				
(u_2)	CUSTOM PRODUCTION — highest levels of unit cost — lowest levels of output		CUSTOM-SMALL BATCH PRODUCTION — highest levels of unit cost — mid-range levels of output			EMPTY — highest levels of unit cost — highest levels of output	
MIDDLE (u_3) Value	Region R3		Region R4			Region R5	
ADDED PER POUND OF METAL	CUSTOM-SMALL BATCH PRODUCTION — mid-range levels of unit cost — lowest levels of output		mid-BATCH PRODUCTION — mid-range levels of unit cost — mid-range levels of output			LARGE BATCH PRODUCTION — mid-range levels of unit cost — highest levels of output	
(u_4)							
LOW (u_5)			Region R6			Region R7	
(u_6)	EMPTY — lowest levels of unit cost — lowest levels of output		LARGE BATCH PRODUCTION — lowest levels of unit cost — mid-range levels of output			MASS PRODUCTION — lowest levels of unit cost — highest levels of output	

both adjoin region 7. Observations in regions 2 and 3 are designated as custom or small batch production industries, since they have at least one of the attributes of custom production (either high cost or small volume), and none of the attributes of mass production (neither low cost nor large volume). Analogously, observations in regions 5 and 6 are designated as industries which produce in large batches , since they have at least one of the attributes of high volume production, and none of the attributes of custom production. The industries which fall into each of these cells will be shown in subsequent sections.

For each major group of industries, the boundary points defining each of the regions are shown along the top and down the left side of Table 1, enclosed in parenthesis. For example, in the illustrative table, the mid-range levels of output are defined to lie within the the closed interval $[m_3$ $m_4]$. The minimum output of any industry in falling into region 2, region 4, or region 6 is m_3 pounds, and the maximum output of any industry falling within these same regions is m_4 pounds. The mid-range levels of unit cost are defined to lie within the closed interval $[u_3 \ u_4]$. The maximum unit cost for any industry falling into region 3, region 4, or region 5 is u_3, and the minimum value of unit cost for any industry falling in these same three regions is u_4. The minimum and maximum values of output for all industries included in the group is given by m_1 and m_6 respectively. The maximum and minimum values of unit cost for all industries included in the group is given by u_1 and u_6 respectively.

The crux of the classification problem is to partition the observations of output and of unit cost into low, medium and and high levels. This partitioning is made easier by first subdividing all of the 101 industries in the four digit data set according to each industry's "parent" major industry group (SIC 34, 35, 36, and 37), and forming four groups of industries. While their is no unambiguous boundaries for partitioning the observations, there appears to be rough agreement across all four of the industry groups as to what comprises low, medium and high levels of each variable (Table 2). For example, within each major group, industries which process nearly 10.0 (millions of) pounds of metal per establishment are set off from the "pack" of industries in a scatter plot of value added/pound of metal vs pounds of metal/establishment. Similarly, industries which process less than 0.33 pounds of metal/establishment or less also are also set off from the most of the observations. Based on the way in which industries in the scatter plots are distributed across output levels, the following boundaries are suggested for identifying low, medium and high volume industries.

Low volume: $m/e <= 0.33$

Medium volume: $0.44 <= m/e <= 7.9$

High volume : $m/e >= 9.6$

Table 2: Partitioning Observations of Output and of Unit Cost
into Low, Medium and High Ranges in the Four Major Industry Groups

Partitioning of Pounds of Output/Number of Establishments (m/e)

Major Industry Group (SIC)	Range of Low Values	Range of Medium Values	Range of High Values
34	a)	0.83 - 7.9	10.6 - 35.5
35	0.06 - 0.28	0.40 - 5.3	9.6 - 12.3
36	0.0 - 0.3	0.50 - 6.6	9.9 - 49.8
37	0.0 - 0.155	0.83 - 6.8	11.2 - 49.8

Partitioning of Value Added/Pounds of Metal (va/m)

Major Industry Group (SIC)	Range of Low Values	Range of Medium Values	Range of High Values
34	0.08 - 0.47	0.64 - 1.30	1.83 - 2.27
35	b)	0.59 - 1.40	1.65 - 13.1
36	0.35 - 0.55	0.73 - 1.47	2.52 - 14.13
37	0.35 - 0.47	0.70 - 0.96	2.56 - 13.1

a) Minimum value of m/e is 0.83 in SIC 34

b) Minimum value of va/m is 0.54 in SIC 35

with

$$m/e \ = \ \frac{\text{pounds of metal}}{\text{number of establishments}}$$

Within each of the four major groups, a boundary between medium and high levels of unit cost is apparent, since those industries with value added/pound of metal exceeding 1.65 are set off from the "pack" of observations. The boundary between medium and low cost industries is drawn at 0.55. Perhaps the boundary between medium and low cost industries could be shifted downward to 0.5 or to 0.4. If this were the case, fewer industries would be included in the mass production region. Since one of the desired objectives of this classification is to obtain an upper bound estimate of the proportion of value added that is produced by mass production industries, the more conservative (higher) boundary between low and medium cost production is used. If the boundary line between medium and low level of unit cost were moved downward, fewer industries would be classified as being dominated by mass production. Thus, a large batch producer might be misclassifed as a mass producer, but it is unlikely that a mass producer would be misclassifed as a large batch producer.

Based on the way in which industries in the scatter plots are distributed across levels of unit cost, the following boundaries are suggested for identifying industries with low, medium and high unit cost.

High values of unit cost : $va/m >= 1.65$

Medium values of unit cost: $0.59 <= va/m <= 1.47$

Low values of unit cost: $0.55 <= va/m$

with

$$va/m \ = \ \frac{\text{value added}}{\text{pounds of metal processed}}$$

For each the four major groups of industries, SIC 34, 35, 36 and 37, the grouping of industries according to low, mid-and high levels of value added per pound of metal and according to pounds of metal processed/establishment are shown in histograms (Tables 3, 5, 7 and 9 respectively.) Based on the boundaries specified in Table 2 (which are partly based on the histograms), industries are placed into the seven regions, as shown in Figures 2, 3, 4 and 5. In these figures, each four digit industry is designated by an integer (1, 2, 3, 4,... etc.) A map identifying the SIC code and the name of the industry designated by each integer is given in Tables 4, 6, 8, and 10). These maps are organized by the seven regions described earlier, so an industry's position in the figure will roughly correspond to its position in the map. In the map, industries within a region are ordered according to their measure of value added per pound of metal. The industry with the highest measure of unit cost is listed first. Following the results for the separate groups, the histogram of value added per pound of metal and of pounds of metal/establishment (Table 11) and the scatterplot of value added per pound of metal versus pounds of metal/establishment (Figure 6) are shown for the pooled four digit data set.[4]

The results of the classifications for each of the four major groups are discussed below. The discussion focuses on industries in the plots of value added per pound metal versus pounds of metal which are outliers, and on industries which are classified as being dominated by mass production.

3.1 Classification of Industries in SIC 34: Fabricated Metal Products

Refer to
Histogram : Table 3
Scatterplot: Figure 2
Map: Table 4

Industries falling into the mass production region are

- crowns and closures (point 21),

- iron and steel forgings (point 18)

- metal cans (point 1)

- auto stampings (point 20)

- metal barrels and shipping containers (point 2)

- miscellaneous metal products (point 15)

There is little doubt that the metal can and auto stamping industries are correctly classified as mass producers. The unit cost measure for the miscellaneous metalwork industry (point 15) is substantially lower than for all the other industries. Considering the relatively simplicity of the principle product of this industry, concrete reinforcing bars and joists, and the low grades of material used (this industry pays the least per pound of metal purchased of all industries in SIC 34), it is not surprising that the processing requirements per pound of metal processed are so low. Simple processing requirements performed with low grades of materials also explain why unit cost measure of the metal barrel and container industry (point 2) is so low. Note that the two industries with the lowest levels of unit cost also use materials which have the least expensive purchase price per pound.

---

[4] A separate map showing the identity of all of the industries in Figure 6 is not necessary, since it would be the same as combining all of the maps for SIC 34, 35, 36 and 37.

Table 3: Histograms of Average Cost and of Output for SIC 34

HISTOGRAM OF VALUE ADDED PER POUND OF METAL (IN NATURAL LOG UNITS)

MIDDLE OF INTERVAL	NUMBER OF OBSERVATIONS		INDUSTRIES (SIC CODES)	
-2.500	1	●	3449------------------------|	
-2.250	0				
-2.000	1	●	3412	
-1.750	1	●	3448	
-1.500	2	●●	3441,3465	Low levels
-1.250	4	●●●●	3411,3493,3444,3462,	of unit cost
-1.000	5	●●●●●	3496,3495,3498,3431,3446	
-0.750	5	●●●●●	3469,3499,3466,3443,3452-----|	
-0.500	2	●●	3442,3433-----------------|	
-0.250	2	●●	3451,3429	mid-range levels
0.000	2	●●	3423,3494	of unit cost
0.250	1	●	3432----------------------|	
0.500	1	●	3425 --| Highest levels of	
0.750	1	●	3463 --| cost	

HISTOGRAM OF POUNDS OF METAL/NUMBER OF ESTABLISHMENTS (IN NATURAL LOG UNITS)

MIDDLE OF INTERVAL	NUMBER OF OBSERVATIONS		INDUSTRIES (SIC CODES)	
-0.250	2	●●	----------|	
0.000	1	●			
0.250	2	●●			
0.500	1	●		mid-range levels
0.750	6	●●●●●●		of output
1.000	0				
1.250	2	●●			
1.500	5	●●●●●	---|	
1.750	0				
2.000	3	●●●	----|	
2.250	1	●			
2.500	2	●●			
2.750	1	●		High levels of
3.000	0			output
3.250	0				
3.500	2	●●----------------		

Figure 2: Value Added Per Pound of Metal Vs Pounds of Metal, SIC 34

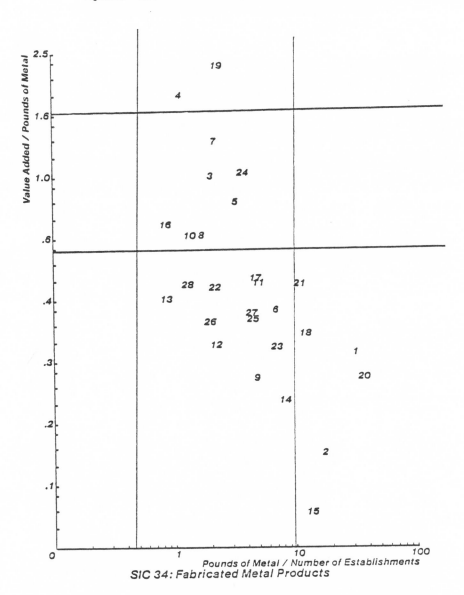

SIC 34: Fabricated Metal Products

Table 4: Clustering of Industries by Mode of Production: SIC 34

POUNDS OF METAL/NUMBER OF ESTABLISHMENTS

(millions)

	LOW		MIDDLE		HIGH	
	(0.0)	(0.0)	(0.83)	(7.9)	(10.6)	(35.5)

HIGH

(2.27)

(1.83)

Region R1	Region R2
	19] 3463 Nonferrous forgings
	4] 3425 Handsaws and saw blades

MIDDLE

(1.3)

VALUE

ADDED

PER

POUND

OF METAL

(0.64)

Region R3	Region R4	Region R5
	7] 3432 Plumbing fittings, brass goods	
	24] 3494 Valves and pipe fittings	
	3] 3423 Hand and edge tools, nec	
	5] 3429 Hardware, nec	
	16] 3451 Screw machine products	
	8] 3433 Heating equipment, except electrical	
	10] 3442 Metal doors, sash and trim	

LOW

(0.47)

Region R6	Region R7
17] 3452 Bolts, nuts, rivets, washers	21] 3466 Crowns and closures
11] 3443 Fabricated platework, boiler shops	
28] 3499 Fabricated metal products, nec	
22] 3469 Metal stampings, nec	
13] 3446 Architectural metalwork	
6] 3431 Metal sanitary ware	
27] 3498 Fabricated pipe and fittings	
25] 3495 Wire springs	
26] 3496 Misc. fabricated wire products	
12] 3444 Sheet metalwork	18] 3462 Iron/steel forgings
23] 3493 Steel springs, except wire	1] 3411 Metal Cans
9] 3441 Fabricated structural metal products	20] 3465 Auto stampings
14] 3448 Prefabricated metal buildings	
	2] 3412 Metal barrels, containers

(0.08)

| | 15] 3449 Misc. metalwork |

Crowns and closures (point 21) and iron and steel forgings (point 18) fall within the mass production region, although they are both close to the designated border between large batch and mass production. It is not know whether these industries are correctly classified, or whether they might be more aptly described as large batch producers.

The nonferrous forging industry (point 19) and handsaws and saw blades (point 4) industries are classified as custom-small batch producers. The two industries in SIC 34 with the highest unit cost also use the materials with the highest price per pound purchased. A major manufacturer of nonferrous forings, who responded to the CMU 1981 Robotics Survey (CMU, 1981) indicated that 60 percent of their products were custom or small batch produced, and that another 40 percent were produced in larger sized batches. This suggests that the nonferrous forging industry is correctly classified. The complex nature of the processing requirements probably explain why unit cost in this industry, and in the handsaw and sawblade industry, are high enough to locate these industries in the custom-small batch production region, as opposed to the medium batch production region.

3.2 Classification of Industries in SIC 35: Machinery, except electrical

Refer to
Histogram : Table 5
Scatterplot: Figure 3
Map: Table 6

None of the industries in SIC 35 are classified as mass producers. Region 7 is empty because the three industries with the lowest values of unit cost, industrial trucks and tractors (point 11), farm machinery and equipment (point 3) and construction equipment (point 5), are designated as having mid-range levels of unit cost. The lowest levels of unit cost in SIC 35 are comparable to the mid range levels of unit cost in the other major groups of industries. One of the largest manufacturers of construction equipment indicated that 90 percent of their products were made in large batches, and the remainder were made in mid-size or small batches. This provides some evidence for the claim that none of the industries in SIC 35 are dominated by mass production.

The turbine and generator set industry (point 1) falls into the top right corner because it has the highest level of unit cost and the highest level of output. This is the only observation in the whole sample that falls into this position. From plant tours, it is known that the turbine and generators set industry is comprised of custom-small batch producers. A plausable explaination for the large level of output in this industry because is the massive size of the products. In this industry, it appears that the large unit weight of the product results in a flagrant misclassification. For purposes of allocating value added by mode of production, the engine and turbine industry is included in custom-small batch production.

Table 5: Histograms of Average Cost and of Output for SIC 35

HISTOGRAM OF VALUE ADDED PER POUND OF METAL (IN NATURAL LOG UNITS)

MIDDLE OF INTERVAL	NUMBER OF OBSERVATIONS		INDUSTRIES (SIC CODES)	
-0.500	2	**	3537,3523------------------|	
-0.400	1	*	3531		
-0.300	3	***	3564,3535,3536		
-0.200	2	**	3581,3582,		mid-range levels
-0.100	2	**	3553,3524		of unit cost
0.000	5	*****	3534,3562,3585,3547,3568	
0.100	4	****	3559,3532,3589,3542	
0.200	4	****	3549,3533,3519,3599	
0.300	3	***	3561,3552,3569-------------|	
0.400	0			
0.500	4	****	3566,3563,3511,3551--|	
0.600	3	***	3567,3554,3576		
0.700	1	*	3544		
0.800	0			
0.900	2	**	3592,3541		
1.000	0			
1.100	1	*	3555		
1.200	0			
1.300	0			
1.400	1	*	3545		Highest levels
1.500	1	*	3566		of unit cost
1.600	0			
1.700	0			
1.800	0			
1.900	0			
2.000	0			
2.100	0			
2.200	0			
2.300	1	*	3573		
2.400	0			
2.500	0			
2.600	1	*	3574----------------|	

Table 5, Continued

HISTOGRAM OF POUNDS OF METAL/NUMBER OF ESTABLISHMENTS (NATURAL LOG UNITS)

```
MIDDLE OF      NUMBER OF        INDUSTRIES
INTERVAL       OBSERVATIONS     (SIC CODES)
 -2.750           1        •      ---|
 -2.500           0                  |
 -2.250           0                  | Lowest levels of
 -2.000           0                  |   output
 -1.750           2        ••        |
 -1.500           0                  |
 -1.250           1        •      ------|
 -1.000           0
 -0.750           2        ••      ---------|
 -0.500           1        •                |
 -0.250           6        ••••••           |
  0.000           1        •                | mid-range levels
  0.250           4        ••••             | of output
  0.500           3        •••              |
  0.750           4        ••••             |
  1.000           4        ••••             |
  1.250           3        •••              |
  1.500           4        ••••             |
  1.750           1        •      -----------|
  2.000           0
  2.250           2        ••      --------| Highest levels
  2.500           2        ••      --------| of output
```

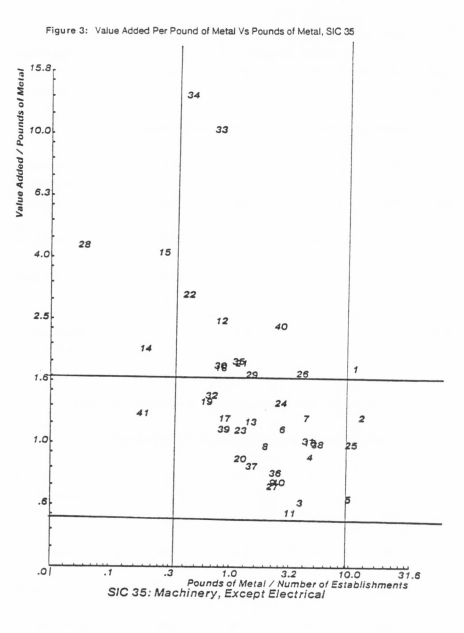

Figure 3: Value Added Per Pound of Metal Vs Pounds of Metal, SIC 35

SIC 35: Machinery, Except Electrical

Table 6: Clustering of Industries by Mode of Production: SIC 35

POUNDS OF METAL / NUMBER OF ESTABLISHMENTS

(millions)

	LOW		MIDDLE		HIGH	
	(0.06)	(0.28)	(0.44)	(5.3)	(9.6)	(12.3)

HIGH

(13.07)

Region R1	Region R2	
	34] 3574 Calculating and accounting machines	
	33] 3573 Electronic computing equipment	
28] 3565 Industrial patterns		
15] 3545 Machine tool accessories	22] 3555 Printing trades machinery	
	12] 3541 Machine tools,metalcutting	
	40] 3592 Carburetors,pistons rings,etc.	
14] 3544 Special tools,dies,jigs,etc.	35] 3576 Scales and balances	
	21] 3554 Paper industry machinery	
	30] 3567 Industrial furnaces and ovens	
	18] 3551 Food products machinery	
	26] 3563 Air and gas compressors	1] 3511 Turbines,generator sets

(1.65)

| | 29] 3566 Speed changers,drives,gears | |

MIDDLE

(1.40)

Region R3	Region R4	Region R5
	32] 3569 General industrial machinery,nec	
	19] 3552 Textile machinery	
41] 3599 Machinery,nec	24] 3561 Pumps and pumping equipment	
	7] 3533 Oilfield machinery	

VALUE

| | 17] 3549 Metalworking machinery, nec | |

ADDED

| | 13] 3542 Machine tools,metal forming | |

PER

| | 39] 3589 Service industry machinery,nec | |

POUND

| | 6] 3532 Mining machinery | |

OF METAL

	23] 3559 Special industry machinery,nec	
	31] 3568 Power transmission equipment	
	16] 3547 Rolling mill machinery	
	38] 3585 Refrigeration and heating equip.	
	8] 3534 Elevators and moving stairways·	

Continued, Next Page

Table 6, Continued

Region R3	Region R4	Region R5
	4] 3524 Lawn and garden machinery	
	20] 3553 Woodworking machinery	
	37] 3582 Commercial laundry equipment	
	36] 3581 Automatic merchandising machines	
	10] 3536 Hoists,cranes,monorails	2] 3519 Internal combustion engines,nec
	9] 3535 Conveyors and conveying equipment	25] 3562 Ball and roller bearings
	27] 3564 Blowers and fans	5] 3531 Construction equipment
	3] 3523 Farm machinery and equip.	
(0.59)	11] 3537 Industrial trucks and tractors	

LOW   Region R6   Region R7

The industries classified as large batch producers are internal combustion engines, nec (point 2), ball and roller bearings (point 25), and construction machinery and equipment (point 5). The output of the ball and roller bearing industry is roughly the same as that of the engine and turbine industry. This is just one of many examples in this data set where industries which produce products which are small in size have a larger output than industries which produce products which are large in size. While differences in the unit weight of products undoubtedly introduces some noise into the analysis, and an occasional gross error (point 1), it is clearly *not the case* that differences in the unit weights of products are primarily responsible for the differences in the levels of output across industries.

Machinery, nec. (point 41) is classified as a custom-small batch producer. According to the industry descriptions in the Census of Manufacturers, this industry is primarily comprised of machine "job shops", establishments primarily engaged in producing or repairing machine and equipment parts on a job or order basis. Based on this information, it seems the industry is correctly classified.

The metalcutting machine tool industry (point 12) is classified as custom-small batch production. Most products of this industry are well known as being custom or small batch produced. This is supported by several machine tool producers who responded to the CMU robotics survey. A major producer of automotive transfer line equipment reported that 85 percent of their products are custom produced. A major manufacturer of general purpose milling machines reported that all of their products are made in small batches. The special tool and die industry (point 14) is also classified as custom producer, which is consistent with the Census industry description which states that most of the products of this industry are produced on a job or order basis.

Almost industries in SIC 35 purchase a large amount of processed metal inputs, in addition to basic metal inputs. This is also true of industries in SIC 36 and 37. Two industries in SIC 35, calculating and accounting machines (point 34) and electronic computing equipment (point 33) are distinguished by the fact that processed metal inputs account for practically all of the total metal input, and the ratio of processed metal purchases to basic metal purchases is inordinately high. These are also the two industries with the highest levels of unit cost in SIC 35. While the unit cost levels for these industries should be high, since they both produce complicated products requiring intricate assembly, there is no apparent reason why the level of unit cost should be so much higher than other industries which also produce complicated products requiring intricate assembly. It is possilbe that the extremely high unit costs are due to an underestimate of the pounds of metal processed. The pounds of metal embodied in a dollar's worth of a processed metal input are estimated in a way that is known to understate the weight of the purchase.

3.3 Classification of Industries in SIC 36: Electrical and Electronic Machinery and Equipment

Refer to
Histogram : Table 7
Scatterplot: Figure 4
Map: Table 8

Only 18 of the 36 four digit industries in SIC 36 are included in the analysis. The rest are excluded because the dollar value of basic and purchased metals does not equal at least 50 percent of the dollar value of total material inputs.

Industries producing noncurrent- carrying wiring devices (point 7), household appliances, nec (point 5) and household laundry equipment (point 4) are classified as mass producers. The main classes of products made in industry SIC 3644, noncurrent carrying wiring devices are electrical transmission line equipment and pole hardware and conduits. The main classes of products made in industry SIC 3639, household appliances, nec, are household water heaters, dishwashers and garbage disposals. Among the industries excluded because of the coverage criteria are household refrigerators and freezers (SIC 3632), electric housewares and fans (SIC 3634), household vacuum cleaners (SIC 3635), and sewing machines (SIC 3636). Based on common observation only, the author guesses that these industries are all dominated by mass or by very large batch production. The household cooking equipment (point 3) industry, which produces electric and gas ranges, is classified as a large batch producer.

The radio and TV communication equipment industry (point 12), is classified as a custom producer. Products made in this industry include communication equipment, television broadcasting equipment, search, detection and navigation equipment, high energy particle accelerator systems, and high energy particle electronic systems. With little doubt, these products are all custom produced. The very high level of unit cost is reasonable, considering the highly specialized and complex nature of the products. There is most likely some understatement in the estimate of pounds of metal processed, since the ratio of processed metal costs to basic metal costs is high, but not nearly as high as the extreme cases in SIC 35.

The radio and TV receiving set industry (point 11), also has a very high level of unit cost. Televisions receivers are the primary product of this industry, and hi-fi systems and speakers are the next largest product. It is puzzling that this industry falls within the custom-small batch production region. Most radios and televisions are consumer goods which are produced in large volumes. However, one must keep in mind that in the year 1977, almost 90 percent of all black and white televisions, 30 percent of all color televisions, and probably a very large percentage of radios were

Table 7: Histograms of Average Cost and of Output for SIC 36

HISTOGRAM OF VALUE ADDED PER POUND OF METAL (IN NATURAL LOG UNITS)

MIDDLE OF INTERVAL	NUMBER OF OBSERVATIONS		INDUSTRIES (SIC CODES)
-1.000	1	•	3644--| Lowest levels
-0.750	1	•	3639	of unit cost
-0.500	1	•	3633--|
-0.250	3	•••	3646,3631,3612--|
0.000	3	•••	3677,3621,3699	mid-range levels
0.250	2	••	3648,3645	of unit cost
0.500	1	•	3694-----------|
0.750	0		
1.000	2	••	3643,3675--|
1.250	0			
1.500	0			
1.750	1	•	3678	
2.000	2	••	3651,3676	Highest levels
2.250	0			of unit cost
2.500	0			
2.750	1	•	3662-----	

HISTOGRAM OF POUNDS OF METAL/NUMBER OF ESTABLISHMENTS (NATURAL LOG UNITS)

MIDDLE OF INTERVAL	NUMBER OF OBSERVATIONS		INDUSTRIES (SIC CODES)
-1.000	1	•	---- Lowest levels of output
-0.750	1	•	----|
-0.500	3	•••		
-0.250	0			
0.000	4	••••		mid-range levels
0.250	0			of output
0.500	1	•		
0.750	0			
1.000	1	•		
1.250	1	•		
1.500	1	•		
1.750	0			
2.000	1	•	----|
2.250	2	••	-------|
2.500	1	•		
2.750	0			
3.000	0			Highest levels
3.250	0			of output
3.500	0			
3.750	0			
4.000	1	•	-----|

Figure 4: Value Added Per Pound of Metal Vs Pounds of Metal, SIC 36

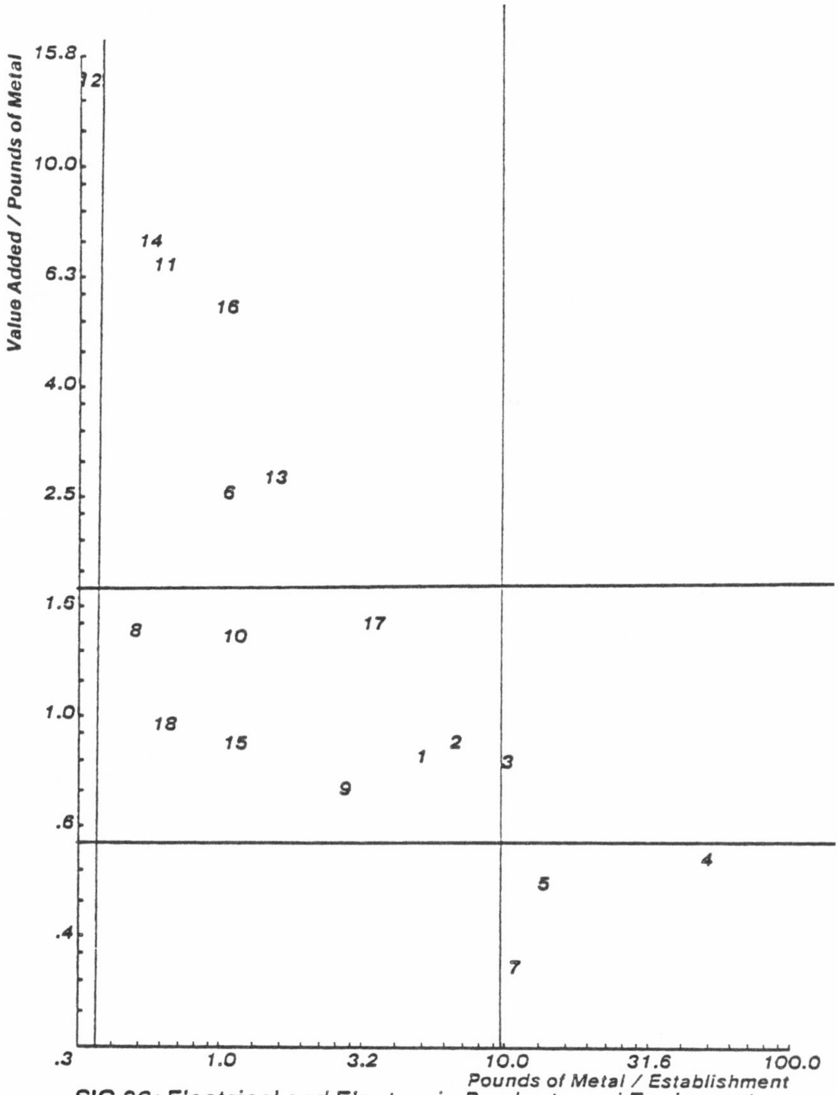

SIC 36: Electrical and Electronic Products and Equipment

Table 8: Clustering of Industries by Mode of Production: SIC 36

POUNDS OF METAL/NUMBER OF ESTABLISHMENTS

(millions)

	LOW	MIDDLE	HIGH	
	(0.0)	(0.33) (0.49)	(6.6) (9.9)	(49.8)
HIGH	Region R1	Region R2		
(14.13)	12] 3662 Radio and TV communication equipment			
		14] 3676 Electronic resistors		
		11] 3651 Radio and TV receiving sets		
		16] 3678 Electronic connectors		
		13] 3675 Electronic capacitors		
(2.52)		6] 3643 Current carrying wiring devices		
MIDDLE	Region R3	Region R4	Region R5	
		17] 3694 Electrical equipment for internal combustion engines		
(1.47)		8] 3645 Residential lighting fixtures		
VALUE		10] 3648 Lighting fixtures, nec		
ADDED		18] 3699 Electronic equipment and supplies, nec		
PER		2] 3621 Motors and generators		
		15] 3677 Electronic coils, transformers, inductors		
POUND		1] 3612 Transformers		
OF METAL		9] 3646 Commercial lighting fixtures	3] 3631 Household cooking equipment	
(0.74)				
LOW		Region R6	Region R7	
(0.54)			4] 3633 Household laundry equipment	
			5] 3639 Household appliances, nec	
(0.35)			7] 3644 Non-current carrying wiring devices	

sold in the U.S. were imported. This leads one to wonder what types of televisions and radios were made by the domestic companies.

The industries which produce electronic components and accessories, electronic resistors (point 14), electronic connectors (point 16) and electronic capacitors (point 13), are all located in the custom-small batch production regions. Two respondents to the CMU Robotics survey who are classified in SIC 367, electronic components and accessories, indicated that almost all of their products custom or small batch produced, which suggests that these industries are correctly classified.

3.4 Classification of Industries in SIC 37: Transportation Equipment

Refer to
Histogram : Table 9
Scatterplot: Figure 5
Map: Table 10

There is no doubt that the industries producing motor vehicle parts and accessories (point 3) and assembling motor vehicles and car bodies (point 1) are correctly classified as mass producers. This is supported by the results of the several plants producing motor vehicle components who responded to the CMU Robotics Survey, as well as by the author's tours of several auto assembly plants. The motor vehicle parts industry (SIC 3714) processes about the same quantity of pounds of metal as the motor vehicle assembly industry (SIC 3711), but has eight times as many establishments. This is why pounds of metal processed/number of establishments is so much larger for SIC 3711 than for SIC 3714. The railroad equipment industry (point 11) also falls into the mass production region. This might be another example where the industry is misclassified because of the very large unit weight of the product. The motor home industry (point 5) also falls into the mass production region. Truck and bus bodies (point 2) and truck trailers (point 4) and transportation equipment, nec (point 14) are classified as large batch producers.

The guided missile and space vehicle industry (point 12) has the highest unit cost in SIC 37. Unquestionably, the products of this industry are custom-small batch produced. This is supported by the author's tour of a missile plant. The output/establishment of the the missile and space vehicle industry seems relatively large, since it is about the same as that of the aircraft industry. This is due to the fact that this the missile/spacecraft industry has only a few establishment producing very large products. All of the aircraft related industries, aircraft and parts (point 6), aircraft engines and parts (point 7) and aircraft parts and auxiliary equipment (point 8) are classified as custom-small batch producers. All three of the major manufacturers of aircraft who responded to the CMU Robotics survey indicated that almost all of their products were custom produced.

Table 9: Histograms of Average Cost and of Output for SIC 37

HISTOGRAM OF VALUE ADDED PER POUND OF METAL (IN NATURAL LOG UNITS)

MIDDLE OF INTERVAL	NUMBER OF OBSERVATIONS		INDUSTRIES (SIC CODES)
-1.000	5	●●●●●	3716,3715,3799,3714,3711--| Lowest levels
-0.750	2	●●	3743,3713----------------| of unit cost
-0.500	0		
-0.250	1	●	3792--| mid-range levels
0.000	1	●	3731--| of unit cost
0.250	0		
0.500	0		
0.750	0		
1.000	1	●	3732--|
1.250	1	●	3728 |
1.500	1	●	3724 | Highest levels
1.750	0		| of unit cost
2.000	1	●	3721 |
2.250	0		|
2.500	1	●	3761--|

HISTOGRAM OF POUNDS OF METAL/NUMBER OF ESTABLISHMENTS (IN NATURAL LOG UNITS)

MIDDLE OF INTERVAL	NUMBER OF OBSERVATIONS		INDUSTRIES (SIC CODES)
-1.750	1	●	---- Lowest levels of unit cost
-1.500	0		
-1.250	0		
-1.000	0		
-0.750	0		
-0.500	0		
-0.250	1	●	------|
0.000	1	●	|
0.250	0		|
0.500	1	●	| mid-range levels
0.750	0		| of output
1.000	2	●●	|
1.250	0		|
1.500	1	●	|
1.750	1	●	|
2.000	2	●●	--------|
2.250	0		
2.500	1	●	----------|
2.750	1	●	|
3.000	1	●	|
3.250	0		|
3.500	0		|
3.750	0		|
4.000	0		|
4.250	0		|
4.500	0		|
4.750	0		|
5.000	1	●	------------|

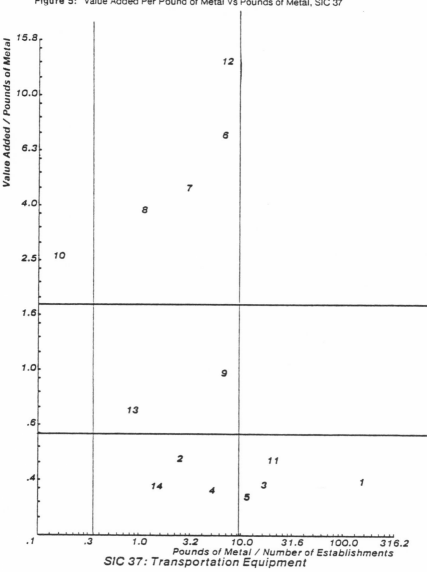

Figure 5: Value Added Per Pound of Metal Vs Pounds of Metal, SIC 37

SIC 37: Transportation Equipment

Table 10: Clustering of Industries by Mode of Production: SIC 37

	POUNDS OF METAL					
	(millions)					
	LOW		MIDDLE		HIGH	
	(0.0)	(0.155)	(0.84)	(6.8)	(11.2)	(148.05)

	LOW		MIDDLE		HIGH	
HIGH (13.10)	Region R1		Region R2			
			12] 3781 Guided missiles and space vehicles			
			6] 3721 Aircraft and Parts			
			7] 3724 Aircraft engines and parts			
			8] 3728 Aircraft parts and auxiliary equipment			
(2.56)	10] 3632 Boat building and repair					
MIDDLE (0.96)	Region R3		Region R4		Region R5	
Value ADDED PER POUND OF METAL (0.70)			9] 3731 Ship building and repair			
			13] 3792 Travel trailers and campers			
LOW (0.47)			Region R6		Region R7	
			2] 3713 Truck and bus bodies			
					11] 3743 Railroad equipment	
			14] 3799 Transportation equipment.nec		1] 3711 Motor vehicles and car bodies	
			4] 3715 Truck Trailers		3] 3714 Motor vehicles parts and accessories	
(0.35)					5] 3716 Motor homes	

The small boat industry is classified as custom production and ship building and repair are classifed as mid-batch production. While ships are in fact custom produced, they are large enough so that many of the components, such as side panels, can be fabricated and assembled in batches. This observation is supported by the author's tour of a large shipyard producting military craft. The very large unit weight of the ships probably results in the industry falling into the batch production region, as opposed to the custom production region.

3.5 The Pooled 4-Digit SIC Data Set

Refer to
Histogram : Table 11
Scatterplot: Figure 6
Map: None given

The industries located in the lowest and highest values of unit cost and of output are designated in Table 11. The ten industries with the lowest levels of unit cost are all in SIC 34, fabricated metal products. For the most part, these ten industries shape and/or join inexpensive types of of steel (with the exception of 3411, which uses a large amount of aluminum). None of these industries purchase considerable amounts of processed metal products. The motor vehicle industries in SIC 371 are ranked between 10 and 20 with respect to industries with the lowest unit cost. All of the industries in 371 purchase sizable quantities of processed metal components. The five industries with the highest unit cost are

3662	Radio and TV communication equipment
3574	Calculating and accounting equipment
3761	Guided missiles and space vehicles
3573	Electronic computing equipment
3676	Electronic resistors

Four of these industries also have the largest ratios of processed metal costs to basic metal costs, which indicates that they are mostly involved in the assembly of components. This suggests that that unit costs is strongly influenced by the nature of the processing requirements. This is explored in greater detail in the next chapter.

The five industries which process the most pounds of metal per establishment are:

3711	Motor vehicles and car bodies
3633	Household laundry equipment
3465	Automobile stampings
3411	Metal cans
3743	Railroad equipment

The five industries which process the smallest amounts of metal per establishment are:

Table 11: Histograms of Average Cost and of Output for
Pooled 4-Digit Data Set, SIC 34-37

HISTOGRAM OF VALUE ADDED PER POUND OF METAL (IN NATURAL LOG UNITS)

MIDDLE OF INTERVAL	NUMBER OF OBSERVATIONS		INDUSTRIES (SIC CODES)
-2.500	1	*	3449
-2.250	0		
-2.000	1	*	3412
-1.750	1	*	3448
-1.500	2	**	3441,3465
-1.250	4	****	3411,3493,3444,3462
-1.000	11	***********	3496,3716,3495,3644,3498,3715,3431,3799,3714,3711,3446
-0.750	8	********	
-0.500	6	******	
-0.250	12	************	
0.000	15	***************	
0.250	11	***********	
0.500	9	*********	
0.750	3	***	3544,3463,3592
1.000	5	*****	3541,3643,3732,3675,3555
1.250	1	*	3728
1.500	3	***	3545,3565,3724
1.750	1	*	3678
2.000	3	***	3651,3721,3676
2.250	1	*	3573
2.500	2	**	3761,3574
2.750	1	*	3662

Table 11, Continued

HISTOGRAM OF POUNDS OF METAL\NUMBER OF ESTABLISHMENTS (IN NATURAL LOG UNITS)

MIDDLE OF INTERVAL	NUMBER OF OBSERVATIONS		INDUSTRIES (SIC CODES)
-2.750	1	*	3565
-2.500	0		
-2.250	0		
-2.000	0		
-1.750	3	***	3732, 3599, 3544
-1.500	0		
-1.250	1	*	3546
-1.000	1	*	3662
-0.750	3	***	3555,3574,3645
-0.500	4	****	3676,3651,3699,3552
-0.250	9	*********	3569,3573,3567,3551,3541,3451,3446,3792,3589
0.000	7	*******	
0.250	6	******	
0.500	6	******	
0.750	10	**********	
1.000	7	*******	
1.250	6	******	
1.500	11	***********	
1.750	2	**	
2.000	6	******	
2.250	5	*****	3531,3631,3562,3644,3466
2.500	6	******	3511,3716,3462,3519,3449,3639
2.750	2	**	3714,3412
3.000	1	*	3743
3.250	0		
3.500	2	**	3411,3465
3.750	0		
4.000	1	*	3633
4.250	0		
4.500	0		
4.750	0		
5.000	1	*	3711

Figure 6: Value Added Per Pound of Metal Vs Pounds of Metal, Pooled 4-Digit Data Set

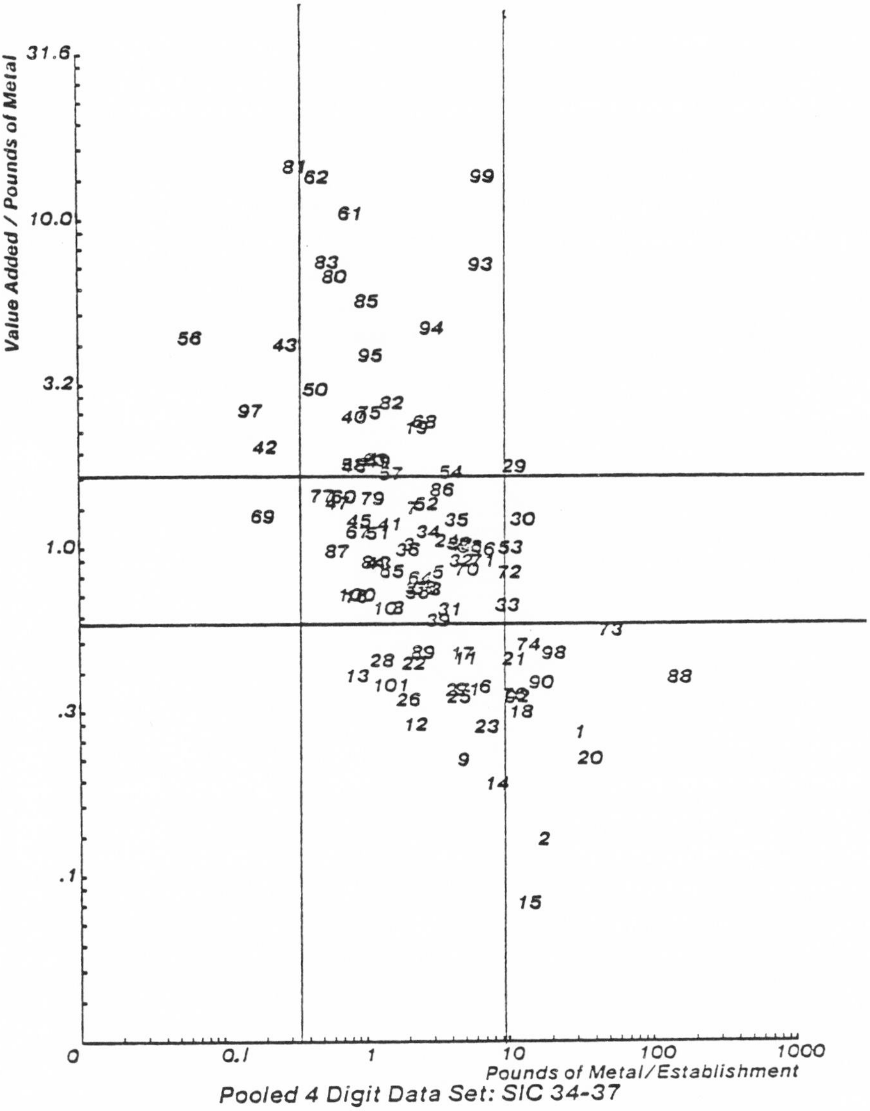

Pooled 4 Digit Data Set: SIC 34-37

3565	Industrial patterns
3732	Boat building and repair
3599	Machine shops
3544	Special tools and dies
3545	Machine tool accessories

It is important to note that, with the exception of railroad equipment, the industries which make the largest, heaviest metalworking products, such as boilers, turbines, and structural metal for buildings, are not among the five industries with the largest output per establishment. Also, industries which produce some of the smallest, lightest products, such as screws, bolts and nuts, and electronic components are not among the five industries with the smallest output per establishment. While there are several notable cases where the large unit weight of an industry's product probably resulted in the industry being misclassified (most notably with engines and turbines), it should be evident that differences in unit weights among products is not dominating the whole analysis.

While it is true that the sources of noise in this classification procedure are potentially troublesome, it is still believed that most of the industries are correctly classified. For example, there is no doubt that the auto related industries--automotive stampings, motor vehicles and car bodies, and motor vehicle parts and accessories-- are all dominated by mass producers, and that industries that produce aircraft, computers, metal cutting machine tools, and special tools and dies are all dominated by custom or small batch producers. It appears that most of those industries where the dominant mode of production is known are correctly classified, and that in most other cases, the classifications are not inconsistent with what is generally known about which types of products are generally custom, batch, and mass produced.

Of special importance, it is believed that the analysis identifies all of the industries in the sample dominated by mass production. The boundaries between regions were established to err on the side of locating some industries which are probably large batch producers in the mass production region, as opposed to the other way around. For example, in SIC 34, crowns and closures (SIC 3466) and iron and steel forgings (SIC 3462) fall within Region 7 and are classified as mass production industries, even though these are believed to be dominated by large batch production. In SIC 37, locomotives (SIC 3743) and motor homes (SIC 3716) are classified as mass production industries, even though they are believed to be dominated by large batch production. However, if the boundaries between medium and high values of unit cost and output were reestablished so as to place these industries in the region of large batch production, it would not be possible to have the same boundary points apply across all four industry groups. The estimate distribution of value added by the mode of production derived from this analysis can be regarded as giving an upper bound on the proportion of value added coming from mass production industries (or correspondingly, as giving a lower bound on the proportion coming from batch production industries).

4 The Distribution of Value Added By Mode of Production

The distribution of value added is estimated in Table 12, based on the classification of 4 digit industries in each of the four major metalworking groups. The value added within each of the regions is also summed across the four major groups to estimate the distribution by mode of production for metalworking as a whole (SIC 34-37). Since industries within each major group with low material coverage are omitted, the combined value added of industries in the sample is less than the combined value added of all industries in each major group. The coverage of the sample is given by the ratio of the combined value added of the industries in the sample to the combined value added of all industries in the major group. The industries included in the sample account practically all of the value added in major groups SIC 34, 35 and 37, but for only half of the value added and output in SIC 36. Across the four groups, the industries in the sample account for 86 percent of total value added of all of the industries in SIC 34-37.

The regions are aggregated together in Table 12, since noise in the data introduces the possibility of misclassification, and since there is some degree of arbitrariness in specifying the boundaries between regions. Observations in regions 1, 2, and 3 are classified as custom-small batch production industries. Observations in region 4 are classified as mid-batch production industries. Observations in regions 5 and 6 are are classified as large batch production industries. Observations in region 7 are classified as mass production industries. Construing batch production broadly to mean everything that is not mass production, regions 1 through 6 would cover the full spectrum of batch sizes, ranging from very small batch (custom) to large batch production. The industries in regions 1 through 6 account for just under 75 percent of the value added in the sample. Stated another way, the industries with the lowest levels of unit cost and with the highest levels of output, which are assumed to be the mass production industries, account for about 25 percent of the value added in the sample. The proportions accounted for by mass production might be slightly understated since several industries in SIC 36 which manufacture household appliances and which are probably dominated by mass production, are excluded from the sample. Included in this group of excluded industries are household refrigerators and freezers (SIC 3632), electric housewares and fans (SIC 3634), household vacuum cleaners (SIC 3635), sewing machines (3636), and telephone and telegraph apparatus (3661). If the value added of these excluded industries were added to region 7, mass production would still only account for 28 percent of the value added in the sample (and for only 25 percent of total value added in SIC 34-37, included all industries). An upper limit on the the proportion of value added and of output accounted for by high volume production, include both large batch and mass production, is derived by considering all of the industries in regions 5, 6, and 7 as high volume producers. The industries in these regions account for just over 42 percent of the value added in the sample.

This analysis corroborates the claim that most of the value added in the metalworking sector is accounted for by products which are batch produced.

The distribution of the total value of output by mode of production is given in the top of Table 13. Total value of output includes material cost as well as value added. The mass production industries account for a higher percentage of the total value of output in the sample than of value added because they process more material per dollar of value added than the batch production industries. Nonetheless, the batch production industries (those in regions 1-6) still account for almost two thirds of the total value of ouput in the sample as compared to three fourths of the value added.

The distribution of the number of production workers by mode of production is given in the bottom of Table 13. The mass production industries account for a smaller percentage of the number of production workers in the sample than of value added because they are relatively less labor intensive than the batch production industries. The batch production industries (those in regions 1-6) account for almost 78 percent of the employment of proudction workers as compared to 75 percent of the value added.

Table 12: Distribution of Value Added by Mode of Production, SIC 34-37

Region and Mode of Production	Major SIC Groups				Total, 34-37
	34	35	36	37	
PERCENT OF VALUE ADDED FOR INDUSTRIES IN SAMPLE					
R1 + R2 + R3 (Custom and Small Batch)	1.1	41.0	58.5	30.6	31.5
R4 (mid-Batch)	28.3	42.5	30.2	6.9	26.4
R5 + R6 (Large Batch)	45.5	16.5	3.1	2.9	16.4
R7 (Mass)	25.1	0.0	8.2	59.6	25.7
Sample coverage of value added in all industries	94.2	95.8	51.2	97.4	86.0

Table 13: Distribution of Output and of Employment by Mode of Production, SIC 34-37

Region and Mode of Production	Major SIC Groups 34	35	36	37	Total, 34-37
PERCENT OF TOTAL OUTPUT FOR INDUSTRIES IN SAMPLE					
R1 + R2 + R3 (Custom and Small Batch)	1.0	36.3	55.1	20.1	24.8
R4 (mid-Batch)	25.3	44.3	31.4	4.9	23.3
R5 + R6 (Large Batch)	45.1	19.4	3.8	2.8	19.9
R7 (Mass)	28.6	0.0	9.7	72.2	35.7
Sample coverage of output in all industries	95.1	96.1	51.0	98.0	88.0
PERCENT OF PRODUCTON WORKER EMPLOYMENT FOR INDUSTRIES IN SAMPLE					
R1 + R2 + R3 (Custom and Small Batch)	0.9	41.8	51.5	24.9	27.8
R4 (mid-Batch)	27.6	42.1	37.9	12.9	29.4
R5 + R6 (Large Batch)	51.9	16.1	3.1	4.6	20.2
R7 (Mass)	19.6	0.0	7.5	57.6	22.6
Sample coverage of production worker employment in all industries	94.5	95.9	52.5	97.3	85.8

References

[Census of Manufacturers 81a]
>Bureau of the Census, Industry Division, U.S. Department of Commerce.
>*Industry Statistics, Part 3, SIC Major Groups 35-39*. Volume II: *1977 Census of Manufacturers*.
>Government Printing Office, Washington, D.C., 1981.

[Census of Manufacturers 81b]
>Bureau of the Census, Industry Division, U.S. Department of Commerce.
>*Industry Statistics, Part 2, SIC Major Groups 27-34*. Volume II: *1977 Census of Manufacturers*.
>Government Printing Office, Washington, D.C., 1981.

[Census of Manufacturers 81c]
>Bureau of the Census, Industry Division, U.S. Dept. of Commerce.
>*Selected Materials Consumed, MC77-SR-11: 1977 Census of Manufacturers*.
>Government Printing Office, Washington, D.C., 1981.

[Census of Manufacturers 81d]
>Bureau of the Census, Industry Division, U.S. Dept. of Commerce.
>*1977 Census of Manufacturers, MC77-SR-1: General Summary*.
>Government Printing Office, 1981.

6. Summary of the Industrial Technology Institute Computer Conference on the Ayres and Miller Papers

BACKGROUND

The computer-conferencing facilities of the Center for Social and Economic Issues of the Industrial Technology Institute, Ann Arbor, Michigan, were used to review the four papers by Ayres and Miller. During 1984, the center had an ongoing computer conference referred to as the Forum on Integrating Research in Socio-Technical Systems. In April, 1984, the computer conference organizer, Patrick Sweet, asked the Forum to focus its efforts on reviewing the Exploratory Assessment of Second Generation Robotics and Sensor-Based Systems. The review started in mid-April and continued through mid-June, 1984.

A "proceeding" of the computer conference follows. All participants are identified at the beginning, followed by the body of comments. For ease of presentation, we have grouped the comments into 15 subject categories. These subject headings were identified after the close of the conference. Within each subject heading, comments are presented in chronological order.

Participants in The Teleconferencing Review of the Ayres/Miller Paper

Robert Ayres   Vary Coates   Faye Duchin
John Ettlie   Eileen Havener   Jan Helling
Bill Hetzner   Timothy Hunt   Dorothy Leonard-Barton
Leonard Lecht   Leonard Lynn   Joe Martino
Steve Miller   David Morrison   Dave Roitman
Gerry Ross   Patrick Sweet   Bobbie Turniansky
Peter Unterweger

Robert Ayres   412-784-8369
I am Professor of Engineering and Public Policy at C-MU. Also co-director of a new interdisciplinary MS program in manufacturing. Also co-author, with Steve Miller of a recent (1983) book entitled Robotics: Applications and Social Implications; and a book due momentarily, entitled The Next Industrial Revolution. I am interested in robots and flexible CIM, but more generally in the processes of technological innovation and change.

Vary Coates   202-966-9307
At present I am running a project on office automation and productivity for the Office of Technology Assessment, U.S.Congress. I'm putting together a panel on impacts of robotics for World Future Society's June Meeting.

Faye Duchin   212-598-3414

John Ettlie   312-321-7803
My most pressing passion is a study funded by NSF on the implementation of programmable manufacturing systems like FMS and robotic cells. I have just finished writing a draft of phase I of this four year project, which is available for review on request. It eventually will be a chapter in a book published by Josey-Bass and is edited by Don Davis based on our conference in Norfolk last Nov(83).

Eileen Havener   412-337-5827
I am a member of Alcoa's Technology Planning Department. Our objective is to investigate technologies of interest to Alcoa – existing as well as emerging technologies. My current assignment is Flexible Manufacturing.

Jan Helling   46-520-85830
Organization consultant within SAAB-SCANIA AB, SAAB Car Division in Sweden. I am involved in organizational change and development projects in various functional areas of our business. I am also member of our Research Contact Team which serves as an environmental scanning function.

Bill Hetzner   202-357-9804
I am a program manager for the Productivity Improvement Research Section of the NSF. I am interested primarily in implementation of advanced manufacturing technology.

Timothy Hunt   616-343-5541

I am a Senior Research Economist at the W. E. Upjohn Institute for Employment Research in Kalamazoo, Michigan. I am co-author of a recent book (1983) titled 'Human Resource Implications of Robotics'. I have a continuing research interest in the employment implications of robotics and technological change generally.

Dorothy Leonard-Barton   617-495-6607
I am a member of the Production Operations Management faculty at Harvard Business School. My research focuses on the development and implementation of new manufacturing technologies such as Artificial Intelligence, robotics, and CAD/CAM. I am particularly interested in the management of organizational interfaces and the introduction of change.

Leonard Lynn   412-578-3571
I am an assistant professor of sociology at Carnegie-Mellon University. My research interests are in the relationships between technology and society. I am now involved in comparative studies of office automation and robotics in Japan and the U.S.

Joe Martino   513-229-3036
I'm a Senior Research Scientist at the University of Dayton Research Institute. I also teach a course in Technological Forecasting in the UD Engineering Management Department. My primary research interests are in measurement of technology, with an emphasis on finding measures which can be forecast.

Steve Miller   412-578-2291
Ass't Prof. at Carnegie-Mellon Univ. I study the applications and implications of flexible automation and machine intelligence in the workplace. Current interests: costs and benefits of these technologies and tools to guide strategic selection of alternatives. Also impacts on job and skill requirements.

David Morrison   313-763-4212
Mid-point in sabbatical leave year looking at implications of robotics and automated manufacturing developments for U.S. foreign policy and for third world economic development. Associated with ITI and the Univ. of Michigan's Institute of Public Policy Studies.

Dave Roitman   313-763-9203
I recently joined the Industrial Technology Institute as a researcher in the Center for Social and Economic Issues. My PhD. is in Psychology, but I don't care for that pigeon-hole; I thrive on multi-disciplinary work. I'm starting up a set of case studies focusing on automotive suppliers who have implemented advanced manufacturing technologies. Looking forward to meeting you.

Gerry Ross   313-662-8050
I am at ITI and ISR and am interested in the managerial and organizational implications of advanced programmable technologies.

Patrick Sweet   313-763-0572
I am interested in the implications of technological change in a variety of settings. More specifically, change processes and introduction techniques, and there implications for altering organizational roles/ managerial responsibilities, are of interst to me and concern to many others experiencing competitive pressure

to adopt new technologies.

Bobbie Turniansky   313-764-6188

I'm a doctoral student in Organizational Psychology at U of M. I work at ITI and am trying to plan my dissertation. My main interests as of now are participative management, employee ownership, and the management of change.

Peter Unterweger   313-863-4702

I am a technology analyst for the UAW, and I work in their research department in Detroit. My academic background is in electrical engineering but now I work mostly in economics, collective bargaining, and computer applications.

Leonard Lecht   212-751-3019
I am an economic consultant concentrating in the human resources area. Previously, I was director of special projects at the Conference Board and director of the Center for Priority Analysis at the National Planning Association. Currently, I'm an adjunct professor at the City University of New York.

OTHER CRITICISMS NEC

Item 50* Leonard Lecht: I have read the manuscript mainly with an eye to the employment and social implications. One question keeps popping up as I read: Why haven't robots taken on more rapidly in the U.S. in the past five years? Given the fact that Level I robots have many limitations, they also have many uses. American industry has spent large sums indeed on plant and equipment, and industry has been quick to adopt electronic technology. Why not the same, or a similar pace, with robots?

The paper, especially the Miller paper, says a great deal about labor costs that is interesting. But it is done under highly restrictive or speculative assumptions. I suspect the real cases lie in-between. As Ayres points out in his essay, widespread use of robots would change the size and location of manufacturing establishments before the year 2000. This may be overly optimistic, but to the extent it is true it is unlikely that robots would be used simply as a substitute for direct labor without otherwise affecting the organization of production and the technology used. In these cases, there must be substantial capital costs involved as well as the cost of the robot itself. Where part of the existing plant or equipment is no longer needed because of the use of robots, the savings from the use of the robots must cover not only the cost of the machines but the loss of value of the old plant and equipment that is no longer used. It would make economic sense for firms to delay the introduction of robots until their existing plants had been depreciated away or the savings in labor costs covered the loss.

The discussion of new industries in which robots could create employment was interesting. It may be that the potential of robots is greater in new fields such as mining the ocean beds than in depending on high price elasticities for existing products to lead to output increases that more than offset the labor displacement caused by robots. There are two elasticities here. One is the elasticity of demand for the product. The other is the elasticity of substitution between robots and blue collar workers. The discussion of labor displacement potentials in large plants, especially with Level II robots, presupposes that you expect the elasticity of substitution to be high. If it were not, there would be no serious employment problem.

The emphasis on quality and performance rather than price competition appears to be overdone. There are tradeoffs here. For most noncritical uses people will put up with mediocre quality if the price is low enough. The Japanese have made their inroads in the U.S. in autos, or consumer electronics, because they produced good quality products and products that were priced lower in most instances than their domestic competitors.

Ayres' discussion of alternative sources of employment for displaced workers, say in building or rebuilding the infrastructure, is worth more attention. He seems to have something in mind that sounds like a SuperMAC, or at least an industrial policy. It might be feasible to use something like the old goals approach to project the jobs that would be created by serving needs that are currently unmet by the marketplace or by government, say in infrastructure, mental health, child care, etc. Otherwise, much of the discussion of alternatives for displaced workers leads to the point that so many of the alternatives offer limited opportunities in terms of numbers of jobs or income. For a few minor nitpicks – Since when have most jobs in the U.S. become interesting and interactive? I doubt it. The U.S. Dept. of Labor's projections indicate that the two occupations with the largest estimated employment growth in the next ten or fifteen years are janitors and sextons, number 1, and nurses aides and orderlies, number 2. Neither sounds very exciting. I haven't seen the most recent data, but I doubt if the female dominated fields – teaching, nursing, secretarial work – have been increasing relatively in terms of income. College educated women have been dealing with this problem by going into other fields in larger numbers. Pay for all women tends to stay stuck at about what it has been in the past, about three-fifths of men's pay. I trust Ayres is right about the waiters.

Item 11* Leonard Lynn: The four papers in the Ayres/Miller NSF Report address several issues of considerable importance that have not generally been approached by a systematic analysis. In the first paper Ayres convincingly argues that humanoid robots are not likely to appear in the near future (if ever). There is no particular reason for anyone to develop such devices. If is far more likely that producers of robots will find a better market for specialized devices. Ayres also argues that the actual displacement of workers by robots is likely to be relatively small in the next two generations or so. This is a reasonably comforting response to fears that may have been engendered in the science fiction literature. Ayres says that many blue collar workers who are now well-paid members of the middle class are 'likely to slip back in the socio-economic hierarchy.' He makes a good argument that the answer for these people is not retraining, but rather upgraded education to allow them to assume new positions. But what happens to the people who have not received the upgraded education? Do we give it to them in middle age, write them off, or ignore them? A useful addition to this discussion would be some sense (perhaps based on Miller's data in the later papers) of about how many people are involved, and when the advances are going to occur that cause them to be displaced. Much of the work Miller and Ayres have done seems relevant to this -- they may well be as qualified as anyone working in this area to give us estimates. As the paper now stands it gives an impression of glossing over real potential problems. On the bright side, for example, it argues that female dominated occupations such as teaching will rise in status and salary. No reason is given for this belief. Ayres speculates that Americans will be more appreciative of the skills of those providing services, but doesn't supply us with much reason for thinking so. I think these arguments should either be developed or dropped.

11.2-6* Peter Unterweger: First of all Bob Ayres deserves recognition for venturing into an area that is necessarily strewn with difficulties. There is much in this paper that is insightful and convincing, especially the analysis which shows that humans are superior generalists while machines excel at specific tasks. In addition, many of the 'predictions', from the decline of the manufacturing workforce, to the increasing importance of education (as opposed to training) are right on target.

The functioning of the business cycle in conjunction with new technology, as outlined in the paper has already been demonstrated in the auto industry. The downturn lays workers off, but it is new technology (together with overtime) which prevents their return to work.

The paper could, however, be improved in several ways. First of all, I felt there was a lack of continuity; the post-2000 scenario flows from first section but not from the second. Second, there is a certain lack of balance; eg. the problem of declining manufacturing employment and the painful adjustments that it will necessitate receive less space than the vacuum cleaner robot. Finally, I agree with Leonard Lynn's comment that the ideas in the last section need some work.

It is not so much that I find fault with the ideas themselves than with the manner in which they are presented. They are insightful speculations, and they are useful as such, but they lend but weak support to the comforting inferences (no declining middle, etc) with which they are linked.

The paper underemphasizes the displacement problem; not so much by erroneous estimates of magnitudes, but by not dealing with the economic system's inability to provide sufficient employment as things stand. It can be argued that there already is a longterm trend of rising unemployment upon which the robotics displacement will be superimposed.

There are references to unions, which touch on deeper problems and which either deserve more detailed discussion, or should be eliminated. all unions do not blindly champion protectionism. The UAW, for example, calls for Local Content, not quotas. Their position on this issue as in their call for an Industrial Policy rests on an analysis which indicates that our complicated world (and national) economy is not well served by laissez faire economics, totally unregulated trade etc. These positions can be disputed, but it is unwise, and perhaps unfair to simply gloss over the debate.

These are complicated issues; some trade regulation might lead to higher prices, on the other hand, import invasions, or too rapid technological change can result in social dislocations that may be even more costly. Retraining, better education, infrastructure projects, all require higher government expenditures, which ultimately mean higher taxes. This is not to argue against any of these positions, but simply to show that they deserve a fuller airing than would be appropriate to the main purpose of this paper.

In the concluding paragraphs Ayres states that were it not for unions, some blue collar workers would not be as well paid as they are. Would that be a desirable development? Is the lowering of decent standards of living part of the strategy for America's future?, or will the additional output due to new technology be used to raise the standards of those on the lower end of the scale? If the purpose is merely to say that weakening industrial unions may have difficulty defending their members' incomes, why not just say that? The actual formulation suggests that they don't deserve what they get, which is again a much broader question. Do the wealthy and powerful in this society deserve their privileges? Again, I think that a fair treatment of this topic goes beyond the scope of the paper. There are still other points, but this comment has become quite long as it is.

11.7–8* Robert Ayres: I want to respond very briefly here, mainly to let you know that I am paying attention. Perhaps a more detailed response will be forthcoming later. Regarding Leonard Lynn's question about why the salaries of teachers (and some other female dominated job categories) can be expected to rise in relative terms, I think the answer is simple: demand for teaching will increase as it is increasingly recognised that education is the key to higher income. Peter Unterweger's comments were too extensive to respond to in full (until I can get hard copy), but the points that stick in my mind relate to two issues: (1) will there be a significant increase in gross national income resulting from the introduction of advanced manufacturing technologies and (2) how will it be distributed. Both are clearly dependent on macroeconomic policy and social policy and, therefore, are way beyond the scope of the paper. Perhaps we should write a book. The first question would be relatively easy to answer if (and only if) the U.S. were not competing in a world market. But we are, and while I can see some arguments in favor of protecting certain domestic industries, I also see strong arguments against it. Frankly the whole subject is so complex -- notwithstanding the oversimplifications introduced from both sides of the debate -- that I honestly don't know where I stand on it. Except, that I am convinced that we (the U.S.) will be a lot worse off if we go slow on robotics than if we don't. With regard to distribution (social policy) again the complexities are awesome. I'm sympathetic with the complaints of minorities and disadvantaged persons who have long had incomes at the low end of the spectrum. The have not had a "decent standard of living" in the past, primarily because they have not had the kind of clout a strong industrial union provides. Now the unions are losing their clout, and some of their members will be in the same situation as others who lack skills or education. I don't suggest they "deserve" this, because I don't have the right to make moral judgments. But those who haven't been so fortunate in the past possibly didn't deserve their fate either. What do we do about the displacees? That is a better question. We discussed the matter at some length in our book (Robotics) and in other papers. I guess we didn't want to repeat all this old material in the recent paper.

11.9* Peter Unterweger: I think Bob is at least in part confusing my list of key issues (entered as a response to item 4 with the intention of stimulating debate) with my discussion of his paper. My comments did not urge a fullscale debate of macro or social policy, but suggested that it was better to not stray into these areas than to give them short shrift. The same goes for the question of who 'deserves' what. I did not argue against robotization, but I must admit that I am not so certain that full speed ahead is the best social policy; for one thing it might lead to more resistance than a more controlled approach that allows time for attrition, and retraining.

11.10* Vary Coates: I felt that Bob had set up a straw man, rather unnecessarily, by again delving into the question of differences between humans and robots. But once you were there, why not mention the great advantage of being able to conform physically -- that is be shaped and sized -- to fit the job (humans are pretty standard in this regard); and the disadvantage that robots have of being non-affective and thereby unmotivatable? change More importantly, the lack of integration between the chapters now is troubling; and Steve's chapters would be very much more useful and interesting if each had some interpetive analysis, i.e., at least a final or several final paragraphs of 'so what does this all mean?'.

11.11* Peter Unterweger: I agree that a summary and interpretation would be very useful in Steve's papers. Leonard Lynn raised similar concerns. The purpose of the papers, actually I should have said audience, is not clear. I don't think technical or engineering audiences and social science people can be satisfied within the same paper. The second paper should be split into two parts. Technical categorization and description in one part, and a summary of likely applications and their implications as a second part. The third paper could benefit from summary and interpretation as indicated above.

Item 12* Leonard Lynn: The Miller papers are of considerable interest, but also some frustration for a non-technical reader (like me). The first gives an impressive account of the developments in robotics technology over the past few years. I would very much like to see Miller go one step further with it and help me more clearly see the forest (as opposed to the trees). A table indicating the advances and their industrial implications, together with dates when other advances are expected might be very helpful. I think this would make the paper accessible and readable to a far wider audience than it now is. It would also set more explicit ties with some of the conclusions Ayres reaches. The second paper also promises much that might be delivered with a little more explanation. It takes up the extent to which unit costs and labor requirements might be reduced in manufacturing if robots and FMS were mor widely used. There is a lot of very interesting analysis here, but I would like to know more about the data on which it is based. Mention is made, for example, of surveys of workers within certain occupations that could be replaced by robots in manufacturing, but no information is provided on the sample of 22 establishments surveyed. I think there is a real need to make such estimates as transparent as possible so that others can test or evaluate the underlying assumptions and the possible biases embedded in them. Perhaps in a revised draft Miller could move in this direction. In general I think the volume has the promise to make major contributions. The authors combine uniquely an understanding of the technology with a sensitivity to the social and economic issues involved.

12.1* Steve Miller: Leonard, thanks for being the first to jump into the criticism of the reports and to come forth with some challanging questions. A few comments on your comments: 1) Making the technology piece more accessible to "less technical" people: At this point, it is not technical enough to be of any use of people doing robotics research. It is interesting that you note it is almost too technical for people doing policy related work. To this day, I do not know how to write in a way that satisfies both the depth requirements of the technology community and the breadth requirements of the policy community. 2) On supporting the data presented in the impacts piece: the details are provided in gory detail in my Ph.D. thesis, which I will be glad to forward to you, if necessary. To what extent do they belong in a piece that summarizes conclusions? Does that paper attempt to do too much?

12.2* Leonard Lynn: Steve, with regard to 1: All I can suggest is that you spell things out in terms of their implications. That's why I suggested the table. I suspect that there is a wide audience that might be reached if you led them step by step from the technology to the social impacts. Re 2: I was reacting to the paper as it stands and what I imagined its purposes to be. For me it covers a lot of ground very quickly without providing me with detail to sink my sociological teeth into. More generally I wonder if you should be trying to satisfy both the depth requirements of the technology community and the breadth requirements of the policy community. I suspect the needs of the policy community are greater for the kind of work you do. I would very much like to hear what others think about this.

12.3-4* Steve Miller: Leonard wants the discussion of the impacts of new robotic technologies summarized, with a statement of their broader industrial implications, and finally, with a step-by-step progression starting with the technology and leading from the social impacts. I have some comments: 1) When I started to write about sensor based systems, I had a long list of general topics to cover, including visual, tactile and force sensing. Then within each area, such as visual, there are many applications, such as parts acquisition, relative positioning and inspection. I started out with the intention of making taxonomies of different technologies, and their capabilities. After working closely with respected research scientists in the CMU Robotics Institute, I found out that it was incredibly difficult to find cute ways of summarizing new developments, like machine vision, in ways that did not have serious flaws to the expert. (Granted, the typical social scientist would never know the difference.) Since this was my first real crack at writing about sensor based systems, I concluded the only way to get a feel for what is going on is to work a particular problem in detail. This is why the section on sensors concentrates on the problem of parts acquisition. Only by focusing on one problem domain can one make any meaningful statements about what types of tasks are hard and easy for the technology.

Perhaps the level of detail needed to explain how the technology works is in itself a clue to the reader of the difficulty of using the technology. Granted, the policy community wants accurate and reliable statements that can lead them from the technology to the impacts. Unfortunately, it takes years for people with the courage to straddle both sides of the fence in a serious manner to develop enough insight to make statements that are not embarrassing. Today, the people who are willing to make these types of statements do a small service, but I think, for the most part, they are not trusted (and to some extent even discounted) by the knowledgeable technical community, who themselevs have trouble generalizing about the capabilities and impacts of the 'leading edges' of these new technologies.

Item 14* Joe Martino: Overall, I think the paper by Ayres and Miller is good. In particular, I agree that 'general–purpose' robots are unlikely. Specialized machines are more likely. In the household, for instance, we don't use a general–purpose machine, robot or otherwise, to wash the dishes and mop the floor. Instead we use a specialized machine for washing the dishes, and we use wall–to–wall carpet or some other means of designing the mopping job out of the house. Industrial practice will be the same. An all–purpose robot makes no more sense than does an all–purpose machine tool.

Now for some specific comments on the Ayres' paper:

Page 12 – retrofitting of robots into existing factories may not be the most common route. My studies of the diffusion of innovations into industry show that one of the most critical factors in rate of adoption is degree of compatibility with the existing work force. Incompatible machines go into new factories rather than being retrofitted.

Page 21 – the shift toward capital intensive techniques may be slowed by high interest rates. The real interest rate is well above historical levels and shows no signs of dropping.

Item 26* Timothy Hunt: I sympathize with Steve Miller on the problems of revealing information about the firms in a sample. Cooperation of firms often can only be obtained with guarantees of confidentiality, yet objective research requires the kind of information mentioned in Lynn's earlier comment (Item 12). Does anyone know of a solution? How can the estimates be made transparent as Lynn suggests yet maintain confidentiality?

26.1* **Steve Miller**: I clearly identify the firms in the sample in the full PH.D dissertation which I believe both you and Leonard have seen. The reason for not including the list in the summary paper is solely out of laziness and a desire for to be brief.

26.2* **Leonard Lynn**: Steve, I don't think you need to repeat your dissertation to at least partially respond to our sense that you should give more information about your sources. In the case of your table 1 on page 4 of your second paper, for example, you could add such information as the sponse rate, perhaps the organizational position of the person responding, etc. It would also be useful for the reader to know how you happened to choose these particular firms. You could do quite a lot in this paper (if it is to stand as an independent document and not just a teaser to get people to read your dissertation) to satisfy the critical reader without listing all the firms in your sample.

Item 42* **Gerry Ross**: Sorry to be late getting into the action but I wanted to say what a thought provoking document the report was. I agree with Leonard that it is a mouthful for us non-technical types but also sympathasize with Steve concerning writing for multiple audiences. I thus shall leave the style debate alone for the moment.

I would like to make some comments on each of the papers to add to the discussion. The remarks I shall make here are mainly editorial. These are the 'sins of comission', if you will. In a leter item, I will suggest what I think the 'sins of omission' are, mainly dealing with the second and third order consequences of technology which are entirely missed in this report (perhaps it was designed to omit them!).

Briefly, my reactions are as follows:

(1) 'THE ROBOTICS SOCIETY'

p14: I like the table of 'increasing difficulty' which is powerful conceptually. However, the diagram is awfully busy and could use some clustering. Perhaps a two-by-two or three-by-three matrix could be employed using such dimensions as 'routine', 'intricacy', etc.

p.22: There is a bit of a gap between the last two paragraphs. The first explains how industries in the East can out-strip us in mass production. The second then makes a leap to say that the 'only' viable strategy for the U.S. is to shift to flexible batch production. I happen to agree with this overall statement but I think more explanation is needed to back up the statement that no other way is viable.

pp. 28-30: The many examples are informative but ramble on a bit. I must say I am looking forward to the 'button operated drink server though!

(2) "RECENT DEVELOPMENTS...."

Although I don't even qualify as an armchair engineer, I will hazard a few comments:

pp. 51-52: It seems to me that humans not only enjoy great flexibility in 'processing' information but in creating in- formation to process (e.g. in complex pattern recognition, such as in the use of analogies). At the extreme, humans have the capacity to 'define a problem' as well as solve it. The former, of course, then helps define what information is required.

pp. 82-86: The economics here are stunning but where's the 'bad' news. There are a number of FMS around the country where this potential has not been realised (i.e. they were downright failures).

(3) "IMPACTS OF ROBOTICS......"

This analysis, particularly in the latter part of the paper, represents a Herculean task and I am impressed by the thoroughness of the analysis. I do not have alternate data to contest the results but would like to hazard a few remarks.

pp. 126-127. It seems that labor 'displacement' is not the only issue here. The new generation of workers dealing with program- mable automation may well be much more mobile. Their skills may not merely be different, they may be more readily transferrable. Workers will know, for example, how to process information using computers, how to indulge in effective group problem solving, how to coordinate their own activities with fellow workers, etc.

pp. 126-127 cont'd: Lou Ferman told the story the other day of a fellow in a Ford plant whose job it was to stick his hand in a bucket of paint to test its viscosity. He would wiggle it around and then would pronounce what needed to be added. He was, of course, extremely valuable in that capacity but his skills were useless once robotic paint sprayers were introduced.

pp. 126-127 cont'd: In short, this new generation of workers, unlike their forebearers, would probably be just as much at home working operating a kidney dialysis machine in a hospital as a paint sprayer in a plant!

pp.140-141:The idea that several older plants might typically be consolidated into a single modern facility suggests the need for a strong regional dimension in the analysis of future projects, particularly from the policy point of view.

P.S. With a high level of integration, what about the impact on management?

(4) "CUSTOM, BATCH AND..."

pp.146-147.I admire the gymnastics in getting the required information from census data but wonder if it wouldn't have been easier some other way (e.g. using financial statement data – giving cost of goods sold, fixed vs. variable costs, etc. – on a selected sample of firms). As a (reformed) accountant, I am suspicious of data based on such a 'card house' of assumptions.

p 161: As a footnote to the above, such an eleborate procedure is used to determine batch vs. mass production and yet an anecdote about a tour through a turbine plant is used to shoot holes through the methodology (shot in the foot?). All this said, I am not a methodologist and applaud the paper overall.

pp 146–175: This paper sounds like it might have had little reworking since its previous incarnation, undoubtedly as a chapter in Steve's dissertation. Almost 30 pages are spent on the ins and outs of the classification system to make the simple point that "most of the value added in the metalworking sector is accounted for by products which are batch produced". Could these first 30 pages be put into an appendix and the rest expanded upon?

42.1* **Steve Miller**: You casually commented that with the new technology, a worker should be as comfortable operating a kidney dialysis machine as a paint spraying machine. I think this is a very important point, and the possibilities of relocating people to jobs which have similar informaiton processing requirements, but very different surroundings and job titles has not really been explored. As your comment suggests, computer intensive factories are getting to be more like office workplaces, and capital intensive or office workplaces are getting to be more like production oriented shops. I wonder if there is a hospital in the country who would hire an ex machinist over a young person out of a two year medical technology program on the basis that the machinist has a more fundamental knowledge of how to moniter and control a machine. Interesting proposition.

Item 43* **Eileen Havener**: For the most part, I thought the report was very enlightening. In view of my present assignment with Alcoa (ie, technology planning), I'd like to comment specifically on the technology content of the Ayres/Miller report.

I recognize that this paper deals with the social and economic issues resulting from technology development and implementation. However, we're finding within Alcoa that systematically understanding the rate of change (past and future) of the underlying technologies enables better strategic business planning to occur.

(1) We in technology planning are concentrating on 'understanding' rates of technological change not 'understating' (i.e., underestimating) them. In fact, all too often business strategy is actually based on an understatement of the benefits provided by technology -- and we feel that this is in error. We are trying to help Alcoa strategists better understand the rapid rate at which technologies such as computers and flexible manufacturing are actually moving so that they can make better proactive decisions (rather than play catch-up with others later on). By 'understand the technology', we look at all areas in which it is found -- emerging technology in academia and labs as well as in-place in industry.

(2) My main concern with your report is that although you begin with a robotic focus, discussion quickly swings to flexible manufacturing without much consideration of the other technologies that are enabling it to occur. I have no trouble with your focused robot discussions -- In fact, I think they are quite good!

I see some of this reasoning with regard to robots and sensors in the report, but, what about the other technologies which are making flexible manufacturing possible -- advanced process control, statistical quality control, database management, optimisation techniques, artificial intelligence, etc.?

I also noticed that when future robotic and sensor technology is discussed in the report, it is basically an extrapolation of the past and present; in fact, some of the "future" robot applications discussed are actually implementations of current technology. How far will science permit this and other relevant technologies to go in the next decade or so?

(Notice that saying that my car now gets 20 mpg and I think my new 1990 car will get 30 mpg is very different from asking how far science will permit car mpg to go -- probably upwards of 80 mpg. Then, I can ask about the engineering barriers which must be overcome to approach this limit. Now, I am ready to consider the social and economic issues arising from overcoming certain of these barriers and not being able to overcome others.)

I don't mean to imply that Ayres or Miller should bury the reader in technology buzzwords or detail. I do understand Leonard Lynn's frustration as a non-technical reader of this report.

I am, however, suggesting that it is possible to understand the rate of technological change in the language of science and then converting/ communicating this information to the non-technical person. The six of us in technology planning at Alcoa have been doing this during the last two years; and, we have seen several major strategic shifts as a result of the business managers being able to understand the potential offered by technological change.

In summary, I do suggest that better understanding of the social and economic implications arising from the emergence of flexible manufacturing is possible through better understanding of all the underlying technologies.

Item 45* John Ettlie: My initial reaction to this report as a whole (not individual chapters) is "what's new?" Overall, most of the material has been expressed elsewhere in perhaps more fragmentary form, but, its there in papers and books, etc., generally available. Now my specific comments.

1) I have no comments on chapter chapter 1. My experience with technological forecasts and assessments has been limited, but the ones done on the diffusion of NC machine tools have been wrong for 20 years.

2) The second chapter is not prospective enough in my opinion. Although prospective material might be better in the next chapter. Examples are what is meant by 'flexible' in referring to FMS, FMC(cell), etc. We (myself and people on my NSF project on implementation) feel this is one of the key issues for the 80's and 90's. Also, there are some very exotic systems now being planned and installed in the U.S., that would have been nice to describe (if possible); the foreign perspective (in detail) would have been nice to include here or in the next chapter.

3) on p. 13 of chapter 3, "Respondents to these survey (sic) over- whelmingly ranked

efforts to reduced labor costs as their main motiviation for installing robots.' First of all, the justification for these advanced systems is not a standardised procedure—so most of these respondents don't know what they are talking about. Example, in the spray robotic application, about 25% of the savings comes in the material savings area. Also, many of my robotics respondents say that quality is drastically improved. In the FMS area, not a single respondent disagreed with the statement that 'FMS cannot be justified on the basis of labor savings alone,' so I conclude that labor savings is not as important as it has been cracked up to be.

4) more on chapter 3,   I believe an important research question, that Gene Merchant is often quoted as saying, is the resistance to adoption of these systems. I believe the Rosenthal and Graham study should do much to address this issue.

5) on chapter four, the recurrent question that comes to mind is why the machine tool builders have not adopted their own technology? (I have a few hypotheses, and several came up at our recent steering committee meeting of the Great Lakes Gov. Council meeting at MIT on the machine tool industry.).Overall, the comments I made earlier, without the benefit of the report still stand. I like the report and hope it will be published.

45.1* Steve Miller: John, the survey responses were collected at the very beginning of 1981 and applied specifically to robots. People were very clear in their approch to cost justification. Maybe we forget that perspectives have changed a bit in the past few years, both as the technology has become more familiar, as more ambitious systems probjects have been planned, and as economic activity has picked up. Also, bear in mind that NSF did not fund us to do research on the topic. It was a relatively very small grant basically designed to summarise existing work.

ROBOT/FMS TECHNOLOGY

Item 16* Steve Miller: I state that the range of tasks that can be integrated into flexible manufacturing systems will expand as the technologies of AI and sensor-based-systems improve and become a part of FMS technology. Do people have specific examples of where this is happening? Examples of whether this is happening slowly or quickly?

16.1* John Ettlie: I have examples of 'super cells' which combine these various technologies and believe this general type of configuration is a good guess about what will be happening in the next five years or more. Even some retrofited systems qualify under this type of supercell design with no limit on the type of control systems that are integrated. This integration process is a major problem in a number of the implementation cases I have studied.

Item 28* Timothy Hunt: Will FMS be used in a wide range of applications in a few years? Westinghouse did not adopt their experimental system for assembly of electric motors. You may also wish to check whether IBM's FMS assembly system is actually operational. To what degree are these systems really flexible anyway?

Is AI being applied in any FMS settings currently? Operationally? AI expert systems appear to require so much investment that I doubt diffusion will occur rapidly.

28.1* Steve Miller: Regarding technology, either I spend too much time with the robotics people and industry people who do this type of thing and lose the ability to distinguish what goes on in research and what goes on in the world generally, or you do not spend nearly enough time in factories and in planning offices of the Fortune 500. I am inclined to say there is a weighted average, with the larger weight on the second factor. Westinghouse has not adopted the APAS system per se, but since that time they have installed a number of robotic assembly cells for electronics manufacture. IBM's designs of FMS for assembly are very real and very impressive. Unlike GE, who likes to advertise everything they do, IBM has a tightlipped policy about process automation, so knowledge of their accomplishments are limited to people who go to technical conferences and 'insiders'. The comment that expert systems are too expensive as to be prohibitive? This is tricky. But there is clearly a major effort by Digital and by many other small companies to commercialize such things. There is even a simple expert system program released for a PC, I am told. The question is what constitutes an expert system. Any bloody hackjob with if-then rules, or something that does a darn impressive job of capturing real organizational and technical knowledge?

28.2* John Ettlie: I have seen a number of expert systems in the field, and this concept has been around for awhile but not called that. I agree that cost is an interesting and maybe 'tricky' issue. it is not as easy as 'you get what you pay for,' but in general I see a trend toward more applications of expert systems in materials handling automation augmented with robots.

28.3* Dorothy Leonard-Barton: John, are you at liberty to disclose where you have seen expert systems applied in the field? I know only of DEC's, and one for oil exploration. The problem with costing out the expert systems I am familiar with is that their benefits can be more easily measured in terms of improved quality than, say, reduction in labor.

28.4* John Ettlie: Dorothy, it might be negotiated -- that is release of information -- but I am currently working with the firm that supplied the two cases on expert systems on two projects, so it will have to wait, I'm afraid.

COST-BENEFITS OF ROBOTICS, FMS, ETC

Item 16* Steve Miller: On pages 114-117 (of The Impacts of Robotics...) robot costs are discussed. The point is made that the robot per se is only a fraction of total installation cost. However, estimates of just how much larger the total installation cost is with respect to robot cost varies widely, as shown by the data in the report. I give examples where robots are 20 to 33 percent of the cost. The table taken from the Tech Tran report gives examples where robots are 40 to 70 percent of the cost. Any comments on which figures seem more realistic?

How do these percentages change as the second, third, ... and 50th robots are installed?

16.1* John Ettlie: I have some data on what factors affect the percent of total cost the robot represents in various flexible systems. However, I would have to get clearance from respondents to release this information.

Item 21* Faye Duchin: Costing-out of robots in specific settings should be done for new installations not retrofits. It is more likely to be cost-effective. Cost effectiveness should be evaluated relative to other technological alternatives for new plant.

21.1* John Ettlie: When interest rates go up, payback periods go down. I have some cases in the file where the required payback on a robotics system was 8 months, others were 2-3 years.

21.2* Steve Miller: John, could you give some details on your payback data. 8 months? Was this the first installation of one robot? What type of application? You seem to indicate that 2-3 years is the longest? Again, is this the first or the nth installation? I have seen a number of cases where the payback is over about 5 years.

IMPORTANCE OF FLEXIBILITY IN FMS/ROBOTICS IMPLEMENTATION— MASS PRODUCTION

Item 44* David Morrison: Comments on the Robotics Society Paper:

I would agree that the batch production mode will be an increasingly important factor in U.S. manufacturing competitiveness but this will not be a dramatic change since the bulk of U.S. manufacturing already is in the batch mode. I would not share the view, however, that this is the only viable long-term strategy for U.S.-based firms. The other arguments in the Ayres paper about the growing importance of transportation costs and proximity to markets as labor costs decline in importance are persuasive - - and applicable with equal force to batch or mass produced goods.

It might be helpful to the analysis to include a review of the underlying theory of comparative advantage (i.e., the United States or Michigan will exchange goods with Japan or California only if there are significant differences in relative efficiencies in producing such goods). If Michigan is twice as efficient as California in all types of production, there will be no trade. Californians will produce their own goods and enjoy half the return to their combined inputs of labor and capital compared with the return enjoyed by owners of labor and capital inputs in Michigan (or, if mobile, labor and capital will migrate). But if Michigan is three times as efficient as California in producing cars but only twice as efficient in producing computers, we will sell cars and buy computers - - unless transportation costs or other barriers are sufficiently high to negate the comparative advantage in production efficiency.

As Robert Lawrence of the Brookings Institution and many other economists noted, trade is a matter of finding niches in which product design, efficiency of production or marketing or some other attribute serves to secure a market share. Many of our competitors abroad (e.g., SAAB) have used batch production very effectively to secure niches in international trade. Still others have done better in mass production areas. There is no reason that I can find in the robotics/ FMS developments which would argue that our comparative advantage, taking into account transport costs, lies solely in batch production. Many of the sensing, control and hardware developments now emerging from laboratory design will have wide applicability in batch and mass production facilities. Ironically, much of the U.S. utilization of robots and other reprogrammable equipment is in mass production auto plants.

Item 46* Peter Unterweger: —Item 3. It seems to me that the trend to more batch production that many observers predict, is based not so much on the nature of the hardware as on the future structure of markets. It is argued that in the auto industry there is a trend towards customization, because that is what the consumer wants, and because that way the manufacturer will be able to carve out a niche and charge a higher price. Also fluctuations in demand for particular features of the product will be easier to meet. It seems to me that the more automated Japanese will try to use their flexible capacity in this way, which would then put great pressure on other producers to follow suit.

There is another reason why I see programmable automation as winning increasing diffusion in mass production industries. Their "mass character" does not seem to be a barrier to the use of CAD, manufacturing resource planning, automated storage and retrieval, group technology, etc. It seems to me that once several of such systems are in place there will be pressure to integrate manufacturing with the rest of the system.

Finally, automation of some assembly tasks in mass production is projected to grow strongly during the remainder of the decade. It seems to me that these will be flexible systems (how complicated an assembly operation could a "hard" automation system perform). Once the flexible systems are in place why not use them to fragment the mass market in order to enjoy the competitive advantage this holds?

46.1* John Ettlie: Peter, on your examples of flexible automation vs. mass, I would argue that not all of these technologies necessarily make an organization more flexible. It depends on the technology and how it is applied. For example, Bob Jacobs and one of his graduate students at IU have some preliminary findings from a plant that the installation of group technology (GT) has decreased flexibility.

46.2* Peter Unterweger: I think one of our problems is that perhaps we are not all using the term "flexible" in a very exact sense. In what I wrote it seems that I was using it interchangeably with programmable. At any rate I would be interested in how group technology decreased flexibility.

STRUCTURE OF FMS/ROBOTICS INDUSTRY (OLIG. VS. COMPET)
STRUCTURE OF USER INDUSTRIES

14.1* John Ettlie: The robotics suppliers have been specialized in their supply of these technologies, but I see a movement toward generalization and diversification in their strategies...one burning issue: how (or what type of firm is capable) of supply of the 'system' of the future?

Item 17* Steve Miller: The conclusion is that based on the narrow justification criteria mentioned, the economic incentives are much greater for establishments which use many robots (10, 20, and more). Thus, it is assumed that must robots will continue to be concentrated in larger establishments, as has been the case for the past 20 years. Any comments?

17.2* Jan Helling: There seems to be a claim that robots and FMS-systems are more common in large companies. (Robots/worker). In Automotive Industry research it is found that this is not necessarily true in Japan and Sweden. Here we find as much or more high tech equipment per worker in manufacturing systems among first or second tier supplier companies. They need as much or more flexible manufacturing equipment than the larger OEM-company. Besides — both in Sweden and Japan the amount of robots per worker is higher than in US. How do you explain this?

17.3* Bob Jacobs, IU Bloomington: Jan, We found this to be true in Japan recently. Here there was some tremendous problems with availability of space. Their Flexible Cells were used by a number of small suppliers who had very restrictive space limitations and great flexibility requirements. A beautiful example was a company called Heirai. Here they made extensive use of welding robots with automatic loading and unloading with queueing in the loading system. There is nothing in the open literature on this. One interesting thing about this company is that employees with basically high school educations were programming the cells including the welding robot patterns (no high level languages were being used). Very interesting!

17.4* Steve Miller: Last year I spent some time looking at the big fat book of Japanese robot statistics distributed in the U.S. by Prentice Hall (I think). (You cannot miss this one. It sells for about $500) Anyway, from the available data on the robot population, it seemed clear that robots were more evenly distributed among smaller, medium sized, and larger establishments than in the US. I do not know why this is so. But considering that materials costs account for about 50 to 60 percent of the total cost of output in many industries, and that the 'second tier' people supply much of the materials, the benefits of getting robots into these small manufacturers seems to be very large.

17.8* Peter Unterweger: I was interested in the analysis which showed that large firms have a greater incentive to install robots, etc. than small. A recently completed study for the Joint Economic Committee of Congress by Reese, Briggs and Hicks confirms this for the metalworking industry. Large plants and old plants both showed consistently higher rates of innovation adoption (NC, CNC, computers in design and manufacturing, and programmable handling equipment including robots). The findings were based on a survey of 628 plants (none in auto). Auto as the largest single user of robots, and with large plants, supports the finding as well.

I also find the argument that forms the basis of Steve's second question sound. Cost is undoubtedly an important factor; there are others, however, and in sum, they might well outweigh the cost factor. These include: capital availability, competition, the availability of trained personnel, and organizational resistance to change. I have gained a healthy respect especially for the latter. I think all of these contribute importantly to the gap between potential applications and actual as well as forecast installations.

17.12* Leonard Lynn: If I may briefly jump back to the question of why smaller firms in Japan so frequently use robots.... One reason is simnply that the Japanese government has a number of policies to encourage this. Concern over the low productivity of the small and medium sized firms has long been an issue. Thus smaller firms can get loans to buy robots. In many instances larger firms will also lend money to smaller firms among their suppliers to buy robots (and threaten firms which are reluctant to buy them). The industry associations (with some government help) also lease robots (primarily to smaller firms) at low rates. Some of these policies are summarized in a paper I did in the ANNALS issue on Robotics (November 1983).

17.13* Steve Miller: I would be interested to know to what extent systematic attempts have been made in this country to spread the use of robotic technology to 'second tier' manufacturers, or why the policies mentioned by Leonard in the last item would not (or maybe would) work here.

17.14* Peter Unterweger: It seems to me that the real problem is not the dearth of workable policies or even models but the attempt to in fact implement social and economic policies. I think these policies have worked in Japan and in Sweden, and with appropriate modifications they would most likely work here too.

17.15* Peter Unterweger: Back to Steve's point that there are incentives for multiple robot installation. The IG Metall study bears this out; not for robots but for NC/CNC machine tools. What I mean is that there are no data for robots, but there are some for machine tools. The survey responses show that plans to expand NC/CNC capacity was greater for plants that already had many than for plants with none or one.

SOCIAL ORGANIZATION FOR PRODUCTION
(LOW SKILL, HIGH CONTROL VS. HIGH SKILL, LOW CONTROL
=CENTRALIZED VS. DECENTRALIZED)

Item 29* Jan Helling: In our Body welding operations at SAAB we have had a lot of welding robots and organised work groups. Peter Unterweger had some questions regarding our experiences and suggested an open item reply – which is this. Peter wanted to know in the first place about the groupleaders. (1) The best practise we have found (among some others) is to organise groups (production teams) with about 10 people and a rotating teamleader. A supervisor in our Body shop supervises 3 – 6 production teams. Depending on what kind of manufacturing the teams take on. The teamleader is a blue collar union member. If the members of a group are willing and prepared (trained) to take on the responsibilities to manage themselves in specific areas (decided on in negotiations between management and unions) all members of the team get an increased wage level (As they all will take on the responsibility for a week). This rotating teamleader institution keeps rotating because the freedom is so desired. The teamleader does not take part in day to day operations if not necessary because of to many absent for instance. In the same fashion is the groups able to increase their wages if they are willing to take on other duties (and trained to do so) like materials handling, maintenance, quality cntrol, personnel matters, house keeping etc. For every step of new   duties they are able to take responsibility for they all in the group/team get a higher wagelevel up to an agreed maximum – even if not everyone in the team is able or willing to take those job enriched duties that belong to the group/team. What is needed here is of course a socio-technical inspired change process in which the technical systems are changed in a way that the workers can be more and more free from the technical system. This is in most cases achieved by new automated equipments. But here are many more things to talk about., but my entry will in that case be too long. I am hovever willing to answer questions about for instance –Effects on various service functions in the environment –The matrixform of group workorganization when the production teams take care of their own robots –The boundaries and banks needed between teams –etc.

34.5* Steve Miller: Is it really possible to generalise about the skilling vs deskilling issue without getting into particulars of how the technology is implemented? Reading I have done suggests that it is not the technology per se, but the way in which work is designed that affects skill requirements.

34.6* John Ettlie: Steve: That is a key point–it is not technological determinism, rather, job design is an orgaizational alternative. See item 40 on a related issue.

34.7* Jan Helling: It is also an issue of whose skill level will increase or decline. When you put FMS and Robots into the workplace you can perceive them as tools. But whose tools? The tool owner and user will undoubtedly increase his skill level adding handling of this new equipment to the tasks. You have at least two different options: (1) Get rid of as many unskilled blue collar workers as possible. Keep a small number that can stand easy, repetitive and boring jobs left in between machine systems. Operations too expensive or complicated to automate. Make the manufacturing engineers or industrial engineers the "owner" of the new machines. Help them be even more trained and skilled. (2) Train and educate the blue collar workforce to use and maintain their own tools — new and more advanced. Make them more skilled along with their enriched and enlarged jobs and new responsibilities for their workgroup/production team. Reduce the number of indirect employees in staff and service functions. We have both options and I know that companies for various reasons take on different strategies when redesigning their organization structure and organizational development programs. In our case — even different production units within the company take different routes. Why not find some companies or units that use different strategies in this respect and find out the effects of their actions on this more or less skilled issue?

34.11* **Steve Miller**: Going back to Jan Helling's remark: I think it is a very interesting observation that one can either 'stretch' the jobs of the floor people, and make them more 'managerial' or stretch the jobs of the industrial or manufacturing engineers in the office and make them more 'floor' oriented in the sense of managing the actual transformation activity as well as planning functions. Jan, you said there are examples of each strategy being followed. What are your general opinions/anecdotes about advantages/disadvantages in each case? It is interesting that in one of your scenarios, it is the exempt worker, the industrial engineer, who could have his job 'deskilled' by the non-exempt worker. That would offer an interesting twist.

34.12* **Peter Unterweger**: I think the question is whether we construct organizations that distribute responsibility, and the exercise of creativity fairly evenly, or whether we construct tightly controlled (automated) organizations. American management, at least in the auto industry, believes in the latter. In Japan plant managers report directly report to the CEO without elaborate intervening staff and control structures. Subordinates are told what is expected and will then be left to carry it out. They are trusted to do the right thing. The same principle applies all the way to the shop floor. US management believes in substituting external control for motivation. FMS technology can be used to implement either management philosophy.

34.13–14* **Jan Helling**: If I go deeper into my (1) and (2) cases my perception is that it is in practice both a policy/strategy and a power issue. In case (2) we have a strong leader and his strong steering team. They have a rather broad and 'wholeness' business perspective. They believe in job enlargment from bottom up as best strategy for increased manpower utilization and productivity increase when they get along with automation and robotization. When the blue collar production teams now (after many years of experience) increase skills and capacity they take over bit by bit tasks that earlier belonged to supervisors and nearby staff/service units. When they lose some of their low level/status tasks they need to increase their own level by striving for tasks that belong to higher managerial levels. The higher in the hierachy this striving reaches into middle management, levels the stronger the resistance. In functional areas with relatively weak leaders and steering teams and with narrowminded specialist–oriented leaders it becomes more a power issue with hidden strategies rather than a logical policy to use new manufacturing technology in order to upgrade the entire sociotechnical organization.

Remarks on Case (1). Here relatively weak leaders support various specialists and their units which have a strong and at times too strong power and influence over development processes and projects. Blue collar union representatives with an ambition to increase skill levels among shop floor people with FMS/robotics are confronted with suspicious and adversarial management – someone is after their privileges. Better keep blue collar skill level as low as reasonable. They had better do what they ought to be best at. Another factor of importance here for the power and strategy issue is the composition of skill and status levels in the blue collar population in the plant.

34.15* John Ettlie: 1) I agree that the most resistance comes from middle management, 2) I would like to comment that the first line supervisor is more a key than has been indicated in the U.S. cases I have looked at first hand. One of the persistant complaints that I get from middle managers in manufacturing is that there is a shortage of qualified people to promote to first line supervison. Are more people the answer or more sociotech to relieve the pressure on existing supervisors in the factory of the future?

40.4,6* Jan Helling: In Sweden we have several companies that measure productivity increase by sociotechnical changes. We can not claim socio-tech inspired changes are all over the place in Swedish industry. Still we envision them as enlighted islands here and there. But they are definitely increasing in numbers. We have research done on this and many papers – in Swedish though. But I know most of the researchers – and they are all able to speak and write in English. If you wish I can give you some names for references. I will myself now be involved in research and development projects here at SAAB when we will go further in our socio-tech efforts in two areas: body welding and final assembly of cars. These two projects will be sponsored by a 'Swedish National Joint Management Union Science Foundation'.

Volvo Kalmar plant is 10 years old by now! Many new steps have been taken both there and at other companies as well. By ' new' technologies I believe you mean not only new technical systems for manufacturing but also integrated changes of social and administrative systems as well. I see sociotech changes as integration of them all at once in projects aimed at increased productivity and better work conditions. We have many examples of new technical sytems along with the rest. I could give you a list of interesting companies and plans in Sweden but give you here instead some sociotech researchers as reference persons: Ulf Karlsson, Chalmers School of Technology, 412 96 Gothenburg, Phone 46 31 810100 Sven Eric Andersson, The Swedish Institute of Production Engineering Research, 412 85 Gothenburg, Phone 46 31 81 01 80 Christian Berggren, Royal Institute of Technology, 10044 Stockholm Phone 46 8 7870120, Torbjorn Stjernberg, Stockholm School of Economics, 113 83 Stockholm Phone 46 8 7360120 Stefan Aguren, Swedish Management Group, 10330 Stockholm, Phone 46 8 7360 Phone 46 8 235410

Item 46* Peter Unterweger: —Item 1. On its own the market and social institutions are giving us a mix of the two alternatives. It seems to me that a society based on the latter approach would be more productive (see what I have argued in the education discussion elsewhere), and that therefore we should develop social policies and economic incentives to further the wide diffusion of technology related knowledge and skills.

FACTORS LIMITING RATE OF TECHNOLOGICAL CHANGE

17.5* Jan Helling: As is mentioned even in a response to item 28 — I think an important impact issue is the distance between research labs and the shopfloor linemanager. Maybe we have some reasons here why we have more robots per worker in Sweden and Japan. Maybe the environment for the installation process is better? It is important to delimit the geographic and mental distance between high tech engineers at research labs and universities and the shopfloor linemanager with his ability to motivate and increase interest among subordinates for better manufacturing tools and other equipment to make the workplace better and more efficient. From our experiences nothing really happens in the shop until the manager is prepared and encouraged to take the next step to bring in new manufacturing tools from wherever he is. To find out where the people are and their need for next step ahead to remain competitive is essential for the development process.

17.6* Steve Miller: My everyday experience with airline clerks, store clerks, etc. is making me think that complicated information processing systems (including factory automation projects) will not work unless the people lowest in the chain broaden their view of the system and take responsibility for making things work. My, how things might change if this were to happen!

17.7* Peter Unterweger: I think Steve is right, but part of the problem is that not everyone is agreed on the desirability of this. The more farsighted managers want to broaden responsibility, but it does threaten the kind of control Taylorism has propagated since the turn of the century. Workers also have some reservations; it means change, learning, perhaps working harder. The paper Jan Hellig sent to me shows that workers were not that thrilled with more responsibility. The technology lends itself to both more distributed responsibility but also to more centralization. In many offices new technology is replicating the factory of the past.

34.13214* Jan Helling: If I go deeper into my (1) and (2) cases my perception is that it is in practice both a policy/strategy and a power issue. In case (2) we have a strong leader and his strong steering team. They have a rather broad and 'wholeness' business perspective. They believe in job enlargment from bottom up as best strategy for increased manpower utilization and productivity increase when they get along with automation and robotization. When the blue collar production teams now (after many years of experience) increase skills and capacity they take over bit by bit tasks that earlier belonged to supervisors and nearby staff/service units. When they lose some of their low level/status tasks they need to increase their own level by striving for tasks that belong to higher managerial levels. The higher in the hierachy this striving reaches into middle management levels, the stronger the resistance. In functional areas with relatively weak leaders and steering teams and with narrowminded specialist—oriented leaders it becomes more a power issue with hidden strategies rather than a logical policy to use new manufacturing technology in order to upgrade the entire sociotechnical organisation.

Remarks on Case (1). Here relatively weak leaders support various specialists and their units which have a strong and at times too strong power and influence over development processes and projects. Blue collar union representatives with an ambition to increase skill levels among shop floor people with FMS/robotics are confronted with suspicious and adversarial management – someone is after their privileges. Better keep blue collar skill level as low as reasonable. They had better do what they ought to be best at. Another factor of importance here for the power and strategy issue is the composition of skill and status levels in the blue collar population in the plant.

34.15* John Ettlie: 1) I agree that the most resistance comes from middle management, 2) I would like to comment that the first line supervisor is more a key than has been indicated in the U.S. cases I have looked at first hand. One of the persistant complaints that I get from middle managers in manufacturing is that there is a shortage of qualified people to promote to first line supervison. Are more people the answer or more sociotech to relieve the pressure on existing supervisors in the factory of the future?

38.10* Peter Unterweger: I wonder whether education, or training, or a combination is really capable of resolving the kinds of issues Patrick uses as an example. No doubt, well educated negotiators, thoroughly familiar with the likely outcomes, are always to be preferred to those less prepared, but institutions, constituencies, and I would argue differing world views are more important in determining outcomes in such situations. Education I would argue is a necessary, but not a sufficient condition for progress. I was really thinking of a narrower set of issues, such as arise when workers who previously operated traditional machine tools, move from NC, to CNC, to FMS, etc. I would argue that teaching the fundamentals of numercal control, computer operation, simple programming skills, and other principles that govern the operation of the entire system, would facilitate the adaptation to any particular system, and of course greatly enhance the workers ability to keep the system functioning.

38.11* Patrick Sweet: I agree with you, Peter, for the most part, and certainly appreciate your attempt at focusing on a narrower set of issues. However, your focus begs the question of "is it simply retraining or the lack thereof that is the locus of resistance to new technologies?" I think not, for most workers understand that there is a steadily decreasing ratio of "new jobs–to–people available". They have learned in life that retraining today means retraining tomorrow; ask any welder or toolcutter.

Management faces the same problem from another perspective. If a company is to put money into retraining but knows that there is going to be a lesser need for employees with requisite skills than its move toward higher technologies is making available through displacement, how does it choose/discriminate among employees available?

My point is only to raise these questions and not to attempt to resolve them. I also believe that if we focus on retraining alone there will be a goodly number of individuals left out. Since our economy depends on the purchasing power of all and not just a few it seems important to transcend retraining and displacement issues. But maybe this conference is not the best place to do so.

38.12* Peter Unterweger: I agree that my comments did not address the question of where the locus of resistance to new technologies lies. My point was simply that more broadly qualified people allow sophisticated systems to function more efficiently. I think that a key element of the resistance to new technology is resistance to change, which affects managers as well as production and office workers. A key question is why is our society producing so many individuals who crave the security of the status quo. For those on the low end of the income distribution a big part though by no means all is everpresent material insecurity. I think both the need to retrain, with all the attendant unknowns, as well as the shortage of suitable training/education are inhibitors to change. I don't understand your point about the importance of the income of all (with which I agree for both social as well as economic reasons), and how it makes the transcendence of training and displacement issues necessary. Economically everyones income is important because there is more than a grain of truth to 'underconsumptionist' explanations of the stagnation afflicting our economy. How training fits in, except through underutilization of human resources, I do not immediately see. Perhaps you could explain.

40.4* Jan Helling: In Sweden we have several companies that measure productivity increase by sociotechnical changes. We can not claim socio–tech inspired changes are all over the place in Swedish industry. Still we envision them as enlighted islands here and there. But they are definitely increasing in numbers. We have research done on this and many papers – in Swedish though. But I know most of the researchers – and they are all able to speak and write in English. If you wish I can give you some names for references. I will myself now be involved in research and development projects here at SAAB when we will go further in our socio–tech efforts in two areas: body welding and final assembly of cars. These two projects will be sponsored by a "Swedish National Joint Management Union Science Foundation".

40.5* John Ettlie: Jan, I think that's great, but honestly, are there really 'new' technologies involved? I know the Kalmar Plant's mobile assembly platforms make a good exampe -- maybe a model -- but are there really a whole lot of others?

40.6* Jan Helling: Volvo Kalmar plant is 10 years old by now! Many new steps have been taken both there and at other companies as well. By ' new' technologies I believe you mean not only new technical systems for manufacturing but also integrated changes of social and administrative systems as well. I see sociotech changes as integration of them all at once in projects aimed at increased productivity and better work conditions. We have many examples of new technical systems along with the rest. I could give you a list of interesting companies and plans in Sweden but give you here instead some sociotech researchers as reference persons: Ulf Karlsson, Chalmers School of Technology, 412 96 Gothenburg, Phone 46 31 810100 Sven Eric Andersson, The Swedish Institute of Production Engineering Research, 412 85 Gothenburg, Phone 46 31 81 01 80 Christian Berggren, Royal Institute of Technology, 10044 Stockholm Phone 46 8 7870120 Torbjorn Stjernberg, Stockholm School of Economics, 113 83 Stockholm Phone 46 8 7360120 Stefan Aguren, Swedish Management Group, 10330 Stockholm, Phone 46 8 7360 Phone 46 8 235410

43.3* John Ettlie: On the 'catch up' issue. I found in my recent qualititative study programmable manufacturing in the U.S. that 'leap frog' is not a game that can be successfully played by most organizations--that is, in order to participate in the factory of the future, you have to understand the technology of the factory of today. It is very difficult to master the next generation of a technology if you don't understand the current generation, according to the vast majority of my respondents. As one manager put it "you have to creep before you crawl."

Item 44* David Morrison: Regarding the Social Questions, it may be necessary to ask in what manner the new manufacturing technology and the resulting benefits will diffuse -- both within the United States and internationally. There could be a highly competitive diffusion of technology (as has occurred with computers) resulting in many producers and widely enjoyed benefits for consumers in the form of lower prices and improved quality. Or there could be oligopoly situations in which the gains in manufacturing technology are converted to economic rents shared (as in other oligopoly circumstances) between shareholders and wage/salary earners at the expense of consumers. FMS is coming on track as a large scale technology -- notwithstanding its batch capabilities but so did the computer at an earlier stage of commercial development. FMS potentially could be downscaled and replicated at low cost for a vast variety of production uses if the hardware and installation costs drop sharply in an expanding, competitive market. A bit more attention to the large scale/small scale issue and to the importance of the macro-economic framework in which change is likely to occur would enlarge the public policy setting in which the paper is framed.

44.1* Steve Miller: A minor comment to your well thought out responses: -Be careful in talking about how easy it would be to downscale FMS technology and replicate it at a low cost in a wide variety of production applications. It seems there is a surprisingly large requirement to tailor systems for the nasty particulars of each and every process, and this makes across industry or across process transfers a little slower than one would like. But this is a minor point. I agree that technological displacement is swamped by the business cycle, but I think there are interesting relationships. The auto industry reduced its employment because of the business cycle. However, now that demand is picking up I would speculate that the Big 3 are restraing rehiring in order to have the freedom to automate without directly displacing people. Does this count as displacement?

SKILLS (UP OR DOWN?)

Item 34* Robert Ayres: Theme 2. Will skill levels rise or decline as a result of the introduction of flexible automation and robotics? (Unterweger). It seems to me that the answer to this is inextricably linked to the answer to the question posed above. If we ignore the total employment (in the U.S.) issue, the answer seems to me to be fairly straightforward: flexible automation will tend to replace unskilled workers -- who are often called "semiskilled" -- whose jobs are basically dependent on eye-hand coordination and sensory abilities they were born with. This will be true even when those workers make simple decisions, such as whether a change in the operating conditions justifies turning off the machine and calling the supervisor or the maintenance man.

By the way, it seems to me that this is in marked contrast to the overall impact of mechanization over the past hundred years. In general, with some exceptions, of course, the trend has been to substitute bigger and bigger, more and more sophisticated machines for the use of skilled workers. Since the net result is that fewer workers are needed, and their role has ben largely reduced to 'baby-sitting' machines -- at least in mass production -- it has been said that the labor force is being deskilled'.Any comments on the difference, as I perceive it, between this pattern and the new one which will leave the more skilled workers largely unaffected? Further, on this topic, all of the above must be modified in view of the growing importance of international trade in the U.S. economy. Substitution of robots and computer-controllers for workers within U.S. factories is probably much less significant in the employment picture as a whole than the impact of wholesale plant closings and removals to overseas locations. In this context, robotics -- in the broad sense -- could help keep some U.S. plants competitive and thus retain employment opportunities at all skill levels that would otherwise be lost. Any comments?

34.2* Joe Martino: I dispute that the work force as a whole has been "deskilled." Particular jobs may now require less skill than the equivalent jobs of a generation ago, but the workers capable of learning skills have simply moved into other jobs. Thirty years ago the clerks in a dime store knew their merchandise well. Now they don't. What has happened is that the people who were basically capable of being knowledgeable clerks have moved into other jobs demanding their skills and native talents. The clerking jobs have been simplified so they can be handled by people who couldn't have held the job a generation ago. Look at the cash registers in a fast-food restaurant. The keys don't have numbers on them, they have pictures of hamburgers and so on. That's so they can be handled by people who can't handle numbers. Those who can handle numbers get better jobs. One of the ways mechanization and automation have multiplied worker productivity is by allowing who weren't competent to handle skilled jobs to do the work which once required a skilled worker. The skilled worker hasn't disappeared, though. He (or she, in the case of the dime store clerk) has moved to a more demanding job. I expect this same pattern to continue with the introduction of robots.

34.3* John Ettlie: 1) a number of U.S. maufacturers (Rockwell is one) have closed their domestic plants that were unproductive and did not move any production off shore; 2) the majority of my data (going back to 1974 studies of NC and CNC) says that jobs increase in challenge ??? and not deskilled.

34.4* **Patrick Sweet:** John, that is an important point. Could you elaborate?

34.5* **Steve Miller:** Is it really possible to generalize about the skilling vs deskilling issue without getting into particulars of how the technology is implemented? Reading I have done suggests that it is not the technology per se, but the way in which work is designed that affects skill requirements.

34.6* **John Ettlie:** Steve: That is a key point—it is not technological determinism, rather, job design is an organizational alternative. See item 40 on a related issue.

34.7* **Jan Helling:** It is also an issue of whose skill level will increase or decline. When you put FMS and Robots into the workplace you can perceive them as tools. But whose tools? The tool owner and user will undoubtedly increase his skill level adding handling of this new equipment to the tasks. You have at least two different options: (1) Get rid of as many unskilled blue collar workers as possible. Keep a small number that can stand easy, repetitious and boring jobs left in between machine systems. Operations too expensive or complicated to automate. Make the manufacturing engineers or industrial engineers the "owner" of the new machines. Help them be even more trained and skilled. (2) Train and educate the blue collar workforce to use and maintain their own tools – new and more advanced. Make them more skilled along with their enriched and enlarged jobs and new responsibilities for their workgroup/production team. Reduce the number of indirect employees in staff and service functions. We have both options and I know that companies for various reasons take on different strategies when redesigning their organization structure and organizational development programs. In our case – even different production units within the company take different routes. Why not find some companies or units that use different strategies in this respect and find out the effects of their actions on this more or less skilled issue?

34.8* **Leonard Lynn:** I wonder if we really have an adequate basis for concluding that FMS and robotics will require higher levels of skill/training than past automation technologies. It seems to me that in the past technologies have frequently followed a pattern in which they initially were more demanding of operatives -- but that deskilling took place later as management sought to reduce its dependence on skilled labor. In this context I think the points raised by Steve and John above are most relevant.

34.9* **John Ettlie:** We have a lot of "de's" to cope with here: deskilling, depersonalization, etc. Maybe we should start over with better concepts.

34.10* **Peter Unterweger:** I think the IG METALL study, a summary of which I will introduce as a separate item has some interesting things to say about the points raised by both Jan and Leonard. I think both are valid: There definitely are different ways of implementing technology. How else to explain that 32% of the respondents reported lowering of skills, and 42% increasing of skills? Or similar large positive and negative effects on responsibility? Although the trend to declining skill demands (Leonard) may be there clearly much depends which concept, Jan's (1) or (2), the implementation is based on. I am not sure what John's questioning of the "de's" means. These are very real. I don't think semantic redefinitions can do away with them.

34.11* Steve Miller: Going back to Jan Helling's remark: I think it is a very interesting observation that one can either 'stretch' the jobs of the floor people, and make them more 'managerial' or stretch the jobs of the industrial or manufacturing engineers in the office and make them more 'floor' oriented in the sense of managing the actual transformation activity as well as planning functions. Jan, you said there are examples of each strategy being followed. What are your general opinions/anecdotes about advantages/disadvantages in each case? It is interesting that in one of your scenarios, it is the exempt worker, the industrial engineer, who could have his job 'deskilled' by the non-exempt worker. That would offer an interesting twist.

34.13-14* Jan Helling: If I go deeper into my (1) and (2) cases my perception is that it is in practice both a policy/strategy and a power issue. In case (2) we have a strong leader and his strong steering team. They have a rather broad and 'wholeness' business peyspective. They believe in job enlargment from bottom up as best strategy for increased manpower utilization and productivity increase when they get along with automation and robotization. When the blue collar production teams now (after many years of experience) increase skills and capacity they take over bit by bit tasks that earlier belonged to supervisors and nearby staff/service units. When they lose some of their low level/status tasks they need to increase their own level by striving for tasks that belong to higher managerial levels. The higher in the hierachy this striving reaches into middle management, levels the stronger the resistance. In functional areas with relatively weak leaders and steering teams and with narrowminded specialist-oriented leaders it becomes more a power issue with hidden strategies rather than a logical policy to use new manufacturing technology in order to upgrade the entire sociotechnical organization.

Remarks on Case (1). Here relatively weak leaders support various specialists and their units which have a strong and at times too strong power and influence over development processes and projects. Blue collar union representatives with an ambition to increase skill levels among shop floor people with FMS/robotics are confronted with suspicious and adversarial management – someone is after their privileges. Better keep blue collar skill level as low as reasonable. They had better do what they ought to be best at. Another factor of importance here for the power and strategy issue is the composition of skill and status levels in the blue collar population in the plant.

34.15* John Ettlie: 1) I agree that the most resistance comes from middle management, 2) I would like to comment that the first line supervisor is more a key than has been indicated in the U.S. cases I have looked at first hand. One of the persistant complaints that I get from middle managers in manufacturing is that there is a shortage of qualified people to promote to first line supervison. Are more people the answer or more sociotech to relieve the pressure on existing supervisors in the factory of the future?

38.7* Bobbie Turniansky: I also agree with Steve. I think the question of upgrading or downgrading the skills begs the question of what skills are necessary and/or desired. Are the people being given the opportunity to use and develop the skills they have right now? Perhaps we can take a broader look and see how transferable some of those skills are. Maybe they have to be conceptualized differently.

38.8* Peter Unterweger: Skills is too narrow a concept. Computers, FMS, all are governed by some general principles, which, once understood can make dealing with knowledge, as opposed to factual memorization, or concrete manual skills, a lot is transferrable. Education is to teach people more of the structure of the world, so that when new situations are encountered they can be dealt with based on the principles that have been learned. It is education in this sense as opposed to broadly "learning how to live" which will be more important in the more rapidly changing world.

38.20* Peter Unterweger: David's comments agree with what I have read about the computer training of "displaced" middle managers, engineers and other white collar workers. What is not clear, however, is whether they have an especially good aptitude for mastering computer systems, or whether their general understanding of the company's business, management or manufacturing practices, etc makle them much more "valuable than recent graduates. I suspect it is the latter, and that in fact this makes up for any reluctance to get involved with computers.

38.21* David Morrison: The Union–Carbide official I spoke with put emphasis on the general knowledge of company operations but also argued that the older engineers brought with them to the company a broader educational experience than do the new, more specialized engineers now being graduated. He pointed to skills in communication as an area where the older engineers have a marked advantage -- in part because of experience but also because of a better general education.

42.1* Steve Miller: Gerry, you casually commented that with the new technology, a worker should be as comfortable operating a kidney dialysis machine as a paint spraying machine I think this is a very important point, and the possibilities of relocating people to jobs which have similar informaiton processing requirements, but very different surroundings and job titles has not really been explored. As your comment suggests, computer intensive factories are getting to be more like office workplaces, and capital intensive or office workplaces are getting to be more like production oriented shops. I wonder if there is a hospital in the country who would hire an ex machinist over a young person out of a two year medical technology program on the basis that the machinist has a more fundamental knowledge of how to moniter and control a machine. Interesting proposition.

Item 46* Peter Unterweger: –Item 2. How interchangeable is work in the control room of an oil refinery and a powerstation? Intuitively it seems to me that there is wide transferrability of knowledge and skills. As standard programming languages evolve, and machine control systems become more electronic, programming and maintenance tasks on different types of programmable equipment should come to resemble each other more? Schooling could concentrate on the general knowledge that these systems have in common, and on the job training could then provide the specific skills needed to operate the particular equipment installed in a given plant.

EDUCATION VS. TRAINING

Item 14* Joe Martino: Specific comments on Ayres paper   — I agree education is necessary, but we may be on the way to 'de-schooling' education. One suggestion made here in Ohio (but not yet implemented) is that companies be given tax breaks for operating retraining centers for their displaced employees, or the people whom they hire who were displaced earlier.

14.2* Robert Ayres: I find myself frequently making the point that education and training are very different, really. Training is almost always very narrow and specific to a particular machine or system. If the machine is obsolete, the training is obsolete, too. Education, on the contrary, gives people the capacity to be trained more efficiently. Above all, it gives people the use of language, without which all other training is impossible.

14.3* Joe Martino: I agree on the distinction between education and training. I'm not sure education is appropriate for displaced workers. However, training is something the worker gets time and again through his career, one way or another. We probably should emphasize that instead of education.

14.4* Peter Unterweger: Why would education not be appropriate? The argument run that more rapidly changing technology will necessitate more frequent retraining. Better educated workers would be easier to retrain thus reducing retraining costs.

14.6* Joe Martino: Is there real evidence that increased education reduces retraining time? That is, beyond such basic things as education in literacy and numeracy? What kind of education would be beneficial to workers in speeding their retraining for new technology?

14.7* Peter Unterweger: For the evidence we will have to go to the experts. Here are some considerations that lead me to this conclusion. Companies have a relatively easy time retraining skilled workers (say electricians) to do robot maintenance; imagine the problems in retraining janitors for such jobs. Similarly middle level managers and professionals are retraining for computer skills; their relatively high educational level clearly facilitates this.

14.8* Joe Martino: Peter's examples still do not make the case for education. When one retrains electricians for robot maintenance, is the success of the retraining due to some generalized education the electricians already had, or is it due to the transferability of their specialized knowledge about electricity, about repair techniques, and about the use of tools? When middle level managers are retrained for computer skills, how much of the ease in transition is due to ther basic 'smarts,' independent of education; how much is due to their knowledge of organizational precedures, which are now being computerized; how much is due to their demonstrated facility with language (after all, learning programming is simply learning a new language with its own grammar, vocabulary and syntax); and finally, how much of it is due to generalized education? Before we go off on an educational kick, I think we need to take a good hard look at just how much benefit would be gained from it.

Item 38* Robert Ayres: Theme 4. Education vs. training. Specifically, is there any evidence that education makes training easier/cheaper? I have never seen the question put this way before, and it deserves serious attention. I wish we could get more -- and more thoughtful -- responses on this.

38.1* Joe Martino: I'll restate my earlier views, just for the record. I have seen no evidence that generalized education in any way eases retraining. I believe it does not. However, I admit my case isn't based on solid research, but only on anecdotal data. I would like to see the matter settled one way or the other. For instance, I have seen an electronics assembly plant in Bangkok staffed solely by young Thai girls, just out of the paddy fields, who do a very competitive job of 'solder-slinging.' With a fairly short period of training, they learn to stick components in the right holes in the circuit boards, and solder them in place. While 12 years of education is supposed to be mandatory in Thailand, most places don't yet have the school facilities to carry it out, and many of these girls hadn't finished high school. I find it hard to believe that a motivated steel worker couldn't couldn't learn that same skill in about the same length of time, regardless of his level of education. The primary block to his retraining is going to be the notion that he 'deserves' $30,000 a year instead of the $10,000 the electronic assembly job pays. All the generalized education in the world isn't going to change that.

38.2* Patrick Sweet: It seems to me that Joe has a point; motivation is a key factor in learning. However, the very task he uses for his example is one ripe for being automated in the near future if not immediately. Faced with the choice of earning $10.000 on a job that promises to be eliminated rather soon or not working at all, the steel worker will probably give retraining a good hard look. The question remains though 'where will he find the opportunity and resources to pursue it?

38.4* Leonard Lynn: It strikes me that this question is intertwined with the one of whether the levels of skill (know-how) required by flexible automation are going to be higher or not. The Thai example and that of the soon-to-be-replaced job both relate to previous technologies where generalized education may even have been a handicap (assuming better educated people find simple jobs to be more frustrating than less educated people would). It also strikes me that we make a mistake in making education some sort of black box. Education in the classics would presumably have a different impact in preparing a worker than education stressing computer literacy.

38.5* Joe Martino: I don't want to come across as an enemy of education, so I want to expand on my earlier comment. An English teacher I had in high school used to tell us repeatedly, 'A technical education teaches you how to earn a living; a liberal education teaches you how to live.' I still think there's a great deal of truth in that. Generalized education can have a lot to do with improving your quality of life generally. What I don't believe is that it has much if anything to do with learning or relearning skills on the job. General education is valuable, but not for learning how to relearn skills. If we want people to learn new skills, that's a different task entirely from learning how to appreciate good music and good literature, and learning how the present came to be the way it is.

38.6* **Steve Miller**: Joe seems to have a rather narrow concept or workplace 'skills'. The examples cited previously focus on physical dexterity and muscular control and maybe a little bit of memorisation. Right now I am in residence at a computer company (Burroughs) and I see that a major skill requirement of white collar staff people (even at low levels) is to know where to go in the system to get things done and how to approach people in ways that they will cooperate with requests as opposed to ingore them or even block them. It seems that in the many aspects of work that do not deal with sticking a peg in a hole (or a similar task) that general education might in fact be a vary important issue. I suppose I am guilty of expanding the problem of skills beyond the domain originally mentioned (produciton work on a factory floor), but I think the point is still worth making.

38.7* **Bobbie Turniansky**: I also agree with Steve. I think the question of upgrading or downgrading the skills begs the question of what skills are necessary and/or desired. Are the people being given the opportunity to use and develop the skills they have right now? Perhaps we can take a broader look and see how transferable some of those skills are. Maybe they have to be conceptualised differently.

38.8* **Peter Unterweger**: Skills is too narrow a concept. Computers, FMS, all are governed by some general principles, which, once understood can make dealing with knowledge, as opposed to factual memorisation, or concrete manual skills, a lot is transferrable. Education is to teach people more of the structure of the world, so that when new situations are encountered they can be dealt with based on the principles that have been learned. It is education in this sense as opposed to broadly 'learning how to live' which will be more important in the more rapidly changing world.

38.9* **Patrick Sweet**: When we talk of training vs education in the context of adopting new technologies and forms of work organization in order to assist a country or company in becoming more able to compete and survive in our world economy, it seems clear to me that training nor education should be emphasised at the expense of the other. I'm sure that Joe and Peter would agree that the most productive contract negotiations, whether between vendor and supplier or labor and management, occur when all parties involved are well educated as to the future implications of the agreements they are about to make. The relationship between labor and management in the U.S. has been rigidified in the past due to both camps being 'trained' to negotiate and do business today rather than looking down the road a bit further (this is a slight overstatement). Education alone will not avoid this though. Neither will training in business schools or trade schools as they 'teach' today. It is the combination of the two (education and training) and its application to real world experiences that holds the most promise for aiding organizations and ecomomies in understanding and adopting new technologies in the optimum manner, I believe.

38.10* **Peter Unterweger**: I wonder whether education, or training, or a combination is really capable of resolving the kinds of issues Patrick uses as an example. No doubt, well educated negotiators, thoroughly familiar with the likely outcomes, are always to be preferred to those less prepared, but institutions, constituencies, and I would argue differing world views are more important in determining outcomes in such situations. Education I would argue is a necessary, but not a sufficient condition for progress. I was really thinking of a narrower set of issues, such as arise when workers who previously operated traditional machine tools, move from NC, to CNC, to FMS, etc. I would argue that teaching the fundamentals of numerical control, computer operation, simple programming skills, and other principles that govern the operation of the entire system, would facilitate the adaptation to any particular system, and, of course, greatly enhance the workers' ability to keep the system functioning.

38.11* Patrick Sweet: I agree with you, Peter, for the most part, and certainly appreciate your attempt at focusing on a narrower set of issues. However, your focus begs the question of "is it simply retraining or the lack thereof that is the locus of resistance to new technologies?" I think not, for most workers understand that there is a steadily decreasing ratio of 'new jobs-to-people available'. They have learned in life that retraining today means retraining tomorrow; ask any welder or toolcutter.

Management faces the same problem from another perspective. If a company is to put money into retraining but knows that there is going to be a lesser need for employees with requisite skills than its move toward higher technologies is making available through displacement, how does it choose/discriminate among employees available?

My point is only to raise these questions and not to attempt to resolve them. I also believe that if we focus on retraining alone there will be a goodly number of individuals left out. Since our economy depends on the purchasing power of all and not just a few it seems important to transcend retraining and displacement issues. But maybe this conference is not the best place to do so.

38.12* Peter Unterweger: I agree that my comments did not address the question of where the locus of resistance to new technologies lies. My point was simply that more broadly qualified people allow sophisticated systems to function more efficiently. I think that a key element of the resistance to new technology is resistance to change, which affects managers as well as production and office workers. A key question is why is our society producing so many individuals who crave the security of the status quo. For those on the low end of the income distribution a big part though by no means all is everpresent material insecurity. I think both the need to retrain, with all the attendant unknowns, as well as the shortage of suitable training/education are inhibitors to change. I don't understand your point about the importance of the income of all (with which I agree for both social as well as economic reasons), and how it makes the transcendence of training and displacement issues necessary. Economically everyones income is important because there is more than a grain of truth to 'underconsumptionist' explanations of the stagnation afflicting our economy. How training fits in, except through underutilization of human resources, I do not immediately see. Perhaps you could explain.

38.13* Patrick Sweet: It is unquestionable that more broadly qualified people run sophisticated systems more efficiently, and that a key element of the resistance to new technologies is a passion for the status quo. I believe we are in agreement but have missed each other somehow. My point about transcending retraining issues did not include transcending displacement issues as you suggest I meant to in your response. To restate what I was driving at would be to say that retraining is not sufficient for there will not be a one-to-one ratio of new jobs created for old ones replaced. Thus all who are displaced will be in need of retraining, but the company letting them go will not have a job for each and every person even if they have learned new skills. It is at this juncture in the debate that we must recognise that retraining is not sufficient. I think that it becomes quite important to then look beyond it for obvious economic and social reasons.

38.14* Joe Martino: part of the "retraining" issue may be a false one. The assumption seems to be that unless workers with no-longer-valuable skills are retrained, they will not be employable. I suggest that even with retraining, they may not be employable if there is insufficient demand for the skills they have just acquired. Conversely, if there is a demand for workers, either of two things will happen. (1) Employers will undertake the retraining themselves, simply because they need the workders, or (2) the jobs will be simplified by making the machines "user-friendly" (like the fast food restaurant's cash registers with pictures of food items on the keys). The two alternatives are not exclusive, of course. In short, if the workers are needed, somehow a way will be found to use them. If they are not in demand, no amount of training is going to get jobs for ALL of them.

38.15&16* Peter Unterweger: Well it looks like on this point we are agreed. retraining is not necessarily a solution to the displacement problem. I agree also with Pat's restatement of his views, but I am not sure all of us did before we went through this. To me displacement, and job creation (as long as we do not have other means of distribution) are the most important issues, but a close second is the quality of jobs, and that is related to the education/ training discussion. I think that better educated people have better jobs and therefore a better quality of life; if Patrick is right, and I think he is, we clearly are in the happy position that we can simultaneously improve the efficiency of our productive system, and the quality of workers lives. This is the second of the two possibilities that Jan Helling outlined a while ago. Joe's example of putting pictures instead of numbers on the cash register is of course an example of Jan's first point. Unfortunately, it seems to me that the U.S. as a society has so far preferred the "deskilling" strategy, or perhaps more accurately a strategy of underinvesting in "human capital". That at least is the point made in a study by the New York stock exchange, which cites John Kendrick as estimating the "people factor" to be responsible for 20%-50% of productivity growth. The study then attributes a large measure of Japan's outstanding productivity growth to its outstanding educational system, which produces not a brilliant elite but a high average level of capability. According to the report this system ' is shaping a whole population, workers as well as managers, to a standard inconceivable in the United States,...'.

38.17* John Ettlie: I think Bob Jacobs guest response earlier on the Japanese worker programming a robot agrees with Peter's position. Also, I do think there is often an age factor in retraining. Older employees as one of my respondents puts it 'have seen generation after generation of control systems' and are just not as enthusiastic about learning the next generation as younger employees. So what is old? My respondent's estimate is early to mid 40's for their plant (GM Fisher body, about 3500 employees).

38.18* David Morrison: On the age question, I was told recently by an official of a large U.S. corporation that they have had very good experience in training older engineers in use of computers. He said in many cases that it is much easier to train these employees to use computer applications than it is to train newly graduated engineers with computer knowledge to apply the latter to company needs.

38.19* John Ettlie: that's a good case. The age factor is really not a construct, in my opinion. There's an attitude variable often correlated with age I think. In a study I published a few years ago in Training and Development Journal, I reported finding that the best predictor of vendor training school success for customer participants was an attitude variable; although the explained variance was significant, it was relatively small (R square). We have gone on to try and develop this variable (we started out with an 'intrinsic interest' approach) in two studies. One was published in the Journal of Management Studies (Ettlie and O'Keefe) and one in R&D Management—both within the last couple of years.

38.20* Peter Unterweger: David's comments agree with what I have read about the computer training of 'displaced' middle managers, engineers and other white collar workers. What is not clear, however, is whether they have an especially good aptitude for mastering computer systems, or whether their general understanding of the company's business, management or manufacturing practices, etc make them much more valuable than recent graduates. I suspect it is the latter, and that in fact this makes up for any reluctance to get involved with computers.

38.21* David Morrison: The Union-Carbide official I spoke with put emphasis on the general knowledge of company operations but also argued that the older engineers brought with them to the company a broader educational experience than do the new, more specialized engineers now being graduated. He pointed to skills in communication as an area where the older engineers have a marked advantage -- in part because of experience but also because of a better general education.

38.22* Joe Martino: The engineer experiences may be an example of the value of general education. I'd like to see a study, not just anecdotal evidence. But that may not yet be available. My experience in implementing various computer systems in organizations is that familiarity with the organization's practices is of greater importance, over the long run, than is knowledge of computers. You can take someone who already knows the organization and teach them computers a lot quicker than you can take someone who knows computers and teach them the organization.

38.23* John Ettlie: I agree with Joe, with the hair-splitting distinction that we don't 'take' someone and train or educate them—they come forward, enthusiastically, supporting the philosophy of change in the organization, and to some extent, the particular technological alternative currently selected. I think a good research question is still where these 'process' champions come from (as opposed to product champions). Jim Solberg said at our Dearborn meeting of ITI that they investigated several FMS installations after they were well under way and could not locate the original champion (or even get their names) nor could his research team reconstruct the justification procedure successfully.

LOCATIONAL IMPACTS

Item 14* Joe Martino: Specific Comments on Ayres Paper – Page 18 – It may be cheaper to move industries to where the displaced workers are, since whether the workers or the industries move, the workers will still have to "shape up." The cities where the workers are already tend to be transportation centers, which makes them particularly suitable for new industries.

Page 18 Did MAC really pull NY back from bankruptcy, or was it instead one of the causes?

14.2* Robert Ayres: I agree with Joe that there is much to be said for moving industry to where the workers are, rather than the converse. The social costs of replacing infrastructure in new locations has never been adequately assessed. This is a very large subject, of course. I'd like some other people to comment on it too.

Item 27* Timothy Hunt: Ayres' argument that location costs or decisions thereof will be dominated by transport costs by the year 2000 does not appear to be convincing. Currently transport costs are becoming a smaller and smaller part of total costs. There is also a trend toward more footloose firms. I might argue (speculatively) that the factory of the future or the location thereof will be dominated by the wishes of important managers and other human factors -- R&D, universities, nearness to other firms with similar interests, etc. Why will transport costs be dominant by the year 2000?

27.1* David Morrison: I've not seen the details of the argument on location but it is logical to expect transport costs to become relatively more costly than would be the costs of production as the impact of automated manufacturing is felt in the future. Also, with labor costs likely to decline as an element of production costs, the historical pattern of firms looking for locations with low cost labor supplies also may be eroded. Not on this same point but of interest perhaps is the possible shift in terms of trade between agricultural and industrial producers if advances in technology bring more rapid gains in manufacturing production bringing a substantial drop in prices of manufactured goods.

27.2* John Ettlie: There is a tendency toward ??? OUTSOURCING AND FOCUSED ? factory that is smaller and, therefore, causes decentralization of operations (not necessarily production operations but assembly). The transportation and logistics experts ought to be brought into this discussion if we pursue it further. I have one to suggest: Dr. David Vellenga, Iowa State University, Ames.

27.3* Peter Unterweger: The current interest in just–in–time production methods would support a reconcentration argument. In auto, GM'S Buick City Project envisions supplier firms locating closer to their assembly facilities. OECD studies that I have seen say that the final verdict on "world car" vs kanban is not yet in.

27.4* John Ettlie: Peter, that is true in auto, but the scale of operations still goes down regardless, and the relationship beteen supplier and customer becomes more crucial: see, for example, Macmillan, Keith, and Farmer, David, "REDEFINING

THE BOUNDARIES OF THE FIRM,' The Journal of Industrial Economics, March 1979.

27.5* Joe Martino: The success of 'minimills' in the steel industry at least suggests that locating close to customers (and in this case, sources of scrap) is more important than costs of production per se. Of course steel is a bulk commodity, with low value added per pound; even small transportation costs could end up being significant. The same might not be true for high value-added products like pharmaceuticals or microcircuits. The current issue of DATAMATION listed a price of $250 in the 'black market' for a new but scarce 16-bit chip.

27.6* John Ettlie: Joe: I show a tape in my org. design class on minimills and one of my students who was working for Inland at the time was quick to point out that the success of these small operations was very much determined by the availability of cheap scarp created by the 'less efficient' larger producers in the industry. I had to agree with him.

27.7* Peter Unterweger: There is an article in the most recent Scientific American that makes the point that one of the reasons minimills are low cost is that they have a relatively limited, and low grade product line ,compared to the major producers.

27.8* John Ettlie: Also a point my student made; however, these minimills are also well managed, based on my information to date. They are relatively low on union/management conflict and progressive in incentive plans.

Item 39* Robert Ayres: Theme 5. Centralization vs. decentralization. On this topic I admit to being rather speculative -- perhaps too much so. But perhaps a worthwhile research issue has surfaced: does the advent of flexible automation tend to influence locational decisions? If so, how? I am fairly clear on the trend toward smaller, more decentralized manufacturing operations in several industries, but there may be a number of plausible explanations. It is not necessarily true that flexibility has any influence on location. I was mainly making the point that plants will not need to be located near large labor pools, since they will be largely unmanned. It does seem logical that, in the absence of this consideration, location would be determined by other factors.

39.1* John Ettlie: One of the questions I asked all respondents in my recent survey was how they felt about the advantages and disadvantages of 'green field' factory of the future locations -- that is, building a plant out in a 'green field' substantially isolated from existing operations. The advantages and disadvantages notwithstanding, the majority of my respondents said that technology is not a significant determinant of plant location decisons.

39.2* Joe Martino: I agree with John, but I think the issue is even broader. The same technology can be either centralizing or decentralizing, and in some cases it depends on the viewer. The solar energy advocates of a decade ago, particularly Amory Lovins, were touting solar energy as a decentralizing technology, by contrast with the present 'centralized' energy system. Yet Franklin D. Roosevelt advocated a public power system which would create a 'sea of electricity' in which individuals could build the kinds of homes and businesses they wanted to, instead of being forced to move to the cities where the electricity was. Was the electrical generating system we now have centralizing or decentralizing? You pays your money and you takes your choice. You can argue either way.

39.3* John Ettlie: Joe: it is my recall that on nuclear power, transmission costs were rather important in locating the plants close to cities rather than in 'safe' remote areas. Also, the Bela Gold group has presented paper after paper show how unit costs of power generation go down with plant size. They cannot answer the question of the 'cost' of an outage, however.

39.4* Joe Martino: Right, John, transmission costs are a big component of the delivered cost of electricity. There has been a trend in the electric power industry toward larger plants, in order to reduce the cost of generation, but this meant the costs of distribution increased. Therefore there has been a parallel trend toward higher distribution voltages, to cut the losses in distribution (and therefore the costs of distribution). My point, though, was that if you focus on the central generating station you can see our present electrical system as centralizing. If you focus on the widespread availability of electrical power, you can see it as decentralizing because you no longer have to go to the cities to get electricity. Therefore the perception of whether a technology is centralizing or decentralizing depends heavily upon the perceiver.

39.5* Steve Miller: I want to go back to Ettlie's earlier comment (R-1) that his survey respondents say that technology does not affect location choices. Something seems very funny. Technology affects the quality and quantity of all input factors used by the plant. I find it interesting, to say the least, that the requirements for the quality and quantity of all inputs has no affect on where a plant is situated.

39.6* Leonard Lynn: I am surprised that no one has picked up on the notion that firms might use FMS in their effort to escape from labor unions and other 'problems' in the existing industrial areas. In this sense FMS could be decentralizing because the technology offers an opportunity (not a mandate) as new plants are built to take advantage of the new technology.

39.7* Peter Unterweger: Labor is quite aware of the runaway phenomenon, but companies don't seem to need FMS to do this. In fact the labor displacing ability of programmable automation, even if no movement out of existing facilities takes place, is probably a much bigger threat than FMS-inspired plant moves.

39.8* John Ettlie: I still stand on my earlier point. technology, as a single variable, is not a priority factor in plant location decisions. It is a factor (the technology of your product vs. your product) in deciding your overall business strategy. For example, you might decide that even though you cannot sell a obsolete product in the U.S., you can sell is to underdeveloped countries and remain price competitive.

ECONOMIC GROWTH VS. DISPLACEMENT

4.9* Peter Unterweger: Still would a definition of the key issues in factory, or office automation not be an appropriate goal of this conference? Perhaps it would be useful if various participants would provide a list of the key concerns technology raises in their areas of expertise. For starters here is a stab at my list:

1. Will automation result in such a cheapening of goods and services that the increase in demand will offset the laborsaving effect of the technology.

2. The relationship between technology and job structures. Is flexible technology more effectively used by broadly skilled workers, or will the expansion of programmable automation lead to a more intensified taylorism? Will skill levels rise or decline?

3. What institutional restructuring might be needed for the optimum implementation of new technologies?

As you can see these seem like 'macro' questions but they actually need answers at various levels.

4.10* Steve Miller: In response to Peter's first item: Will automation result in lowering of prices and increases in demand that will offset labor reductions in industries using automation? I argue no in most cases. See Ayres/Miller Book, 'Robotics: Applications and Social Implications', chapter 6.

4.11* Peter Unterweger: I tend to agree with Steve's point, but would still like to see a soundly argued case for the opposite.

Item 15* Joe Martino: Comments on Miller Paper –   – The issue of displacing workers appears to start from the assumption that an economy exists to provide jobs. On the contrary, the function of an economy is to ECONOMIZE on resource use, including labor. If we all tried to be self-sufficient, obviously each of us would be employed full-time. By engaging in exchange, through division of labor and so on, we do better for ourselves than we could if we all tried to be Robinson Crusoes. But the economy exists to facilitate exchange. It assumes we all start with something to exchange, not that we need the economy for that purpose. Any discussion of the effect of robotics on labor is going to go astray if it starts with the assumption that the economy provides jobs, in the same way an economy provides steel or cement.

Page 20 – How many workers have been 'displaced' by computers? By automation? With the U.S. showing 4 million new jobs last year, it's hard to make the case that anyone has been 'displaced' by either automation or computers. Instead, a lot of people have been 'fired' by the buying public, which no longer wants what they have to offer.

15.1* David Morrison: The issue of job creation vs a focus on production is an important develop ment issue. I have a rough paper suggesting that the reason rural producers in a number of countries produce so little is that urban populations consume without producing goods in exchange for farm goods. Farmers quickly tire of this one-sided 'exchange' and opt for subsistence farming. Hence, there could be a place for highly efficient, automated manufacturing even in countries with high unemployment or underemployment -- if the output were directed at meeting rural demand. Anyone interested in the paper may request it by message, indicating address or asking to receive it by message. About seven pages long.

15.6* Peter Unterweger: On the purpose of the economy. I think the definitions so far are off the mark. Heilbroner once defined the economy as a system for the production, and distribution of needed goods and services. In this sense economies existed long before the advent of market economies. I would suggest that jobs and the associated incomes are concepts most relevant to developed market economies, although the introduction of large scale automation will probably lead to the breaking of the job income link in order to assure more effective distribution. Leontieff, for one, thinks that worktime reduction and transfer payments will have to be increasingly relied upon. 'Economizing' alone seems far too restricted a view of the economy's purpose.

15.7&8* Joe Martino:PURPOSE OF AN ECONOMY. I can't argue with Heilbroner's definition. Clearly the purpose of an economy is to distribute goods and services. But one has to be careful about the meaning of 'distribution'. Socialists like Heilbroner and Leontieff take it for granted that goods are really a 'socially-owned' lump of wealth, and we somehow have to figure out how to distribute that lump among the people. Not everyone accepts that view. For many of us, distribution is the other side of the coin of production. Goods don't have to be distributed, they have to be exchanged. The producer of something owns that thing. He doesn't need to gain a claim on it. He already has a claim on it. He exchanges what he has produced, or acquired through exchange, for what other people have produced or acquired by exchange. The purpose of an economy, market or barter, is to facilitate this exchange. The only reason people engage in exchange is to get things with less effort than it would take to produce them for themselves. It's in that sense that an economy 'economizes' on resources.

The idea of breaking the so-called job-income link is just a hidden version of the original notion that wealth really belongs to society as a whole, and somehow we have to give people a claim on part of it without making participating in producing it a prerequisite. The other view is that wealth belongs to those who produce it; they need no justification for having it. It doesn't need to be distributed to them.

15.9* Steve Miller: I want to respond to Joe's earlier comment that it is hard to argue anyone was displaced by robots or computers given that there were 4 million new jobs created last year? I think one clearly has to distinguish net gains from gross displacement. Over the last few years, many plants in many heavy engineering industries have been consolidated as management committed to invest in improving blue collar productivity. I have no doubts that many new jobs were created. I do not think this is at all a counter argument to the claim that some jobs were displaced. Now I am willing to argue the source of the displacement. I am prepared to believe that robots per se had a marginal effect, if any at all, and basic rationalization and consolidation, which is really motivated by, if not driven by new automation programs, is what really caused any displacement that might have resulted.

15.10* Joe Martino: Responding to Steve, clearly there has been displacement in some industries. The best example is agriculture, but there are others. The problem has always been not just creation of new jobs, which has always taken place, but the transfer of displaced workers to those new jobs. Since there are net new jobs, clearly the transfer is possible from a numbers standpoint. The problem is feasibility in terms of geographic location, new skills, and so on.

Item 17* Steve Miller: I claim with normal rates of attrition over a several year period, a 10 percent displacement of production workers could be absorbed. Any comments?

The conclusion of pages 133-141 is that flexibly automated plants would have a substantial cost advantage over conventionally organised batch production facilities. While the speculative nature of the analysis is acknowledged, the conclusion is that 20 to 50 percent reductions in cost are realistic (See Table 16 on page 139). Any comments?

Another question: The following scenario is outlined. A flexibly automated plant is built (either new, or from extensive modifications of an exisiting facility), and a company consolidates its production, shutting down one or several other facilities and moving their products into the new flexibly automated facility. My feeling is that plant shutdowns and consolidations are the real threat to worker displacement, and not the introduction of a handful of robots into a facility. Any comments?

15.9-10* Peter Unterweger: One of Steve's questions deals with attrition. I have looked at some numbers we estimated for one of the major auto companies; they show attrition averaged only about 3% '76-'80; it appears to have been substantially higher (more than double) in the late sixties and early seventies. I seems that while the auto market and employment were strong, turnover and attrition were high. Now we are in a situation of declining employment and low turnover. I hope to get some more numbers for the early eighties.

While I think that attrition can help protect the jobs of the currently employed, it does not address the problems of the new workforce entrants. I think it will become a growing problem (even with some demographic slowdown); the average unemployment rate has consistently risen since the 1950's; the first half of the 1980's are continuing this trend. Market growth for most of the goods that fueled the postwar expansion is slowing, while competition stimulates the introduction of labor saving automation. These developments are superimposed on the unfavorable unemployment trend. It is these considerations which convince me that the work by Ayres and Miller, by Timothy Hunt, and by Duchin and Leontief is of crucial importance. We need improved social, and economic policy in this area, and that needs a sound basis.

Steve's final question was about FMS as a stimulus to plant shutdowns. I came to this conclusion when I first saw the numbers on metalcutting time for the average machine tool. Clearly there is vast room for improvement, and the thrust of computer control and improved manufacturing organisation is to bring it about. Unless markets grow more strongly this would generate excess capacity, which would lead to the shutdown of existing operations. There is some evidence that this is taking place in the auto industry, at least in the major assembly companies, but I will have to look at it a little more closely.

15.11* Joe Martino: Peter's observations regarding shutdowns and low attrition are supported by an analogy with the airline industry. Under the old regime of regulated airlines, the Civil Aeronautics Board enforced a cartel. However, the primary beneficiaries of the cartel were the airline employees, who enjoyed wages much higher than the market would have supported. Airline stockholders often took losses. With deregulation, market forces are acting at last to bring airline wages into balance with the rest of society. The growth in the airline industry since deregulation has been in the startup firms. The old-line firms, saddled with uneconomic wage schedules, have been hard put to hold their own. (Braniff's bankruptcy is due largely to management bungling; the wage situation is only partly to blame there.) In the absence of "protection" of one kind or another for the firms stuck with low-productivity, high cost labor, such as tariffs, import quotas, government bailouts, etc., these firms will be pressed hard by startup firms which use the new technology to reduce operating costs. These new firms may well also end up giving their employees greater "job security," and ultimately higher wages, than the older firms did, because the employees will be more productive. So long as an employer can make more money by adding workers, he will bid up their wages in order to attract them from other jobs.

Items 31-33* Robert Ayres: Theme 1:Will Growth Compensate for Displacement? The question arises (Unterweger and others): will the introduction of robots and flexible automation result in product costs dropping sufficiently to stimulate demand and actually increase employment in factories? We have argued elsewhere that, in general the answer is likely to be no. Basically, cost-cutting by reducing labor inputs only results in increased employment if the market is extremely responsive to lower prices. In other words, the price elasticity of demand must be much greater than unity (in absolute value) for this to occur. Large price elasticities are normally associated with luxury goods, or with products newly introduced to the market. For most products that are manufactured in the "mature" industries, the opposite is true. Such products are more or less necessities, markets are close to saturation, and price elasticities are fairly low. For such products, lower cost of production would not stimulate proportional increases in demand, or output. Unfortunately, some proponents of robotics, including the author of a paper commissioned by an important committee of Congress (sorry, I haven't got the reference handy) argued precisely the opposite, using the price history of the Model T Ford as an example. Of course, the market being served by the Model T was anything but saturated, and its great success in the market was at least partly due to sharply rising incomes. By comparison, the average disposable income of workers in the U.S. has not increased since the early 1960's. The real cost of cars, and of fuel, has increased quite a lot, on the other hand. While the non-working population -- mainly the elderly -- have enjoyed an increased income in the past decade, these are not the most eager buyers of cars. Thus, I feel safe in suggesting that the use of robotics by auto firms will not, ipso facto, increase demand for autos. On the other hand, it is important to remember that this step might help keep the auto manufacturers in the U.S. competitive. It is not, after all, total employment in an industry that matters to American workers, but employment within this country. I would appreciate your comments.

31.1* Steve Miller: To the extent there there will be an increase in manufacturing employment due to cost reductions/demand increases, my informal guess is that it will not affect the hands on production people. The general industry attitude, I think, is that it is better to have fewer people touch a product in the manufacturing stage for a lot of reason, quality control, overhead cost reduction, in addition to minimizing labor cost. I suspect the increase will be in servicing and supporting the manufactured goods.

31/33.1* Peter Unterweger: In general I agree with your description of the prospects of the auto market. I recently looked at some numbers that showed that there is more than .8 of a vehicle for every person of driving age in this country (cars and trucks). European studies have concluded that saturation occurs below this level. While this may be true for the more densely populated Western Europe, I would guess that the US saturation level is higher. Still, we don't have much further to go which clearly implies slow market growth. For a good summary of industry views see the Wall Street Journal article that appeared earlier this week. In general I also agree that overall employment rather than that in one industry is what matters. The reservations I have are that we are talking about a very big industry (auto), that others are also affected by 'saturation', and that in this country the social legislation framework which allows countries like Sweden to shift employment out of declining industries is mostly missing. Once income support, education, and employment programs are in place workers will feel a lot less reluctant about new technology.

31/33.2* Joe Martino: I'm in agreement on the auto market. It's saturated. However, human demand as a whole is never saturated. Just how many people do you know who don't want more than they already have? Therefore even with the saturation of so-called 'basic' industries, there is still opportunity for more production of wealth. Despite the Keynesians, Says' Law has never been repealed. Every act of production creates a demand for SOMETHING ELSE, to the full exchange value of the things produced. Thus the trick is to move the workers who have been 'fired' by the buying public into industries where the public is still buying. Ultimately the more we all produce, the more we can all have. Cutting productivity in order to save jobs is ultimately cutting our own throats.

38.11* Patrick Sweet: I agree with you, Peter(38.10*), for the most part, and certainly appreciate your attempt at focusing on a narrower set of issues. However, your focus begs the question of "is it simply retraining or the lack thereof that is the locus of resistance to new technologies?" I think not, for most workers understand that there is a steadily decreasing ratio of 'new jobs-to-people available'. They have learned in life that retraining today means retraining tomorrow; ask any welder or toolcutter.

Management faces the same problem from another perspective. If a company is to put money into retraining but knows that there is going to be a lesser need for employees with requisite skills than its move toward higher technologies is making available through displacement, how does it choose/discriminate among employees available?

My point is only to raise these questions and not to attempt to resolve them. I also believe that if we focus on retraining alone there will be a goodly number of individuals left out. Since our economy depends on the purchasing power of all and not just a few it seems important to transcend retraining and displacement issues. But maybe this conference is not the best place to do so.

38.12* Peter Unterweger: I agree that my comments did not address the question of where the locus of resistance to new technologies lies. My point was simply that more broadly qualified people allow sophisticated systems to function more efficiently. I think that a key element of the resistance to new technology is resistance to change, which affects managers as well as production and office workers. A key question is why is our society producing so many individuals who crave the security of the status quo. For those on the low end of the income distribution a big part though by no means all is everpresent material insecurity. I think both the need to retrain, with all the attendant unknowns, as well as the shortage of suitable training/education are inhibitors to change. I don't understand your point about the importance of the income of all (with which I agree for both social as well as economic reasons), and how it makes the transcendence of training and displacement issues necessary. Economically everyones income is important because there is more than a grain of truth to "underconsumptionist" explanations of the stagnation afflicting our economy. How training fits in, except through underutilization of human resources, I do not immediately see. Perhaps you could explain.

38.13* Patrick Sweet: It is unquestionable that more broadly qualified people run sophisticated systems more efficiently, and that a key element of the resistance to new technologies is a passion for the status quo. I believe we are in agreement but have missed each other somehow. My point about transcending retraining issues did not include transcending displacement issues as you suggest I meant to in your response. To restate what I was driving at would be to say that retraining is not sufficient for there will not be a one-to-one ratio of new jobs created for old ones replaced. Thus all who are displaced will be in need of retraining, but the company letting them go will not have a job for each and every person even if they have learned new skills. It is at this juncture in the debate that we must recognize that retraining is not sufficient. I think that it becomes quite important to then look beyond it for obvious economic and social reasons.

38.14* Joe Martino: part of the "retraining" issue may be a false one. The assumption seems to be that unless workers with no-longer-valuable skills are retrained, they will not be employable. I suggest that even with retraining, they may not be employable if there is insufficient demand for the skills they have just acquired. Conversely, if there is a demand for workers, either of two things will happen. (1) Employers will undertake the retraining themselves, simply because they need the workders, or (2) the jobs will be simplified by making the machines "user-friendly" (like the fast food restaurant's cash registers with pictures of food items on the keys). The two alternatives are not exclusive, of course. In short, if the workers are needed, somehow a way will be found to use them. If they are not in demand, no amount of training is going to get jobs for ALL of them.

38.15&16* Peter Unterweger: Well it looks like on this point we are agreed. retraining is not necessarily a solution to the displacement problem. I agree also with Pat's restatement of his views, but I am not sure all of us did before we went through this. To me displacement, and job creation (as long as we do not have other means of distribution) are the most important issues, but a close second is the quality of jobs, and that is related to the education/ training discussion. I think that better educated people have better jobs and therefore a better quality of life; if Patrick is right, and I think he is, we clearly are in the happy position that we can simultaneously improve the efficiency of our productive system, and the quality of workers lives. This is the second of the two possibilities that Jan Helling outlined a while ago. Joe's example of putting pictures instead of numbers on the cash register is of course an example of Jan's first point. Unfortunately, it seems to me that the U.S. as a society has so far preferred the 'deskilling' strategy, or perhaps more accurately a strategy of underinvesting in 'human capital'. That at least is the point made in a study by the New York stock exchange, which cites John Kendrick as estimating the 'people factor' to be responsible for 20%–50% of productivity growth. The study then attributes a large measure of Japan's outstanding productivity growth to its outstanding educational system, which produces not a brilliant elite but a high average level of capability. According to the report this system ' is shaping a whole population, workers as well as managers, to a standard inconceivable in the United States,...'.

Item 41* Peter Unterweger: NEW GERMAN STUDY OF AUTOMATION/WORK– is a report on a survey of 1100 plants with 100 or more employees in the German metalworking industry which was conducted by IG Metall, the German metal workers' union.

A draft questionnaire was tested on 120 plants, and then revised and developed into a 34 page survey instrument, which covered various means of rationalizing production: new technology, product changes, organizational change, personnel policy, and corporate restructuring.

Interviews with 'works councils'(1) were conducted by trained interviewers. Results are not individual opinions, but consensus of the 'works council' leadership.

The plants covered include 1.3 million employees which is more than 35% of the employment in the industries covered. The only thing nearly comparable that I am aware of is a recently completed JEC study.

The following are some of the major findings. Please note that most of the percentages shown below are the result of reading a chart, ie the error is about 2 percentage points:

1. Rationalization measures were reported in all phases of operation:

   Parts production: 69% of reporting plants
   Warehousing: 64%
   Design, Mfg Planning: 62%
   Administration: 62%
   Assembly: 45%
   Transport: 28%
2. A speed–up in the introduction of NC/CNC and robots is reported.
   40% of all NC/CNC machine tools installed as of year–end 1982
       were installed since 1979.
   75% of all robots were installed during the same period.

3. Large plants report more rapid technology introduction:

> NC/CNC: 88% of 2000+ plants vs. 49% of under 500.
> Flexible Systems: 48% vs. 13%.
> Robots: 32% vs. 4%
> Automated Assembly: 23% vs. 5%
> Autom. Inspection: 76% vs. 16%
> Autom. Warehousing: 40% vs. 1%
> CAD: 53% vs. 7%

4. The impacts of NC/CNC. Affected plants reporting negative and positive consequences:

> Employment: 60+% negative, about 5% positive, rest no change.
> Job hourly rate: about 10% neg, 25% positive.
> Performance Monitoring: 40+ % negative, 8% positive.
> Work Pace: 60+% negative, 5% positive.
> Physical Stress: less than 20% negative, more than 35% positive.
> Psychological Stress: 78% negative, 4% positive.
> Variation (in tasks): 55% negative, 12% positive.
> Responsibility: about 40% negative, 30% positive.
> Skill: 32 % negative, 42% positive.

5. Product changes are strongly linked to technical and organizational rationalization, ie. 80% reporting such measures also report product changes.

6. Many plants report the side by side existence of various data collection and processing systems. Islands of automation waiting to be linked.

7. Impact of office automation:

> Shiftwork: 7% report declines, 12% increases.
> Noise: declined 12%, increased 17%.
> Rigid work position: declined 3%, increased 37%.
> Eye stress: declined 1%, increased 68%.
> Work speed: declined 3%, increased 28%.
> Monotony: declined 3%, increased 34%.
> Social Isolation: declined 1%, increased 68%

8. The study divides plants into strongly innovative firms and others, and shows that the results of 4 and 7 are found to an even greater degree in the more innovative firms.

9. Existing skills are endangered. Most of the programming of CNC machines especially when there are more than 5 in a plant is done by others instead of the operator.

10. New jobs and skills.

> Administration: 58% report new kinds of jobs. 40% of these are

skilled, 36% unskilled, with the rest undetermined.
Production: 67% report new kinds of jobs. 63% of these are skilled
11% unskilled.

11. The reported employment effects are very similar in strongly
innovative and technically backward firms. 62% of the former report
employment declines, and 65% of the latter. (Not known are the
relative size of the declines).

If you can read German, I will be happy to send copies of the report
to you.

(1) "Works Councils" are mandated by law (in plants above a certain
size ?). They consist of workers' elected representatives, which
may, but need not be, union members. Most of what we consider
bargaining, labor relations, and more (since there are far more
legally required topics for labor relations) takes place between
management and the works council. There is no single union
bargaining agent, although I am sure that union members hold the
great majority of works council positions.

Item 44* David Morrison: Regarding price elasticity of demand and worker
displacement issues, it may be misleading to approach the issue in terms of broad
percentages. Only a few percentage point gains in either productivity or
displacement of workers is likely in any year as a result of technological change.
However, in the conversion of particular factories or industry segments, the
process is likely to be a series of jumps (or jolts). Thus if GM's Saturn(?)
project for the small car of the 1990's results in a technological breakthrough in
design, materials and automated assembly, you could have a sharp cost reduction,
rapid market expansion (for that product) as a result of purchases by low wage
earners and displacement of competing imports. (Too often overlooked is the fact
that real income can grow either from wage increases or from the availability of
better or cheaper products for purchase with "old" wages. For those concerned with
equity, it is an interesting question whether equity is better served by expanding
(or preserving) the middle wage earning segment of society or, for example,
providing a lower cost car to the minimum wage fast food employee.) But how do you
measure worker displacement?

In the Saturn small car case, the impact would be complex. If there had been no
such project, how many jobs would have been lost to imports. If Saturn succeeds,
how many jobs will have been taken from domestic production of larger or older
models (of either GM or competitor companies) and for new buyers (low income buyers
who might not have bought a car) what jobs have been lost in other product or
service areas because of the shift of buying into the car market. Somewhere also
there is the question of dynamics -- How does the successful launching of a major
new product (or new product version) whether the auto, the aircraft or the computer
impact on national or regional economies through various mechanisms including
consumer and investor confidence in the future. It is pretty complicated even to
try to conceive of the factors involved at a micro-level which would allow anyone
to draw conclusions about the employment consequences of a change of this kind. On
the other hand, the incremental average percentage change in manufacturing

employment attributable to automation is likely to be dominated at any given time by ups and downs in the business cycle.

Another problem in the displacement concern is the industry immobility which results from sharp differences in wage levels among industries and between industries and services.   In terms of public policy goals of facilitating adjustment to change, it might be useful to set as a national goal the raising of real compensation for minimum wage and other lower wage earners to levels closer to those of presently highly paid industrial workers –– thereby lowering the barriers to movement across industries and changing occupations.

INDUSTRIAL POLICY

Item 8* **John Ettlie**: Does anyone have any comments, great thoughts, etc. on Petition 232 (defense protection petition filed by DOC to limit imports in machine tool industry to 17.5%)?

8.1* **Joe Martino**: Limiting machine tool imports will raise the cost of machine tools, and thereby make U.S. exports less competitive in the world market. It's one more example of strangling the successful industries to bolster the unsuccessful ones. 'Lemon Fascism' as distinguished from 'Lemon Socialism' as practiced in England and France.

8.4* **Robert Ayres**: I can't believe the situation is as simple as suggested by Joe Martino. In the first place, the British managed to become the world's low cost iron/steel producers during the 18th and early 19th centuries while they were practicing a policy of high tariffs discriminating against Sweden and (by the way) their American colonies. Without the tariff protection there would have been no iron industry at all in the UK. The Japanese have also consistently protected their 'infant industries' until they were ready to go out and compete on even terms with the rest of the world. Will the U.S. be better off with no domestic machine tool industry -- as we now have no domestic consumer electronic industry?

8.5* **Peter Unterweger**: Bob Ayres has a good point. Also I don't understand the difference between lemon socialism and lemon fascism?

8.6* **John Ettlie**: The Great Lakes governors have already endorsed petition 232. The Japanese will not be affected by it even if it was endorsed by the President (it was recently sent back to DOC (Department of Commerce). Small machine tool builders in this country, for example, are already importing components like lead screws because the U.S. builders cannot find that level quality at that price here. (I have two good cases on that in the file from last fall). We might want to take up the more important issue of the Great Lakes Council in general (they intend to finish with the machine tool industry in July or August and then go on to the automobile industry) as a policy making body to be influenced, confronted, conferred, etc.

FORECASTING METHOD

item 20* Timothy Hunt: We criticized Ayres/Miller in our book HUMAN RESOURCES for the failure to put a time signature on their forecasts. To some extent that criticism remains valid, especially in regards to meaningful social policy. We think it is difficult to speculate about technology 5–10 years into the future, let alone the year 2000 and beyond ... Policymakers, educators, and others may demand such forecasts, but there are some things that are knowable and some that are not. In any event I concur with Lynn that the Ayres paper is too speculative.

20.1* Robert Ayres: For this sort of criticism there is no answer. On the one hand we are damned for failing to put a time signature on our forecasts -- which was deliberate, by the way. We believe that it is possible to say a good deal about the nature of a technology and its likely impacts, without specifying its time path precisely. This we tried to do. On the other hand, you condemn us for being too speculative. What would satisfy you? Incidently, I, with a graduate student, Jeff Funk, am on the point of publishing a more explicit time dependent forecast. It will not change any of our socio–economic conclusions, however.

item 22* Faye Duchin: Employment consequences. My current work involves the assessment of employment and other economic consequences of robots and other forms of automation (Leontief and Duchin, NSF report, March 1984) in the context of an economy–wide input–output model. I feel the technical work on robots or any other technology should specify labor REQUIREMENTS not DISPLACEMENT. Displacement static concept since it is relative to existing staffing patterns. If all alternative technologies are instead described in terms of their labor (and energy, robot, etc) requirements per unit of output, then any displacement due to the use of one set of technologies relative to another will be the outcome of analysis rather than a fixed parameter.

22.2* Steve Miller: Faye, I think I might agree with you, but I need to see this idea explained in more detail.

22.3* Joe Martino: This approach seems to make a great deal more sense than simply looking at the industry from which workers have been "displaced." If there is net increase in employment, then the possiblility of absorbing the displaced workers exists.

item 23* Faye Duchin: Economic Consequences. Examining the prospects for workers displaced by robots is too ambitious for the framework of these papers. It will depend on factors like future demand for American autos and computers, for example. We need to approach these complex questions systematically and do an equally probing job at each level of analysis. The work by Ayres and Miller on robots needs to be combined with similar work on the prospects for the other sectors of the economy into a comprehensive analytic framework before the global economic questions can be addressed analytically. This requires disciplined collaboration and/or overview and co–ordination by, say, NSF.

23.1* John Ettlie: Hear, hear!

23.2* Steve Miller: Here again, I agree with one, although there are some interesting 'technology transfer' problems.

First off, I do not even feel secure in saying that we will know the impacts of robots (more importantly of flexible automation) in the sectors in which they are used, not even considering other sectors. Up to this time, we have simplistically focused on cost reduction, and have not really addressed issues of how these capabilities might give a producer greater ability to respond to market requirements and do more business. Secondly, the I/O or any other aggregate framework must use highly stylized characterizations of technological impacts. While I am basically a fan of I/O, I have yet to be convinced that the framework adaquately embodies what we need to know about the impacts of these technologies on the economics of production.

23.3* Joe Martino: I'm sympathetic with Steve's view here. I/O is useful in analyzing the short-term impacts of major changes in availability of materials, or shifts in demand patterns. It cannot cope with major changes in technology. Some years back Anne Carter did some interesting work in analyzing the 'technologh' of a whole sector, using I/O, but no one, to my knowledge, has ever been able to link her results to finer-grained analysis of technology. Without that finer-grained analysis, I don't think I/O will be particularly useful in telling us how other sectors will respond to the results of increased use of robots in certain sectors.

23.4* John Ettlie: Agree.

23.5* Faye Duchin: I/O has been evolving. We have developed a dynamic (no longer static) framework and are trying to provide a finer-grained analysis (as in our recent study). We have used the work of Ayres and Miller, and that is precisely why I am interested in seeing it elaborated.

23.6* Steve Miller: Faye, I would be glad to look at any recent work where you want to check assumptions about the changing nature of I/O relationships in the capital goods sector. I am especially interested in any analysis that sheds light on the issue of the 'multiplier' effect of cost reduction/productivity improvement in the engineering sectors.

item 24* Faye Duchin: Levels of Analysis. These papers range from detailed descriptions of different types of robots, to more abstract descriptions of metalworking sectors, to estimates of future employment in manufacturing, to speculation on the situation of the middle class in the 21st century. While this is indeed a logical progression, the problem is that the analysis at each level is carried out with a completely new methodology and is related to the preceding mainly through the intuition of the researcher. I personally would like to see more elaboration and integration of the first two levels with particular emphasis on presenting the work in a form which enhances its usefulness in further analytic work by the same or other researchers.

24.1* John Ettlie: Agree.

24.2* Steve Miller: I admit that these several pieces were done at separate times and the only unifying theme is that they are placed under the same report cover. More integration is necessary. At the moment, I am not sure my time is best spend doing broad brush work on societal impacts, though. My preferences are to go deeper and narrower and spend more time in particular manufacturing organizations seeing what is going on in particular plants. Perhaps you can give me more insight into the type of work that we could do that would feed into the macro analysis and economy wide modelling. This might motivate me to still keep an interest in writing for policy people as opposed to business people.

24.3* John Ettlie: Now we're getting some place -- micro and macro people talking.

24.4* Faye Duchin:We are now extending the framework of the current dynamic model so that the economic decision will be made within the model (ie, endogenously) as to what technology should be chosen when a sector is expanding or contemplating modernization. For this we need a detailed description of the inputs (on current account, capital, and labor) for each alternative. This menu of technological alternatives obviously requires the kind of careful, detailed work you are talking about -- but based not only on what IS being done but also on what is on the drawing boards and what is at the state of the art. If this is done systematically, sector by sector, we can hope to be able to realistically evaluate some alternative paths for the economy over the next couple of decades. I will send you our study -- you will see that the model is considerably more realistic than its many predecessors. Let's certainly discuss the prospects further.

24.5* Joe Martino: I'm certainly not objecting to what Duchin is proposing. This is really what I try to get my clients to do, although on a much smaller scale. However, I have to pass on a warning I give my clients. Don't think that any amount of sophisticated analysis is going to allow you to predict what will happen in an industry when the people in that industry, who know a lot more about it than you do, are themselves uncertain about what will happen.

24.6* Steve Miller: Trying to draw together themes from Faye and Joe: I agree with Joe that industry people themselves are uncertain about where they are going but here is an instance of where the distinction between very micro and macro can be helpful. I do not think that the industry people are uncertain about general long term goals such as x percent per year reductions in labor requirements, inventory costs, etc. Many of the uncertainties are in how to achieve it, and in how to time phase the projects from quarter to quarter. Despite my earlier comments for I/O to get very micro, I wonder if there is an appropriate level of absrtaction that will actually be more useful because it ignores very low level types of noise and uncertainties.

24.7* Joe Martino: The analysis can probably be done successfully at a fairly high level of aggregation. In many cases this is where my work is most useful to industrial clients, who know much more about the micro aspects of ther industry than I do, but who don't or won't take the time to look at the macro aspects.

items 35-37* Robert Ayres: Theme 3. We need quantitative, industry-by industry an occupation-by occupation and year by year estimates of likely worker displacement (Hunt, Lynn, Duchin). Duchin goes on to advocate a comprehensive forecasting and modelling framework -- presumably embodying an input-output core -- for doing this. I have mixed feelings about this. On the one hand, it is easy to point out the extraordinary difficulties of such an enterprise. Comments by several of you emphasised some of the problems of capturing in an aggregate-level model the complexities and subtleties of the relationship between human workers and robots or computer-controlled machines in a range of dramatically different applications. In fact, the recent OTA study essentially threw in the towel on this and substituted an essay on the problems for any serious attempt to solve them. (I'm referring to the interim report, which was published a year ago, since the final report is not yet out. But I'll bet the final report avoids any quantitative forecasts too). On the other hand, I do think a serious attempt at a quantitative forecast would be possible, given a sufficient body of analysis of the substitution possibilities between human workers and sensor-driven, computer-controlled machines in a number of key domains. I've been wanting to do this for several years. A forthcoming PhD thesis by Jeff Funk will provide much of this analysis, with a focus on inspection and assembly. But there are other gaps yet to be filled. In short, it would be a major undertaking, but a worthwhile one.

AFTERTHOUGHTS ON TELECONFERENCING

Item 48* Patrick Sweet: Review Cost Overview. There are 2 things I would like to accomplish with this item:

1) I will provide a rough estimate of the costs of the Ayres/Miller review, and

2) I would like elicit some discussion on the felt benefits/drawbacks of running a review via this medium.

First the costs. Appearently our effort has been relatively cost effective. Since the beginning of the review total computer and communication costs incurred are in the range of $1200. Not all of this is review related activity though, i.e., private massages and other items should be subtracted from this figure. I would estimate total electronic costs for the review to be about $1000. Add to this Organizer time, final editing of the transcript, printing, postage, technical assistance and 'real time' phone calls, and I estimate the cost of the review to be in the range $2850 give or take $100 or so.

When comparing this figure to the cost of face-to-face reviews, not to mention time and effort saved in coordinating such a meeting at everyone's convenience, there appear to be some real advantages. However, there are also drawbacks.

I am interested in hearing how others feel about online reviews, and would especially appreciate Bob and Steve sharing with us their sense of what the process offers since they know best what they expected and what they received.

48.1* John Ettlie: I would raise one issue. Although it did not affect (I think) my reviews, I wonder if the typical journal article review process where authors and reviewers are not known to each other is not better -- better criticism and more indepth? My own opinion is that a good working document can be very effectively reviewed in this forum.

48.2* Leonard Lynn: I was impressed by the process in several regards: it seemed much more spontaneous than a collective written review (which would also presumably have taken far more time), yet it allowed more thoughtfulness than a collective oral review. I would have found it easier to give better responses, however, if I had known an easy way to get printouts so that I could have looked at several of the comments of others while composing a response. John Ettlie's suggestion that reviews could easily be kept anonymous also seems a good one. I also doubt if I would have said anything that I did not here, but it is hard to be sure. Finally, and I think this is simply a transitory problem, I suspect that some of us found ourselves concentrating at times less on the content of what we were saying, then on simply trying to get the machine to take it in some readable form. It takes a few days to get used to the system, at least that was my experience.

48.3* Peter Unterweger: I find the conference both useful and interesting. it seems a good medium for the review process, primarily for the reasons Leonard mentioned above. Anonymity is available, I believe to anyone who would care to use it. I doubt that the public nature of the comments inhibited me. At times it seemed that the interaction was slow and less than inspired, but usually someone like Bob Ayres would jump in, summarize points and crystallize the discussion, which then livened up subsequent exchanges. if there is any lesson in this, it would be for a bit more active intervention on the part of the discussion leader.

48.5* **Steve Miller**: It takes a while to catch on to the fact that questions and responses should be kept very focused because it is hard to deal with several screensfull of information. This is partially fixed by easy output device, but this means one needs a pc and printer, instead of just a good old terminal. I find it annoying that I can jump around across items, but not jump around the responses within an item. When the list of responses gets long, and I want to look at a particular person's response, I have to read more than I want to. So there are a bunch of serious user interface issues to attend to so that a person can extract information in an easy way and can review the history of the proceedings without having a super memory. In its current state, I believe that effective conferencing really requires easy access to paper print out, and that it is the paper print out that makes it possible to pull things together. For most of the conference, I did not have printout, and when faced with long, multi-issued responses, I tended to ignore all but one point that struck my fancy at the time.

48.6* **Peter Unterweger**: Having a printer is very useful. I dont think that I could have participated effectively without one. Without hard copy to review it is easy to miss or misinterpret points. I also found it very annoying not to be able to jump around in the responses to an item. In addition, I would find it very helpful if the system would offer one the opportunity to edit responses, just as it is possible to edit messages.

Item 49* **Patrick Sweet**: Future Conferencing Direction. As Peter Unterweger noted, the discussion of the Ayres/Miller paper has raised a number of very interesting issues. Peter suggested I create a summary list of issues for further discussion and poll participants on their views. I am willing to do this but am reluctant to inflict my bias on others. I would like to suggest the following alternative.

Gerry Ross and I have discussed possible next steps for the conference and have concluded that we (mainly I) may be too content conscious when we should be a little more process conscious. If I were to provide us with a list of issues to pursue, the question remains "to what ends?". The detail and thought that went into Peter's suggestion (see item 46) is a clear example of the level and quality of input each of us has to offer. When it comes to formulating our next steps we should benefit greatly from maximizing Peter's brand of insight and effort.

Very early in the conference (item 1) there was some discussion about possible activities and outcomes that could result from our efforts. I think we all have a good sense of some of the benefits and limitations of this technology (and also what we would like to receive from it ourselves), and are therefor better equipped to readdress our conference's focus and future here. One way to do this would be to set up a Steering Committee. The responsibility of its members would be to agree to address the future direction and outcomes (whatever they may be) of conferencing activities. I don't perceive the exact number of members as being that important, but do feel there should be some formal commitment to participate on the part of those who plan to. Committee discussions should be public whenever possible, and noncommittee conference members should be encouraged to contribute. Having stated all these 'shoulds,' I would like some feedback on how all this sounds.

49.1* John Ettlie: probably a good idea. There's nothing wrong with a hierarchy in a computer conference, just as long as its not FIXED.

49.2* Patrick Sweet: I agree. The steering committee is not intended to remain fixed nor will it need to be. The concept underlying it is mainly to bring several of us together more formal and to share responsibility. I do not see the need for minutes or other procedures associated with committees, but would like some camaraderie in setting agenda and formulating objectives. Any takers?

49.3* Peter Unterweger: This steering committee idea is still not quite clear to me. Usually a 'select' body is formed when the size of the group inhibits efficient discussion and decision making. This group, especially the group of active paricipants, does not seem too large from this point of view. If the committees discussions are to be public, and other conference participants would comment on this discussion, how would this discussion differ from the one we are carrying on now? Is the intention to encourage a higher level of participation from the people who would be on the committee? Presumably the more active participants would be the logical candidates. If so, how would this change the current state of affairs. I hope these comments are not interpreted as negative reactions. If the steering committee idea is pursued, I would be willing to give it a try. Unfortunately, it looks as though I will be away during the important initial phase of any such experiment.

49.4* John Ettlie: Patrick, can you give us a frequency distribution of participation (items, responses, participant)? Would this be an empirical way of forming the nominations for the steering committee?

49.5* Patrick Sweet: First a response to Peter. I think that forming an S.C. would have the effect of focusing our efforts more than they are at present. A committee disucssion may not differ very much from what is happening now but, it might, however, offer a greater sense of shared responsibility among those involved for arriving at conference goals and objectives and may therefore bring a clearer focus to our efforts and discussions. As I see it, greater participation in general is not the issue. Every one has contributed to the success of the conference thus far in their own ways. What we, as a network, need to address at present is the direction of the conference over its content. We could continue to discuss issues raised, but there needs to be some basic agreed upon purpose for doing so if we are to attract new participants and maintain the interest of present ones as well. (This is an opinion based on my experiences with other teleconferences and discussions with curious inquirees.) I don't want to be perceived as pushing the S.C. idea for its own sake and run the risk of having us miss the point(s) behind the suggestion. These are to:

1) Focus our attention on the importance of explicitly
     addressing the future direction(s) of this conference/network;
2) Encourage others to become actively involved in this process;
3) Generate commitment to follow through.

If an S.C. is not appropriate or there are other suggestions, I'd sincerely like to discuss them. I'm not normally known for relying on structure rather than letting group process run its course, but this medium (I think) requires some structure for it is not entirely stimulus–driven. Does this give a better idea of the impetus behind the S.C. suggestion?

REFERENCES

7.1* Dave Roitman: Gustavson, P., and Taylor, J.C. Socio–technical Design and New Forms of Work Organization: Integrated Circuit Fabrication. In 'Automation and Work Design: and ILO Project (Butera, F., & Thurman, J.E., eds.) Probably can be obtained from J. Taylor.

7.2* David Morrison: The Department of State Center for the Study of Foreign Affairs is sponsoring a symposium in Washington (actually, Rosslyn) on May 9th. Topic: Robotics/ Automated Manufacturing. The session on U.S. competitiveness (speakers Robert Lawrence of Brookings and George Kuper of the Manufacturing Studies Board of the National Academies of Science will be addressing, respectively, manufactures in general and machine tools in particular. Phone number for registration is (703) 235–8830 or 235–8831. No charge. More details by bulletin.

Item 40* John Ettlie: Socio–technical research dearth. Some results reported in a literature review recently by Pasmore, et al 1982 (Human Relations, Vol. 35, No. 12, 1179–1204) are most relvant to this confernce. I report them now because they agree with a literature review we have begun in a specific area and they are too important to hold back until published. I quote from Pasmore, et al (the title of the article, by the way, is 'Sociotechnical Systems: A North American Reflection on Empirical Studies of the Seventies,').

1) from p 1179, the abstract: 'the number of experiments involving technological inn- ovation or change is relatively small; moreover, from the results achieved in these experiments, it is obvious that we still have much to learn regarding the design of technical systems for joint optimization.'

2) from page 1196, 'We note specifically that technological chage was a factor in only 9% of the cases <134 experiments total reviewed> for which improved productivity data were reported; furthermore, productivity increased in only 60% of the cases in which technological change took place.' I could go on, but the message is obvious -- we have our work cut out for us -- most of the 'sociotechnical' work that has been done has been either socio or technical and not both.

40.1* Bill Hetzner: There have been actual socio and technical interventions, but in general they have not made their way into the literature. The people doing the interventions are either too busy or basically uninterested in reporting on their work. The Center for Quality of Working Life at UCLA at one time was developing a case protocol for reporting on their and others' interventions. It was never done. They have a number of interesting cases, although generally continuous process technologies. Henry Tosi at Florida is doing a case study for us (something we generally don't fund) on a social and technical change. There is data out there, but nobody has found the time or has been clever enough to develop a way of uncovering it.

40.2* Patrick Sweet: A major factor contributing to the lack of publication of such vital data is the often proprietary nature of production processes -- I believe. It can be very difficult to report on changes made in systems and adhere to the confidentiality agreements frequently arranged prior to embarking on an intervention/study.

40.3,4,6* Jan Helling: I can see other factors of importance as well. Like: No resources and time within a company to spend efforts on finding out facts and writing reports. Lack of talent to write papers. Very little of incentives and management support to write papers for external use. Difficulties to be a prophet in your own house Not least maybe the experiences from academic researchers entering the company – finding facts and doing analyses, writing reports and books pretending the innovations are their own and take the credit them selves. Why not at least support inhouse entrepreneurs to take part in the studies and writing and appear as coauthors.? In our company this is one reason why we are restrictive.

In Sweden we have several companies that measure productivity increase by sociotechnical changes. We can not claim socio-tech inspired changes are all over the place in Swedish industry. Still we envision them as enlighted islands here and there. But they are definitely increasing in numbers. We have research done on this and many papers – in Swedish though. But I know most of the researchers – and they are all able to speak and write in English. If you wish I can give you some names for references. I will myself now be involved in research and development projects here at SAAB when we will go further in our socio-tech efforts in two areas: body welding and final assembly of cars. These two projects will be sponsored by a "Swedish National Joint Management Union Science Foundation".

Volvo Kalmar plant is 10 years old by now! Many new steps have been taken both there and at other companies as well. By ' new' technologies I believe you mean not only new technical systems for manufacturing but also integrated changes of social and administrative systems as well. I see sociotech changes as integration of them all at once i n projects aimed at increased productivity and better work conditions. We have many examples of new technical sytems along with the rest. I could give you a list of interesting companies and plans in Sweden but give you here instead some sociotech researchers as reference persons: Ulf Karlsson, Chalmers School of Technology, 412 96 Gothenburg, Phone 46 31 810100 Sven Eric Andersson, The Swedish Institute of Production Engineering Research, 412 85 Gothenburg, Phone 46 31 81 01 80 Christian Berggren, Royal Institute of Technology, 10044 Stockholm Phone 46 8 7870120 Torbjorn Stjernberg, Stockholm School of Economics, 113 83 Stockholm Phone 46 8 7360120 Stefan Aguren, Swedish Management Group, 10330 Stockholm, Phone 46 8 7360 Phone 46 8 235410

Item 41* Peter Unterweger: NEW GERMAN STUDY OF AUTOMATION/WORK-

This is a report on a survey of 1100 plants with 100 or more employees in the German metalworking industry which was conducted by IG Metall, the German metal workers' union.

A draft questionnaire was tested on 120 plants, and then revised and developed into a 34 page survey instrument, which covered various means of rationalizing production: new technology, product changes, organizational change, personnel policy, and corporate restructuring.

Interviews with "works councils"(1) were conducted by trained interviewers. Results are not individual opinions, but consensus of the "works council" leadership.

The plants covered include 1.3 million employees which is more than 35% of the employment in the industries covered. The only thing nearly comparable that I am aware of is a recently completed JEC study.

The following are some of the major findings. Please note that most of the percentages shown below are the result of reading a chart, ie the error is about 2 percentage points:

1. Rationalization measures were reported in all phases of operation:

   Parts production: 69% of reporting plants
   Warehousing: 64%
   Design, Mfg Planning: 62%
   Administration: 62%
   Assembly: 45%
   Transport: 28%
2. A speed-up in the introduction of NC/CNC and robots is reported.
   40% of all NC/CNC machine tools installed as of year-end 1982
       were installed since 1979.
   75% of all robots were installed during the same period.

3. Large plants report more rapid technology introduction:

   NC/CNC: 88% of 2000+ plants vs. 49% of under 500.
   Flexible Systems: 48% vs. 13%.
   Robots: 32% vs. 4%
   Automated Assembly: 23% vs. 5%
   Autom. Inspection: 76% vs. 16%
   Autom. Warehousing: 40% vs. 1%
   CAD: 53% vs. 7%

4. The impacts of NC/CNC. Affected plants reporting negative and positive consequences:

   Employment: 60+% negative, about 5% positive, rest no change.
   Job hourly rate: about 10% neg, 25% positive.
   Performance Monitoring: 40+ % negative, 8% positive.
   Work Pace: 60+% negative, 5% positive.
   Physical Stress: less than 20% negative, more than 35% positive.
   Psychological Stress: 78% negative, 4% positive.
   Variation (in tasks): 55% negative, 12% positive.
   Responsibility: about 40% negative, 30% positive.
   Skill: 32 % negative, 42% positive.

5. Product changes are strongly linked to technical and organizational rationalization, ie. 80% reporting such measures also report product changes.

6. Many plants report the side by side existence of various data collection and processing systems. Islands of automation waiting

to be linked.

7. Impact of office automation:

    Shiftwork: 7% report declines, 12% increases.
    Noise: declined 12%, increased 17%.
    Rigid work position: declined 3%, increased 37%.
    Eye stress: declined 1%, increased 68%.
    Work speed: declined 3%, increased 28%.
    Monotony: declined 3%, increased 34%.
    Social Isolation: declined 1%, increased 68%

8. The study divides plants into strongly innovative firms and
   others, and shows that the results of 4 and 7 are found to an
   even greater degree in the more innovative firms.

9. Existing skills are endangered. Most of the programming of CNC machines
   especially when there are more than 5 in a plant is done by others
   instead of the operator.

10. New jobs and skills.

    Administration: 58% report new kinds of jobs. 46% of these are
                    skilled, 36% unskilled, with the rest undetermined.
    Production: 67% report new kinds of jobs. 63% of these are skilled
                11% unskilled.

11. The reported employment effects are very similar in strongly
    innovative and technically backward firms. 62% of the former report
    employment declines, and 65% of the latter. (Not known are the
    relative size of the declines).

If you can read German, I will be happy to send copies of the report
to you.

(1) "Works Councils" are mandated by law (in plants above a certain
    size ?). They consist of workers' elected representatives, which
    may, but need not be, union members. Most of what we consider
    bargaining, labor relations, and more (since there are far more
    legally required topics for labor relations) takes place between
    management and the works council. There is no single union
    bargaining agent, although I am sure that union members hold the
    great majority of works council positions.

7. Front Wheel Drive Technology with Rear Wheel Drive Management
or
From Flexible Manufacturing to Flexible Management Systems

Dr. Gerald H.B. Ross

Nesbitt Consulting Group
Toronto, Canada

July 1984

Dr. Ross, a participant in the
computer conference, prepared
this paper to follow through
on ideas that were raised
during the conference.

INTRODUCTION

Many industries, in response to intensified competition and rising costs, are reaching for new technological solutions such as robotics and Computer-Aided Manufacturing (CAM). While these developments are still very much in transition, they hold the promise of new flexibility in manufacturing systems, fully integrating the computer with the production process.

The thesis of this paper is that increased flexibility in manufacturing is inevitable, largely due to the unacceptably high cost of adaptation resulting from many decades of hard automation and standardization. Second, it will be suggested that "management systems" must always mirror production systems. As factories have moved from job shop technologies to batch process to continuous process, management has necessarily evolved from an "ad hoc" approach to a highly standardized bureaucracy. Consequently, it will be argued that as flexible manufacturing systems become more pervasive, they will require new kinds of flexible management systems.

DILEMMAS OF MASS PRODUCTION

Almost 200 years ago, Adam Smith Advocated the mass production of the common pin, using highly standardized operations. A century later, Henry Ford began to practice what Adam Smith Had preached and produced the Model T - revolutionizing the industry. Ford put a car in almost every garage; the secret of his success was the standardized product. the roots of our current predicament lie at the heart of this success story.

Prior to the industrial revolution, virtually all production was based on cottage industry or small scale localized production. Today the job shop operation is the descendant of this rudimentary form of production. The job shop, although not terribly efficient, represents extreme flexibility and organizational simplicity. Little production scheduling is needed, inventory can be checked with a brief glance at the back room and only a handful of employees need to be coordinated in the productive effort. In this sense, the blacksmith of yore is not too dissimilar from the custom machine shop of today.

The next type of production system to evolve was batch process. It involves some degree of standardization, at least for the duration of short production runs. It permits limited flexibility, usually within a family of manufactured products, but there are typically unit cost advantages over the job shop.

Finally, continuous process production systems were developed, ultimately using hard automation. These resulted in spectacular reductions in unit costs, providing that high volume could be maintained. However, virtually all flexibility, except perhaps for minor specification changes, had to be eliminated.

With each of these stages of production systems, a trade-off was inevitably involved between flexibility and efficiency. It seemed impossible to enjoy both. Today, of course, there are many examples of each type still in existence. However, the balance has shifted increasingly in favor of standardized, less flexible systems because of cost advantages.

The nature of these cost advantages are presented in Figure 1. In a sense, this diagram represents the continual "cost pressure" to standardize, where possible.

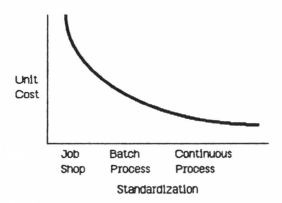

Figure 1. The trade off between flexibility and standardization.

The automobile industry, which is close at hand, illustrates how these costs pressures spurred on standardization.

> The process of standardization followed a
> hierarchy: first came the propulsion choice, then
> overall chassis configuration, and then major
> components were advanced. Finally, once the tech-
> nological change in the components subsided, the
> overall design of the automobile was optimized.
> This trend ended in the 1960's[1].

Perhaps the most serious consequence of this trend toward standardization has been the greatly diminished adaptability of the industry - its ability to respond to the demands of the environment. Highly automated systems became increasingly difficult to change. The end result of this trend over time, both in production and in management, can perhaps best be illustrated by examining adaptation to one critical aspect of the environment - consumer demand.

----

1. Abernathy, 1978, p.20

Abernathy et al[2] traced the costs of a major new model
introduction from the very beginning of the industry in 1900
to the present and compare the costs in terms of the incremental sales stimulated by each model change (Figure 2).

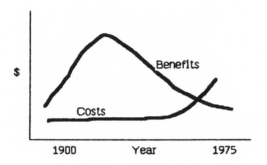

Figure 2. The Costs and Benefits of a Major
New Model Introduction
Inflation Adjusted).

The "Benefits" curve represents incremental sales (i.e. that
increase attributable to the availability of a new, more
desirable model). In the early part of this period, the
industry was characterized by the introduction many new
innovations: four wheel brakes, electric lights and the all
enclosed steel cab. The public responded enthusiastically to
such technological improvements by eagerly consuming the
latest offering. The peak of this trend was in the
mid-1920's.

At the same time, the cost on introducing these new improved
models began to increase, slowly at first and later dramatically. Abernathy notes the relative ease with which Ford
moved the engine in the Model C in 1904 from under the
driver's seat to a more fashionable location at the front of
the car (where the horse used to be). It was simply a matter
of re-routing chain drives and other components. However, to
change from the Model T to the Model A, after almost 20 years
of mass production, Ford had to close his plant for over a
year and permanently its traditional No. 1 spot to General
Motors forever - an enormous cost.

----

2. Ibid

For similar reasons, Volkswagen in the mid-1960's almost went out of business because of the costs of introducing front wheel drive after many decades of producing the "Beetle".   In the 1970's, the U.S. industry had the same kind of experience with "down sizing" to compete with the Japanese. Finally, in the 1980's, GM alone is in the process of spending about $40 billion on new model development - more then has been invested cumulatively in the history of the Company to date. These latter spending plans would seem to indicate that the cost curve in Figure 2 is rising even more dramatically today.

It is appropriate to ask why such a dramatic trend was not reflected in the financial statements? Why wasn't it more evident?".   There are a couple of reasons why this anomaly has occurred. For one thing, every time a new plant or a new facility is built, it is not called an "expense" of the business but an "asset", a term with a much more favorable connotation. The only difference between an "expense" and an "asset" is the assumption about how long the associated benefit is assumed to last. So a business lunch is treated as an expense because the benefit is assumed to last only until the next lunch hour while a new plant is expected to produce a stream of benefits over many years.

This actual "cost" thus only becomes realized as an expense in the profit and loss statement as annual "depreciation" is charged. The latter is a device for smoothing costs over long periods and can obscure the fact that a great deal of money has been irretrievably spent.

The effect of all this accounting wizardry is that "investment" in fixed assets is encouraged, thus trading variable costs (such as labor) for fixed. This is fine as long as the volume of production can be continually increased. The break even point, however, climbs steadily.

The automobile industry is not alone in this predicament. Airplane manufacturers are finding it prohibitively expensive to develop new generations of aircraft. Farmers are finding it increasingly costly to adopt the latest technological advances. Many other industries are similarly affected. This situation is largely due to the gradual loss of adaptability over time as organizations standardize.

The situation presented in Figure 1 represents a kind of "extinction scenario' if nothing were done to change the pattern. Any industry for which the costs of adaptation (in this case to market demand) are greater than the benefits cannot remain viable for long.

These escalating costs (e.g. for new product design, retooling, etc.) represent a substantial increase in the fixed
costs of the enterprise. This has had the effect of pushing
the breakeven point higher and higher, necessitating an ever
greater volume of production to sustain profitably.   Unfortunately, this can leave the organization increasingly vul-
nerable to changing market, regulatory and other conditions.
This escalation in the breakeven point is illustrated in
Figure 3 below.

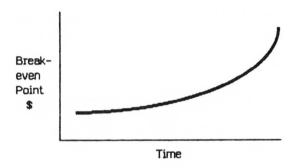

Figure 3. Breakeven point over time.

When the breakeven point is low, a relatively small volume of
sales will generate a profit. In "bad" years, the company
develops strategies, largely through marketing efforts, to
stimulate sales to remedy the situation. However, as automation and other forms of standardization proceed, it becomes
harder and harder to continually expand sales, particularly
as an industry matures.

Ultimately, this can create extraordinary situations where a
company has a best seller but is still losing money. The
Ford Escort/Lynx, for example, is one of the most popular
small cars on the market, with an annual volume of about half
a million units. Unfortunately, the breakeven point is in
the neighborhood of 800,000 units. This is largely due to
the high fixed costs that must be recovered from the initial
new model development and retooling.

This is in stark contrast to Saab which makes a modest profit
on a worldwide volume of about 90,000 units. For example,
there engines are manufactured in small bays of ten people

with the units being shunted off the main line into the
bays. With full production, there are ten people in each bay
and each person rotates jobs periodically. When an engine is
finished, it is sent back onto the main line.

At times of extreme hardship, the lines have been reduced to
as few as one person on the line. Interestingly enough,
although it takes the ten people cumulatively 33 minutes to
manufacture the engine, It also takes one person 33 minutes
working alone. In other words, there is enormous flexibility
designed into the human system as well as into the technology.

The essential dilemma is that the pursuit of productivity
through automation and other forms of standardization over
time has created a situation which threatens the viability of
many organizations. This suggests that some different
strategy is required -ideally one that introduces flexibility
into production processes while retaining the unit cost advantages of continuous process technologies.

Fortunately, the advent of more flexible manufacturing
systems promises a way out of this efficiency versus flexibility dilemma[3]. The versatility of computer controlled
equipment, particularly when integrated with the management
information system, presents the prospect of regaining some
of the flexibility of job shop operations with low unit costs
approaching those for continuous process.

> Such flexible manufacturing systems will one day
> closely approach the high throughput rates and low
> costs of today's specialized transfer line, with
> the added advantage of being able to process a
> wider variety of parts[4].

Figure 4 compares the traditional trade off between flexibility and standardization with an "ideal" situation that
could represent the potential of flexible manufacturing
systems.

----

3. Ayres & Miller, 1981

4. Ibid

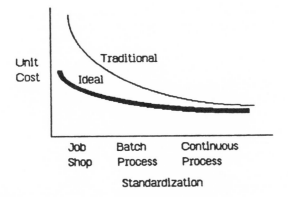

Figure 4.

Although the "ideal" curve may at first seem a remote possibility, there are already some salient examples of where this kind of potential has been realized. For example, Mercedes-Benz can produce truck engines in V-6, V-8 and V-12 configurations on the same production line. The engines are designed so that the front and rear cylinders are the same. The number of cylinders in between are simply varied to produce the desired product.

In contrast, in the early 1970´s, Ford spent millions of dollars and several years down sizing a V-8 plant to produce six cylinder engines - an enormous cost of adaptation.

With a similar philosophy, Toyota has been able to produce the front wheel drive Tercel and rear wheel drive Starlet, in left and right hand drive for the different world markets, on the same assembly line with negligible set-up time. Furthermore, these models are certainly not "custom" made but are price leaders at the low end of the market.

It thus appears that the gains from increased flexibility in manufacturing can be realized without the usual sacrifices in efficiency. While the examples just cited represent a fairly limited degree of flexibility, they are a quantum jump from what went before. Furthermore, they probably provide only a glimpse of the full potential of these new technologies. As such, flexible manufacturing systems may ultimately provide a way out of the dilemma posed by the traditional tradeoff between adaptability and efficiency.

FROM FLEXIBLE MANUFACTURING TO FLEXIBLE MANAGEMENT

As major industries have moved towards standardized, automated production process, management too has undergone a
parallel change. Woodward (1965) and others have demonstrated that the management (and organizational) system must
reflect the nature of the dominant technology.

The pressure to standardize in production technologies is
paralleled in management systems. Routinized decision making
simply costs less than is more specialized attention is required. The term "management by exception" refers to the
former.

In general, managers can make decisions on any of three
bases, according to the degree of flexibility required.
These are analogous to: job shop, batch process and continuous process production.

1. Ad Hoc Management: Where the problem is difficult to
    define and there is high uncertainty, managers must
    interact and coordinate their efforts **intensively** to
    monitor in "real time" the changing situation.

2. Planning: Where the situation is more predictable and
    the various alternatives and consequences are largely
    known, management can plan for defined periods of
    stability (such as a batch run).

3. Management by Exception: Where little change is
    expected, management simply sets up a bureaucratic
    procedure which will deal with a large number of
    similar transactions.

These three approaches represent a hierarchy of costs, with
the first being much the highest. Managers, like manufacturing engineers, will thus try to standardize wherever they
can. Figure 5 illustrates the costs advantage of shifting
towards more standardized management practices, with the
usual penalty in flexibility or adaptability.

Ultimately, management in large concerns, because of the
apparent cost savings, evolves highly standardized practices
and procedures. Purchasing, marketing, production planning,
personnel and other decisions become increasingly routinized,
paralleling the shift in the standardization of production
systems. The principle of "management by exception" exemplifies this process by comparing performance against

predetermined standards. Only exceptions require attention.
Managerial resources are thus "saved". Unfortunately, over
time, this can result in large, unwieldy bureaucratic structures with great inertia. Like the "hard-wired" mass production systems, the management systems become increasingly less
adaptable or responsive. Figure 5 illustrates the tradeoffs
between flexibility and efficiency in management.

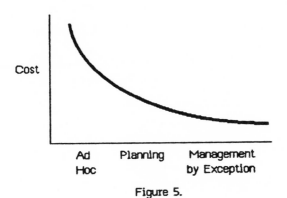

Figure 5.

However, as flexible manufacturing becomes more pervasive,
new kinds of demands are being place on management. In the
manufacturing of electronic chips, for example, where yields
can vary widely, a very different kind of management is often
required. A sudden increase (or decrease) in yield has immediate implications for the scheduling of raw material and
labor inputs down the line. Management must be constantly on
top of such changes in order to be able to respond quickly.
Such variability is in contrast with the more routine management, even within the same organization. In fact, when
managers from traditional continuous process operations are
transferred to such unpredictable environments (and vice
versa), they inevitably encounter great difficulties. Unfortunately, the response of the company is often to fire the
managers concerned for poor performance instead of recognizing the underlying differences in the production
environments.

The new kinds of flexible management systems required by the
advent of flexibility in manufacturing will undoubtedly include the ability of the managers to:

1. Assimilate information in "real time", as it is
    generated by the computer rather that reacting to
    periodic reports produced daily, weekly or monthly.

2. Take full advantage of production flexibilities in
    terms of greater product variety, new kinds of supplier
    relationships (e.g to reduce inventory), etc.

3. Coordinate both human and production resources to
    respond effectively to changing contingencies.

It is thus critically important that such qualities be
integrated by management as flexibility in manufacturing is
introduced. Otherwise, organizations will find themselves
with "front wheel drive technology with rear wheel drive
management"!

A PRESCRIPTION FOR RESEARCH

The case has been made that:

- Decades of standardization have brought our major industries to an impasse whereby the costs of adaptation
  (e.g. as in the introduction of a major new model) may
  now exceed the benefits.

- Management or organizational systems tend to mirror the
  characteristics of technological systems. Thus, the
  transition from continuous processing towards "flexible
  manufacturing systems" would suggest that there must be
  similar changes in human systems which might loosely be
  termed "flexible management systems".

It is thus suggested that future research efforts revolve
around the impact of the new technologies on:

1. Organizational structure: How will the highly rou-
    tinized, "machine bureaucracy" evolve in response to
    the need for a more responsive organizational form?
    Shorter lead times for new products, closer supplier
    coordination and more cooperative labor relations will
    undoubtedly be part of the recipe for future
    organizational success.

2. Training: What sorts of skills will be needed by the
    new "flexible management systems" and to what extent
    will they be acquired by formal training versus on the
    job experience? Furthermore, which of the skills will
    be "portable"?

3. Information systems: How will the requirements for
    management information change in response to the new
    technologies? Why do the Japanese manage with an order
    of magnitude less information than Europeans or
    Americans?

4. Labor relations: What is the new role for labor, re-
    cognizing that the old adversarial role may have to be
    abandoned? How has this transition been handled in
    sectors most strongly affected by flexible automation?

Since the future always exists in microcosms in the past, the
possibility of having "glimpses" of the future exist. It is
hoped that these will provide invaluable incites for those of
us wedded to the present.

PART II

DHR ASSESSMENT

The information in Part II is from *Robotic Technology: An Assessment and Forecast* prepared by James Just, Keith King, Michael Osheroff, George Berke, Peter Spidaliere and Tran Ngoc of DHR, Incorporated for the Aerospace Industrial Modernization Office of the Air Force Systems Command, July 1984.

Acknowledgments

This study was performed by DHR, Incorporated under contract to the Aerospace Industrial Modernization Office of the Air Force Systems Command. The project manager for this effort at the AIM office was Mr. Fredrick C. Brooks.

The DHR project team consisted of:

Dr. Keith King
Mr. Michael Osheroff
Dr. George Berke
Mr. Peter Spidaliere
Dr. Tran Ngoc (Task Manager)
Dr. James Just (Corporate Manager)

An expert panel worked closely with DHR project team and played an instrumental role in the completion of the study:

Dr. Michael Brady, MIT
Mr. Brian Carlisle, Adept Technology
Mr. Dale Hartman, Hughes Aircraft Company
Dr. Robert Kelley, University of Rhode Island
Mr. Charles-Henri Mangin, CEERIS International
Mr. Gene Merchant, Metcut Research Associates, Inc.
Dr. Roger Nagel, Lehigh University
Dr. Charles Rosen, Machine Intelligence Corporation, and
Mr. Jack Thornton, Robot Insider

Executive Summary

Robotics will be a major driving force of the defense industrial base as industrial robots grow and mature into an integral part of next generation computer-integrated manufacturing processes. In this context, the Aerospace Industrial Modernization (AIM) Office of the Air Force Systems Command has been tasked to evaluate robotics technology as a means of improving the manufacturing technology base of the aerospace industry. Consequently, the AIM Office has recently initiated a technology assessment of robotics, of which the results are presented in the present report. The major objectives of this study are to:

o   Perform a critical assessment of the current status of the technology,

o   Review key world-wide R&D activities and discern the principal thrusts and trends in robotics R&D, and

o   Perform a technological forecast addressing future functional capabilities, emerging application areas and future directions of robotics producers and end-users.

CURRENT TECHNOLOGY

Current Functional Capabilities

The functional elements of a generic robot can be grouped into three categories: mechanical, sensing and control. In addition, the performance of a robot can also be judged on two additional dimensions: internal integration of constituent components and external integration with the surrounding environment. Internal integration is normally evaluated by a number of system performance characteristics such as accuracy, repeatability, resolution, reach, working configuration, speed and load capacity. External integration is measured in terms of robot's performance within a work cell or a flexible manufacturing system.

In the following, the current functional capabilities of a generic robot are described by specifying their technical limitations presently encountered in practice:

Mechanical

o   Manipulators, in general, are still clumsy and slow with massive components.

o   Actuators are of three types: pneumatic, hydraulic and electric. Pneumatic actuators are difficult to control while hydraulic actuators are frequently subject to disruption due to failure of precision mechanical components. The major deficiencies of conventional electric actuators consist

265

of low power-to-weight ratio, backlash and lack of rigidity under load due to the use of reduction gearing. In general, no currently available actuators can incorporate control capabilities to modify actuator responses.

o   End-effectors, in general, are crude with no sensor or just binary tactile sensors. They do not have the desired interchangeability in the absence of standardization, and have to be custom-designed for specific applications.

o   Locomotion exists in the form of rail, gantry or wheeled systems and is typically limited to structured environments.

Sensing

o   Vision sensing is limited by poor resolution and difficulty in depth mapping. It is too slow for real-time processing and is hindered by lack of standardization.

o   Tactile sensing capabilities exist in binary form or simple force/torque sensing. Current tactile sensors are not very robust and have a narrow dynamic range.

o   Proximity sensors employ IR, ultrasonic or laser sources. IR and ultrasonic sensors are characterized by poor range resolution and inaccurate location; while laser sensors are furthest developed but expensive to implement.

Control

o   Current controllers can best be characterized as primitive computers.

o   Control software primarily exists at walk-through or teach-pendant levels with some limited off-line programming and crude sensory integration capabilities.

o   Robots generally operate as an "island of automation" and interact with their surroundings via part feeders or fixturings.

Established and Emerging Robotic Applications

A further measure of current robot's capabilities is the degree to which industrial robots have penetrated various application areas. Examination of the application areas penetrated by robots as well as those likely to emerge in the near future helps to identify the present status of robotics technology. Following is a tabulation of robotic applications, separated into established and emerging areas.

Established Applications:

o   Spot welding

o   One-pass, non adaptive arc welding

o   Two-pass arc welding with limited adaptive path control

o   Two-pass arc welding with limited adaptive process control

o   Material handling with parts in known location and orientation
     or transferred from known areas

o   Easy-mating assembly

o   Editable coating/painting with single or multiple robots

o   Inspection of coarse features

o   Sealant application with bead flaw detection

o   Routing or drilling with template

o   Coarse grinding with simple force or torque sensing

o   Investment casting

o   Die casting with simple inspection

o   Forging

o   Plastic injection molding

Emerging Applications:

o   Arc Welding aided by expert systems

o   Material handling with part recognition and acquisition for
     transfer from unstructured supply, e.g. bin picking

o   Fast close-fitting assembly with active compensation

o   Painting/coating with inspection and adaptive process control

o   Routing or drilling without template

o   High-precision grinding

Industrial Usage of Robots

Robots have penetrated several industries to different extents. The automotive industry is the largest user of industrial robots in the U.S., acquiring approximately 50 percent of the total number of installed robots. Other industries that have also made significant use of robots include foundry, electronics, aerospace, non-metal light manufacturing, and heavy-equipment manufacturing. Below is a listing of robotic applications that have established at least a moderate presence in these industries.

o   Automotive: Spot welding, material handling, painting, die casting, arc welding, inspection, assembly and sealant application.

o   Foundry: Material handling, investment casting, die casting, forging and finishing.

o   Electronics: Material handling, assembly, inspection, arc welding, sealant application and plastic molding.

o   Aerospace: Painting/Coating, material handling, inspection, finishing, sealant application and investment casting.

o   Non-metal light manufacturing: Plastic molding, material handling, sealant application and assembly.

o   Heavy-equipment manufacturing: Arc welding, material handling, painting/coating and finishing.

Recent Developments in Robotics Industry

The dynamic, high-tech robotics industry has undergone remarkable changes in the last decade. Since its inception, the industry has evolved through three major stages:

o   Industry definition in the late 1960's and most of the 1970's,

o   Strong initial growth in the 1979-81 period, and

o   Industry consolidation from 1982 to the present.

At the peak of its high-growth period, the robotics industry was characterized by an annual growth rate of about 90% and a large influx of new companies. Many of these companies were financed by venture capital attracted by the industry's high growth potential. This initial growth, however, was relatively short- lived as end-users began to realize the limited capabilities of available industrial robots. In this period the industry was dominated by fewer than six robot producers.

Since 1982, robot sales have slowed considerably. The U.S.-based robot market is becoming highly competitive as a large number of robot vendors (over sixty) vie for a limited market, which is still growing but at a much slower rate, about 30% per year. This leads to a strong belief in the robotics community that an eventual shake-out of the industry is imminent, if indeed it has not already begun. In this environment, the robotics industry is undergoing structural changes which are evident in the following observations:

o   The market-share hierarchy of robot producers has begun to change substantially to reflect the momentum gained by several new start-ups and giant corporations. The current top six robot producers, on the basis of their market shares, are GMF Robotics, Cincinnati Milacron, Automatix, Westinghouse, ASEA and DeVilbiss.

o   Companies are starting to seek out niches by applications, price ranges, targeted customer bases and levels of robot sophistication.

R&D ACTIVITIES

Air Force

Agencies active in the Air Force robotics technology base include the Air Force Office of Scientific Research (AFOSR), the Air Force Systems Command (AFSC) and the Air Force Logistics Command (AFLC). The following table summarizes the key R&D performers and topics sponsored by these Air Force agencies.

AGENCY/PROGRAM	KEY PERFORMERS	R&D TOPICS
AFOSR	o University of Michigan	o High Performance manipulators
	o SRI International	o Sensory control
	o Stanford University	o Microcomputer controller
		o Manufacturing cells

AFSC MANSCIENCE	o Martin Marietta (Subcontractors: ERIM, McDonnell Douglas, RPI, Stanford Univ., University of Mass./ Amherst, VPI)	o Hardware and software design for automated assembly
	o Honeywell (Subcontractors: Adept Technology, Stanford University, SRI)	o Development of multi-arm systems for assembly and inspection
AFSC MANTECH	o McDonnell Douglas	o Machine control language (MCL)
	o RVSI	o Vision sensing
	o Grumman Aircraft	o Drilling/riveting
	o Fairchild Aircraft	o Drilling/trimming
AFLC	o General Electric	o Packaging and Warehousing
	o Georgia Tech	o Inspection of turbine blades

Other Federal Agencies

In addition to the Air Force, significant robotics R&D is also supported by other federal agencies, which include the Navy, the Army, DARPA, NASA, and NSF. Below is a summary of major thrusts of these programs:

Navy:

o   Office of Naval Research (ONR) supports about a dozen university programs to perform basic research on sensory control and advanced sensing techniques.

o   Major projects sponsored by NAVSEA and NAVAIR include autonomous mobility for navigation, welding for ship hull fabrication and deriveting in airplane refurbishing.

Army:

    o   The Army supports a cohesive, directed R&D program to address the question of how applicable will robotics and artificial intelligence be to battlefield situations.

    o   Other efforts supported by various Army commands are focused on improving manufacturing technologies. Areas of emphasis are assembly and inspection. Current commitments in terms of on-going FY84 projects and planned FY85 projects amount to approximately $3.1 million for manufacturing technology.

DARPA:

    o   DARPA's robotics efforts are designed to support the three services in high-risk, high-payoff projects. Its long-term goal is to establish a technology base for non-manufacturing military applications in maintenance logistics and weapons support.

    o   R&D projects sponsored by DARPA are concentrated in two areas: control of specialized manipulators and integration of advanced sensory input for manipulator control and navigation.

NSF:

    o   NSF supports a broad range of basic research projects covering all aspects of robotics.

    o   Total funding for FY83 was about $4.8 million, of which $2.4 million was devoted to sensing, $0.9 million to control, $1.0 million to manipulation, and $0.5 million to system performance.

NASA:

    o   NASA commits about $1.5 million in FY84.

    o   NASA - sponsored projects were focused mainly on vision processing, supervisory control, man-machine interaction and system integration.

Foreign R&D

The U.S. faces a strong technological challenge in robotics from three groups of developed countries, Japan, Western Europe and the Soviet Bloc. In general, technological advances achieved in most of these countries are highly competitive, although the U.S. is still in the forefront in the development of robotics technology. The key features of robotics R&D in these countries are summarized as follows.

Japan:

- o   Robotics R&D in Japan is established as a national policy, which targets those R&D efforts in support of early commercialization and removal of humans from hazardous environments.

- o   This strategy is implemented by two national R&D programs. The first program is aimed at improving those robotic capabilities required for nuclear, undersea and rescue applications. The second national program, also known as the Jupiter Project, is focused on those problems identified as the key technological barriers to robotic commercialization.

- o   The first program was initiated in 1982 with a commitment of about $130 million for the next seven years. The Jupiter Project began in 1983 with an estimated funding in the range of $55-80 million in its entire duration.

Western Europe:

- o   Countries with a concerted, well-supported R&D program in robotics include the United Kingdom, France and West Germany. Also significant, but of a smaller magnitude, are programs in Sweden, Norway and Italy.

- o   Robotics R&D in the U.K. is characterized by close cooperation between government and industry. Funding emphasis is, therefore, placed on immediate payback projects targeted to industrial problems. A major program, which funds most university R&D in robotics, was created in 1980 through the Science and Engineering Research Council and jointly funded by Government and industry.

- o   Robotics R&D in France is mainly represented by a national three-year, $350 million program, starting in 1983. This program is concentrated on R&D in manufacturing technology and also includes training of robotics specialists and promotion of robotic implementation.

- o   At the center of robotics R&D in West Germany are efforts performed at various Fraunhofer Gesellschalt Institutes. Their funding resources are equally contributed from government block grants, industry and specific government contracts. In general, their R&D activities are mainly driven by specific applications.

- o   Other significant robotics R&D programs in Western Europe exist, at a smaller scale, in Sweden, Norway and Italy. Most do not have a cohesive national focus and are normally

Most do not have a cohesive national focus and are normally led by major robot producers such as ASEA of Sweden, Trallfa of Norway and Olivetti of Italy.

Soviet Bloc:

o Among the countries belonging to the Soviet Bloc, with the exception of Yugoslavia, robotics R&D is well coordinated through the Council for Economic Mutual Assistance (CEMA).

o CEMA members are approximately a decade behind the West in robotics technology mainly due to their deficiencies in computer and electronics technologies.

o Notable features of their robotics R&D are advances made in manipulative and sensing technologies and a strong drive to achieve standardization and modularity.

Summary of R&D Activities

Sources	Highlights

U. S. Federal:

o Air Force

o AFOSR committing about $2-3 million for basic research. Multi-year, multi-million dollar commitments to develop robotics technology base provided by AF MANTECH and MANSCIENCE programs.

o Navy

o Basic research supported by ONR and technology development in autonomous mobility, ship hull welding and aircraft de-riveting sponsored by NAVSEA and NAVAIR.

o Army

o Committing about $3 million in FY84 and FY85 mainly to improve manufacturing technologies in assembly and inspection.

o DARPA

o Supporting the three services in high-risk, high-payoff projects.

o NSF

o Sponsoring basic research, totaling about $4.8 million in FY83.

o NASA

o Committing about $1.5 million in FY84 with a strong emphasis on control and integration issues.

Foreign:

o Japan

o Represented by national programs: a seven year, $130 million program that targets nuclear, undersea, and rescue applications and the multi-year Jupiter project that commits $55-80 million to speed robotics commercialization.

o Western Europe

o West Germany, France and the United Kingdom have concerted, well-supported programs. Other significant programs also exist in Italy, Sweden and Norway.

o Soviet Bloc

o Robotics R&D coordinated through CEMA. Robotics development is lagging the Western countries in general, but make significant progress in manipulation, sensing, modularity and standardization.

TECHNOLOGICAL FORECAST

The present study is concluded with a technological forecast, which consists of four parts: projected capabilities, application forecast, industry trends and technological trends.

Projected Capabilities

Mechanical

Manipulator:

Future manipulators will require greater speed, more versatility and enhanced accuracy. In the near term, these needs are being addressed in the development of rigid but lightweight manipulator structures, improved joint and bearing design, parallel linkages and antagonistic drives. In the long term, light, flexible robot arms will become common.

Actuator:

Current actuators suffer from inefficiency, lack of stiffness under load and backlash. In the near future, direct drive electric actuators will alleviate many of these shortcomings. In a longer time frame, tendon drives with high power transmission capabilities will be able to replace conventional actuators for some of the arm joints.

End Effector:

End effectors in use today are generally bulky and lack versatility. The next generation of end effectors will have quick change capability for versatility and will incorporate local sensing. The long term solution to most end effector shortcomings will be the development of a general-purpose dexterous hand, with high resolution force sensing "skin".

Mobility:

Mobile robots in the near term will be descendents of today's computer-controlled parts-transfer carts as used in automated factories. They will run on wheeled suspensions, and improved mechanical registration techniques will allow precise positioning of the robot at work stations. In the long term, mobile robots will make use of active tracked suspensions, legged locomotion and crawling/climbing abilities.

Control

Current work in distributed processing, networking and development of hierarchical software will result in substantially more sophisticated controllers in the near future. Sensing systems will acquire dedicated satellite processors, supplying the controller with sensing results instead of raw data, while software operating systems will manage the housekeeping of distributed processing. The controller will utilize more complex dynamic models to produce better accomodation of workload effects. In the longer term, controllers will tie into local area networks to communicate with surrounding machinery and to receive programs and commands from higher-level supervisory computers. As vertical integration improves, the robot controller will lose much of its identity, becoming just another link in the processing hierarchy.

Sensing

Vision:

The two primary developmental needs for robotic vision are lower cost and increased speed in processing. The near term results of current R&D will be VLSI processors for 2D and 2 1/2D vision that are fast enough to provide real-time results, usable for adaptive control. In the long term, processing speed will be sufficient for real-time results from 3D vision, while signal processing methodologies will be applied to allow use of vision in uncontrolled, visually noisy environments.

Tactile:

In the near future, the VLSI technology that will help vision systems will also enhance tactile sensing. Tactile arrays of modest size and resolution will be packaged with their own dedicated processors, while force/torque sensing will become common. Tactile sensing sophistication will continue to improve, and in the long term will result in high resolution force sensing tactile arrays, capable of acquiring 3D shapes by touching.

Proximity/Ranging:

In the near term, developments in this type of sensing will be driven by application needs, such as eddy current sensing of rivets for aircraft refurbishing and ultrasonic sensing to aid robots in acquiring parts. In the longer term, proximity sensing will become more important due to the needs of mobile robots as an essential component in obstacle avoidance and navigation.

Sonic:

The majority of interest in robotic hearing is focused on speech recognition for command purposes. Today and in the near future, this capability is very limited with respect to vocabulary and reliability. Speech recognition will become a significant capability for robots in the long term with the appearance of artificial intelligence and natural language capability.

Integration

Short Term:

o   Internal integration will improve as some level of communication standardization becomes accepted; sensing systems will be the first well-modularized components.

o   External integration will reduce the robot's dependence on expensive and inflexible fixturings and feeders, replacing them with simpler mechanical systems. Coordination with external computer systems for task assessment and off-line programming with simulation testing will become common for sophisticated installations.

Long Term:

o   Internal integration will become much better, with industry-wide standards for interconnection; a buyer will be able to add to his robot's capabilities by plugging in modules.

o   External integration will connect and coordinate entire production lines, including many robots. CAD/CAM systems will connect with graphics-aided robot programming systems, which will then download the resulting programs to the robot production line. This supervisory system will perform the necessary planning, stock and machine allocation, maintain inventory and maintenance schedules, and support a sophisticated Management Information System.

Application Forecast

With respect to the effect of future developments in robotics, there are three major categories of robotics applications:

Low Growth Applications - Developments in robotic technology will not produce sweeping increases in robotic penetration.

- o Spot Welding
- o Spray Painting and Coating
- o Forging
- o Investment Casting
- o Sealant/Adhesive Application
- o Die Casting

High Growth Applications - As improvements in the laboratory and development stage become commercially available in the near term, these applications will show very rapid increases in robotics penetration.

- o Material Handling
- o Arc Welding
- o Routing, Drilling, Grinding
- o Inspection
- o Assembly

Blue Sky Applications - These applictions require capabilities that are still in early developmental stages. Robotic penetration will be very slow starting, and will not become significant in the near term future.

- o Houskeeping
- o Construction Labor
- o Maintenance by Expert Systems
- o Hazardous Environment Rescue
- o Orbital Construction

Industry Trends

At present, the number of companies in the U.S. producing and marketing robots or robot components is quite high, more than today's market can support. A shake-out is occuring, and many of these companies are likely to withdraw from the market. Small companies that would like to enter the market with a line of components, such as vision systems, are severely hampered by the lack of industry-wide standardization.

During the next several years, the robotics industry is likely to be rather frenetic, characterized by new companies entering the field, some existing. companies withdrawing, and corporate take-overs. However, some trends seem likely to appear:

- o Many larger firms will market flexible manufacturing systems o Suppliers of complete turnkey systems will become more prominent, minimizing the hidden costs of a robot.

o Greater product differentiation and market segmentation will develop, as vendors carve out specific markets.

Technological Trends

The developing trends of robotics technology are:

o Separation of high sophistication robots from simple robots will be established.

o Sensing will become both faster and better, and integration of sensory information will be much more efficient.

o Mobility will be easily available due to improved mechanical and navigation systems.

o Future robots will take advantage of lighter materials and more efficient design.

o Perhaps the greatest change will be in the extent of robot integration. Sophisticated robots will communicate downward to dedicated satellite processors, sideways to adjacent robots, and upward to supervisory control systems.

o Lower cost is going to be a major trend in both sophisticated and simple robots due to improved technology and economies of scale.

o Hybrid robotic/teleoperated devices will become common, leading the way in applications that will eventually be handled by fully autonomous robots.

1. Introduction

Decrease in industrial productivity has been an issue of national concern for the last several years. This concern has permeated the entire U.S. industry and, without exception, strongly affected the aerospace sector. As a consequence, the defense industrial base is faced with a threatening erosion that could entail a significant reduction of the nation's defense capacity. The Aerospace Industrial Modernization (AIM) Office of the Air Force Systems Command has recently been tasked to assess and improve the manufacturing technology base of the aerospace industry. As a part of this effort, the AIM office has initiated an assessment of U.S. and foreign activities associated with robotic technology. This assessment will support the establishment of a full-spectrum, long-range plan for the Air Force robotics implementation program. This plan is intended to guide and accelerate the implementation of robotics technology into the aerospace community in order to reduce the cost of manufacturing, maintaining, repairing and servicing aerospace systems. This report represents the results of the above-mentioned assessment of the current and projected status of U.S. and foreign robotic technology.

Because robotic technology is still at a formative stage, it is necessary in assessing the current technology to establish some working definitions and concepts. The lack of established industrial practices that will be shown to permeate the industry includes even basic definitions. For the purposes of this report, then, the Robot Institute of America definition of a robot as a "reprogrammable multi-functional manipulator" will be adhered to. As a result, the present study will be focused on robots as such and thus will not address in detail related technical and economic issues such as integrated work cells and Flexible Manufacturing Systems (FMS).

Additionally, there are several concepts used in this report that should be clarified here. The first is the concept of "state-of-practice". By state-of-practice we generally mean the level of technology currently in manufacturing use. Clarification of this term will be provided where necessary for each specific example. Secondly, the concept of near and far term is used throughout the discussion of the technological forecast. By near term it is meant to imply a range of several years, generally about two to five years. Similarly, far term is intended to indicate the five to ten year range. The following discussion is devoted to the organization of the report.

The main body of this study begins in Chapter Two with a detailed view of current robotics technology. This presentation is divided into four sections: the robotics industry, current robotic capabilities, current robotic applications, and industrial usage of robots. Each of these topics is discussed separately to highlight the difference between what robots can do today and what they are doing today. Generally the technological capabilities exceed the state-of-practice by at least several years. Additionally, it is realized that the degree of industrial usage of robot and even the structure of the robotics industry play an important ro, in determining the kinds of products and technology that are and will

279

soon become available. From this multi-sided approach to describing the
current robotic technology, an understanding of the driving forces behind
technological developments can be developed. This information is used
later in the report to draw some conclusions about future robot usage.

Chapter Three summarizes a world-wide study of robotics research
and development programs, divided into U.S. and foreign activities. Both
the U.S. activities and the foreign activities will be classified, when
possible, according to the R&D funding community and the R&D performing
community. This distinction is made to draw attention to the individual
government-sponsored R&D projects and overall funding strategies. Based
on an analysis of these project goals and directions, the thrusts of robotics
research are determined in each of the most active research communities,
both domestic and abroad. These individual thrusts are then synthesized
into an overall picture of world-wide robotics research and development.

The fourth chapter of this report consists of a methodical forecast
of robotic technology. The adopted approach includes a study of the anti-
cipated advances from in-progress research programs. Specifically, the
forecast begins with a list of projected functional capabilities, mainly
on the basis of the preceeding analysis of research topics, goals and
directions. The forecast continues with a summary of projected robot
usage by application, based on the conclusions drawn in Chapter Two combined
with the projected functional capabilities. It is believed that this
forecast approach, combined with in-depth consultations with a well-represented
expert panel provides a sound, practical prediction of future robotic
technology. The concluding section of this forecast chapter, Future
Directions, is devoted to a synthesis of the above projections into a
concise, directed forecast of both technological and industry trends.

While the body of this report presents a complete picture of robotic
technology, there are some additional discussions that might enhance the
reader's understanding of several of the topics presented. These discussions
are elaborated at length in the appendices. Appendix A includes a systematic
analysis of key considerations in current robotic manufacturing appli-
catons. An in-depth study of industrial and academic R&D programs is
presented in Appendices B and C, respectively. Finally the references
and personal contacts made during the course of this study are listed
in Appendices D and E.

2. Current Robotic Technology

In this chapter, the current status of robotic technology is reviewed and assessed from several perspectives. In Appendix A, all current robot applications are reviewed in detail. There, in each application area, the involved manufacturing process, basic elements of a typical robot used in such applications, economic motivations and technological constraints in robot usage, and specific examples of the considered application area are reviewed and assessed. In the following, the information contained in Appendix A is highlighted, and at the same time the current technology is assessed from a more general point of view. A quick update of the robotics industry is first provided to inform the reader about the major features of the industry and the latest developments in the field. Section 2.2 then examines a generic version of an industrial robot as it is being used and analyzes its functional capabilities systematically, going through the major components of a robot's subsystems. In Section 2.3, robotic applications are again reviewed with emphasis placed on an overall analysis of robotic applications, economic motivations and technological barriers hindering robot penetration. Section 2.4 is then devoted to assessment of present usage of robots in various industries with regard to industrial operations and response to robot applications. Finally, a composite picture of the present technology is summarized in the last section of this chapter. It is believed that this approach to technology assessment will present a comprehensive understanding of current robotic technology which is balanced and most useful from the perspective of various industrial sectors associated with robotics.

2.1. U.S. Robotics Industry - An Update

The U. S. robotics industry has demonstrated the vitality and dynamism that are typical of a rapidly growing high-technology sector. Sales and production have been expanding at a vigorous annual rate of 30-60% from the 1970's through the early 1980's. As in most high-growth areas, the industry has evolved through several stages: industry definition in most of the 1970's, strong initial growth in the 1979 - 1981 period, and industry consolidation from 1982 to the present.

From the start, the growth pattern of the robotics industry was influenced by two major factors: first, the hourly cost of direct labor was lower than the hourly cost of operating an industrial robot; and second, the benefits expected from a new, unproven technology were still uncertain. As a result, by 1970, ten years after their introduction, only about 200 industrial robots were in use throughout the U. S.

During the 1970's, however, the U. S. economic environment changed significantly. Manufacturing productivity declined steadily and labor cost increased while robot cost did not rise excessively. These trends were taking place at the time that robots became more sophisticated in both manipulative and control/ sensing capabilities. Usage of robots in manufacturing began to increase significantly in the 1970's. Robot population increased from about 200 in 1970 to about 1700 in 1978. This characterized the formative stage of the robotics industry, during which a sizable industry was taking shape from early developmental efforts. In this period, the basic technology solidified and a core group of robot manufacturers established well-defined product lines.

In the next three years, 1979 - 1981, the robotics industry underwent a major development characterized by high growth in sales and considerable penetration into new application areas. This was spurred by the rapidly increasing labor cost coupled with the reassurance of successful applications of robots in automotive and foundry industries. Known for their widespread use in this period were robots designed for spray painting, spot welding, parts transfer and machine loading. Robot population increased at a remarkable rate from about 1700 in 1978 to about 4500 in 1981. This period also witnessed the influential role of venture capital in raising the number of robot producers from under a dozen in 1978 to about 80 by the end of 1981.

Most recently, in the last two years, 1982 and 1983, the robotics industry entered a period of consolidation, which was characterized by a slowdown in growth and entrance into the robotics industry of several powerful giant companies. The initial enthusiasm appeared to have leveled off and end-users began to recognize the limitations as well as capabilities of robots. At the same time, continued infusion of new capital into the industry further increased the number of robot manufacturers and component suppliers. These trends are indicative of an industry that is still growing vigorously but has become highly competitive. This has led to a strong belief in the robotics community that an eventual shake-out of the industry is quite imminent, if indeed it has not already begun.

Presently, there are more than 60 U.S.-based robot manufacturers with indications that this number is still rising. They generally fall into one of the following three categories:

(i)   Pioneeer robot producers that either started in robotics like Unimation, or entered early from their lead in the machine tool area, such as Cincinnati Milacron and Prab Robots;

(ii)   New start-ups financed by venture capital attracted to the field by its high growth potential, including among many others, Automatix and Advanced Robotics; and

(iii)   Major corporations (e.g. General Electric, IBM, and Westinghouse) seeking to parlay their related strengths into robotics and to support their interests in factory automation through robotic developments.

As a result, the early robotics industry was heavily dominated by a small group of pioneer companies. Their strong market positions, however, have slowly been eroded as new companies enter the market. With the industry so dynamic and at such a young stage, it is not surprising to see that relative market shares have undergone great flux in the last decade. A closer look at the market shares of five producers with the largest sales in 1980 (i.e. Unimation, Cincinnati Milacron, Prab Robots, DeVilbiss and ASEA) reveals that their dominance has slipped considerably, from a combined percentage share of 90.9% in 1980 to 42.2% in 1984. This is illustrated in Figure 2-1 below.

	1980	1981	1982	1983	1984
Westinghouse/ Unimation	44.3%	44.0%	32.9%	15.1%	9.2%
Cincinnati Milacron	32.2%	32.3%	16.7%	17.1%	14.1%
Prab Robots	6.1%	5.3%	6.5%	5.0%	4.9%
DeVilbiss	5.5%	4.2%	12.4%	9.1%	7.0%
ASEA	2.5%	5.8%	5.0%	5.4%	7.0%
TOTAL	90.9%	91.6%	73.5%	51.7%	42.2%

Source: Prudential-Bache

Figure 2-1: Combined Market Share of Five Selected Companies
(1980 - 1984)

In their place, there emerged two major producers, GMF Robotics
and Automatix, with several others such as IBM, GE and Cybotech beginning
to show their strength in the robot market.   In general, the market share
held by the long-standing vendors in the industry is declining, giving
way to new entrants supported by venture capital and major corporations.
Within this highly competitive environment, the robotics market is undergoing
structural changes to the effect that:

(1)   The hierarchy of robot producers in terms of their market
position has begun to change substantially to reflect the momentum gained
by several new start-ups and giant corporations; and

(2)   Companies are seeking out niches by application, level
of technological sophistication, price range and targeted customer base.

As a whole, the robotics industry is still characterized by a fairly
vigorous growth despite this increasing competitiveness.   This is apparent
when one examines the sales trends in the last decade of this sector.
Figure 2-2 illustrates the total sales achieved by the robotics industry
since 1975 and the associated growth rate for each year.   In the initial
growth phase, annual growth rate is fluctuating about an unusually high
percentage of 60% during the 1975 - 1981 period.   More recently, this
remarkable growth has slowed down somewhat, varying in the 20%-40% range
in the 1982 - 1984 period.   On the basis of the growth rate of 32%, which
is an average over the last three years, it is estimated that sales will
reach about $470 million in 1985 and about $1880 million in 1990.

In summary, a detailed breakdown of robot sales by U.S.-based vendors
in the last five years is presented in Figure 2-3, where annual sales
figures of the top ten U.S.-based robot producers are tabulated with their
percentage market share included in parentheses.

U.S. - BASED ROBOT SALES

YEAR	75	76	77	78	79	80	81	82	83	84
SALES ($ MILLIONS)	8	15	26	40	60	90	155	190	249 (P)	355 (E)
GROWTH RATE (%)		87	73	54	50	50	72	23	32	42

SOURCES: PRUDENTIAL-BACHE, FROST & SULLIVAN, NORTH & DONAHUE

P = PRELIMINARY

E = ESTIMATED

Figure 2-2: Sales by U.S. Based Vendors

SALES IN $MILLION AND MARKET SHARE (IN PERCENT) OF U.S. - BASED VENDORS: 1980 - 1984

1980		1981		1982		1983 (P)		1984 (E)	
Unimation	40.0 (44.3)	Unimation	68.0 (44.0)	Unimation	63.0 (32.9)	Cincinnati Milacron	42.5 (17.1)	GMF Robotics	60.0 (16.9)
Cincinnati Milacron	29.0 (32.2)	Cincinnati Milacron	50.0 (32.3)	Cincinnati Milacron	32.0 (16.7)	Westinghouse (Unimation)	37.5 (15.1)	Cincinnati Milacron	50.0 (14.1)
Prab Robots	5.5 (6.1)	ASEA	9.0 (5.8)	DeVilbiss	23.7 (12.4)	GMF Robotics	23.5 (9.5)	Automatix	40.0 (11.3)
DeVilbiss	5.0 (5.5)	Prab Robots	8.2 (5.3)	Prab Robots	12.5 (6.5)	DeVilbiss	22.5 (9.1)	Westinghouse Unimation	32.5 (9.2)
ASEA	2.5 (2.8)	DeVilbiss	6.5 (4.2)	ASEA	9.5 (5.0)	Automatix	18.0 (7.2)	DeVilbiss	25.0 (7.0)
Advanced Robotics	1.7 (1.9)	Automatix	3.0 (1.9)	Cybotech	9.0 (4.7)	ASEA	13.5 (5.4)	ASEA	25.0 (7.0)
Copperweld	1.5 (1.7)	Nordson	2.0 (1.3)	Automatix	8.1 (4.2)	Prab Robots	12.5 (5.0)	IBM	17.5 (4.9)
Nordson	0.8 (0.9)	Copperweld	1.5 (1.0)	Advanced Robotics	6.6 (3.4)	IBM	11.0 (4.4)	GE	17.5 (4.9)
Mobot	0.8 (0.9)	Thermwood	1.0 (0.6)	Nordson	4.5 (2.4)	GE	10.5 (4.2)	Prab Robots	17.5 (4.9)
Automatix	0.4 (0.4)	Advanced Robotics	0.8 (0.5)	IBM	4.5 (2.4)	Cybotech	7.0 (2.8)	Cybotech	12.0 (3.4)
Others	3.0 (3.3)	Others	4.6 (3.0)	Others	18.0 (9.4)	Others	50.0 (20.1)	Others	58.0 (16.3)
TOTAL	90.0 (100.0)	TOTAL	155.0 (100.0)	TOTAL	190.0 (100.0)	TOTAL	248.5 (100.0)	TOTAL	355.0 (100.0)

SOURCE: Prudential-Bach Securities

P = Preliminary

E = Estimated

Figure 2-3: Sales and Market Shares of U.S. Based Vendors

2.2 Current Capabilities of a Generic Robot

This section introduces the reader to the components of a generic industrial robot, discusses the interactions of these components (internal integration), and finally considers the interaction of the robot with its surroundings (external integration) as it leads to coordination of work cells and flexible manufacturing systems. The components are described in terms of current state-of-practice; for the sake of clarity, component variations that are rarely used in production may be omitted.

The basic components of a generic state-of-practice industrial robot fall into three major groups:

1) Mechanical - These are the parts that move or produce motion. This group consists of the manipulator, the actuators that power the joints and the end effector that holds the workpiece or tool.

2) Sensing - These are the components that provide the robot with information about its environment. The main types of sensing components in use today are vision systems, tactile or contact systems and proximity systems.

3) Control - These are the components of the controller. Controllers in use today can be as simple as rotating cams that open and close air or hydraulic valves, or can be as complex as a sophisticated computer system. In the latter case, the major components and features can include the processor, I/O and interface units, mass storage, programming language or programming method, and a library of pre-written routines to perform path control and sensory integration. Figure 2-4 presents a taxonomy that illustrates the basic components of a robot. The last unit in that taxonomy, labeled system performance, represents the result of integrating the robot components (internal integration) and characterizing the way in which they perform as a whole.

Mechanical

When one first looks at an industrial robot, the component that dominates the image is the manipulator, the structural framework on which the robot is built. It is composed of rigid links connected by joints. Manipulators can be characterized by types of joints, either rotary or translational, and by the way that they are linked. The two types of joints are illustrated in Figure 2-5. Most manipulators have three degrees of freedom, i.e., three movable joints, with additional joints incorporated in the wrist between the manipulator and the end effector to increase agility. A manipulator that uses only translational joints is referred to as a Cartesian robot because the position of the wrist is specified by the position of each joint along the standard cartesian x, y, and z axes. A schematic drawing of a three joint cartesian manipulator is shown in Figure 2-6. A manipulator that combines one rotational and two translational joints is referred to as a cylindrical coordinate robot, as is illustrated in Figure 2-7. Two perpendicular rotational joints and one translational joint, illustrated in Figure 2-8, results in a spherical

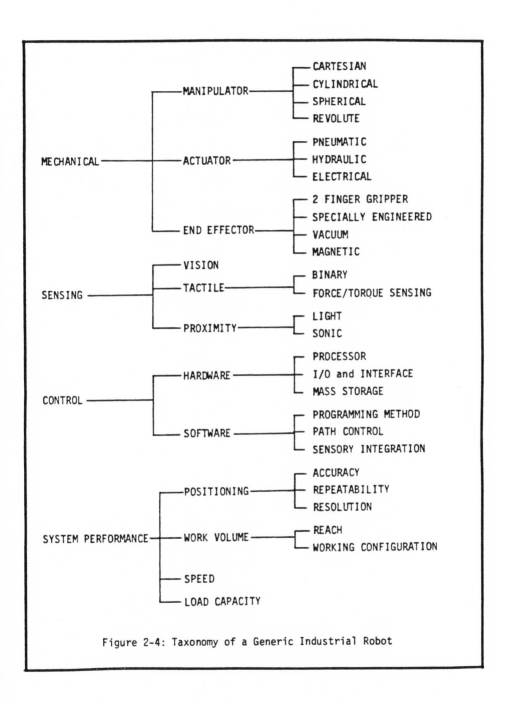

Figure 2-4: Taxonomy of a Generic Industrial Robot

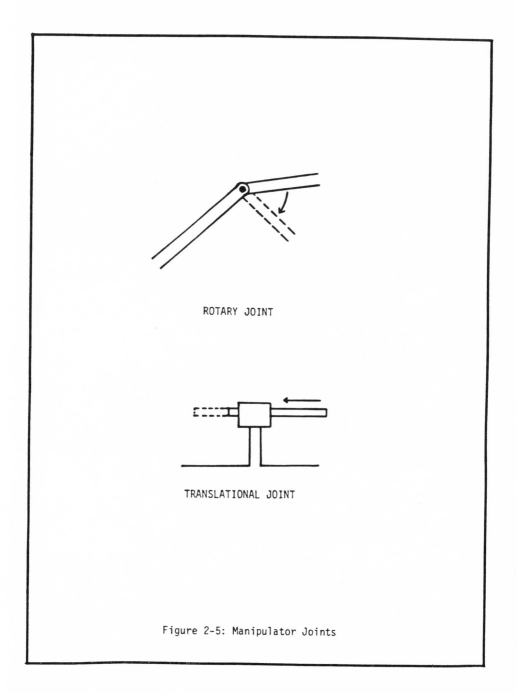

Figure 2-5: Manipulator Joints

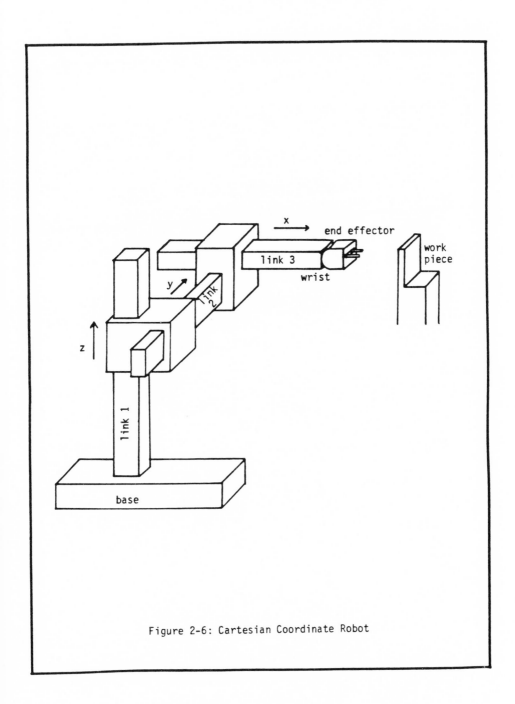

Figure 2-6: Cartesian Coordinate Robot

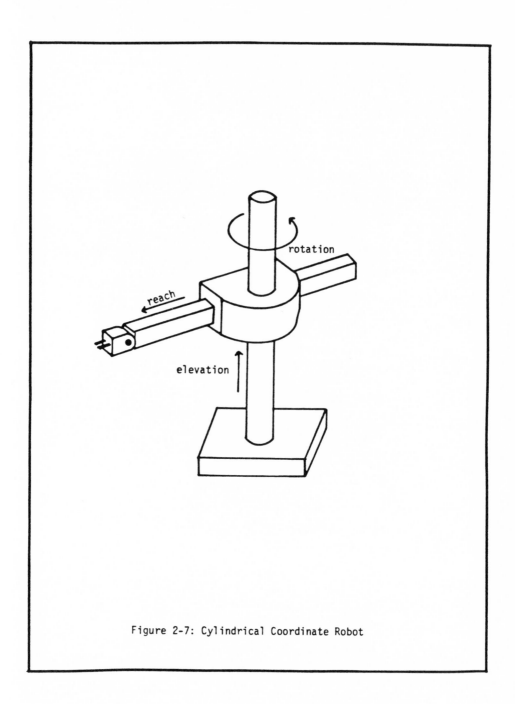

Figure 2-7: Cylindrical Coordinate Robot

Figure 2-8: Spherical Coordinate Robot

coordinate robot. A more agile manipulator configuration consists of three rotational joints, two of them coplanar, as illustrated in Figure 2-9. This is referred to as a revolute or jointed arm; joint 1 is commonly referred to as the shoulder and joint 2 as the elbow by analogy to the human arm.

The actuators generate the force required to operate the joints of the manipulator. Three major types of actuators are in common use today: pneumatic, hydraulic and electric. Pneumatic actuators generally rely on pressurized air to move pistons that produce linear motion or rotational motion via lever arms. Hydraulic actuators draw their energy from high pressure hydraulic fluid, and can be linear actuators or rotary motors, which produce rotation without requiring lever arm linkages. Electric actuators are high performance electric motors that generally produce rotary motion through reduction gearing to provide adequate torque (though direct drive electric motors that eliminate the need for gearing are now appearing) and linear motion by means of ball screw types of converters. End effectors are attached to the end of the manipulator (frequently through a wrist that provides additional degrees of freedom to improve dexterity). They are the components that actually hold the tool or grip the workpiece. Tool-type end effectors in common use today are spot welding guns, welding torches and spray guns. End effectors that grip the workpiece are usually specially designed by the robot user for a particular application. Grippers in use today typically consist of two fingers moving towards each other while the fingers remain parallel. However, gripping of unusually shaped objects with this simple motion has also been demonstrated by re-configuring the finger geometry. Certain types of workpiece can be picked up by end effectors that use vacuum or magnetism.

Sensing

Robotic sensors commonly used today in general fall into three types: vision, tactile (touch sensing and force sensing), and proximity. Robotic vision systems use a video camera to produce an image consisting of a grid of discrete elements called pixels. Typical resolution of the image may range from 100 by 100 up to approximately 400 by 400 such elements. Each element senses the brightness at that position in an analog manner with potentially many levels (gray-scale levels). Most vision systems in use today are called binary systems because the analog output of each pixel is thresholded by external circuitry to yield a binary output, i.e., light levels above the threshold are labeled black, and below the threshold are labeled white, or vice-versa. Many processing algorithms can be applied to interpret these binary images. The major algorithms in use today rely on important goemetrical features to characterize the workpiece. For this reason, present vision systems are more often used for inspection and quality control than for acting as adaptive sensors.

In binary vision systems, careful design and attention must be engineered into the accompanying lighting and optical systems to ensure the reliable acquisition of a high contrast image. For some applications, more sophisticated picture-processing algorithms are used on the original analog (gray-scale) data, enhancing contrast and extracting additional important features. A common method of reducing the complexity of the interpretation problem

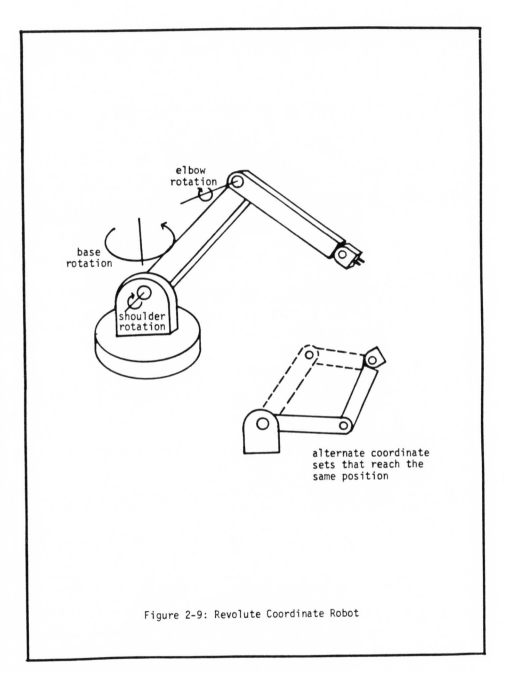

Figure 2-9: Revolute Coordinate Robot

is the use of structured light to illuminate the workpiece. An illustration of the use of structured light applied to a beveled seam to be welded is given in Figure 2-10. An overview is given in Figure 2-10a, showing the seam to be examined; Figure 2-10b shows the arrangement in side view with the camera looking straight down, and the light stripe source illuminating the seam at an angle. The image produced by the video system is shown in Figure 2-10c. The two critical measurements, location of center of seam and depth of the seam are easily and quickly extracted from this image.

Tactile or touch sensing is more common than vision sensing; at its simplest level, an ordinary micro switch located on the end effector responds to presence of an object in the gripper to verify that the robot is holding something. This is called binary contact sensing because the switch only gives a yes/no indication. Simple contact switches are also used for safety purposes: mounted along the manipulator, they can detect contact with unexpected obstacles while the arm is moving.

Another increasingly popular type of tactile sensing is force sensing. It is a step beyond simple contact detection, indicating not just that the robot is touching something, but also how hard it is touching. The amount of force applied can be detected with electronic strain gauges or piezoelectric transducers, which produce an electrical signal indicating how high a force is being exerted. Force sensors in use today are generally limited to sensing along only one axis.

Proximity sensing is commonly used as a substitute for simple contact sensing. For example, instead of an object in the gripper closing a contact switch, it interrupts a light beam to verify that the part is in the gripper. Because there is no physical contact, this type of sensing is not subject to wear, as are contact switches, and the light source commonly is infrared to avoid interference from ambient light.

Control

In today's sophisticated industrial robots, the robot controller is essentially a small computer system. The central component is the processor, which is frequently characterized by the number of bits it uses in parallel for data manipulation. In general, the more bits that are used for the data line, the faster the processing becomes. This is why 16 bit processors are considered more desirable than 8 bit. However, this rule is not absolute; a well implemented 8 bit system can be faster than a poorly implemented 16 bit system.

The ability to interface with other components of the robot is essential in the controller. To direct the motion of the manipulator, the controller must be able to send commands to the actuators, and generally needs to receive information about the position of each joint. Additionally, sensing systems must communicate their information to the controller. The communication of information to the processor is commonly handled through direct connection to the processor's I/O port, while commands from the computer to other components generally require conversion from the low-power signal levels generated by the processor to high power control levels needed

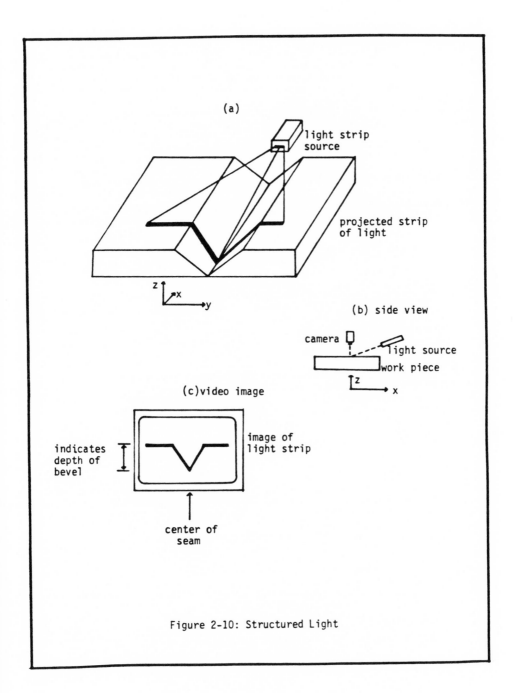

(a)

light strip
source

projected strip
of light

z
x
y

(b) side view

camera   light source
          work piece

z
x

(c)video image

indicates
depth of
bevel

image of
light strip

center of
seam

Figure 2-10: Structured Light

by other components. This conversion is commonly performed by switching transistors or electronic relays.

Some form of mass storage is usually incorporated in today's controllers. Once the robot has been taught to perform a task, a mass storage system is used to retain the commands that perform the task for later retrieval. The most common mass storage unit used in industrial robots today is magnetic tape cassettes. While less common, floppy disk systems are also used, bringing the advantages of faster access to the data and the disadvantages of higher cost and lower tolerance to harsh environments.

The controller components described so far have been, in computer terminology, hardware items. The following software items and the features they make available are equally important for a sophisticated robot.

From the point of view of the controller, there are two fundamentally different approaches to programming industrial robots: walk through or teach pendant in which the robot is "shown" what to do versus keyboard entry of the programs in which the robot is "told" what to do. For a controller to be programmed by the former method, the only software components really required are stored routines to save the joint positions "as shown", to execute the sequence of steps created by the programming and to read from or write to the mass storage device.

In contrast, a controller that is programmed by entering commands from a keyboard needs more software components. For the robot to interpret the program, it must have a compiler or interpreter that converts commands in the programming language to commands that the processor can execute. A system of this type generally requires a more extensive library of stored routines to allow commonly used command sequences to be executed quickly and easily. Normally this library will include robot oriented modules that perform tasks such as receiving information from the sensing system or systems, calculating the joint motion needed to move the end effector to a specific position command, or operating the end effector. Other modules will be program oriented, such as the language compiler, a program editor, and data storage.

Internal Integration

Work volume is simply that region around the robot which can be reached by the tool plate of the manipulator. It is determined by the mechanical constraints of the joints and the lengths of the links. It results from a combination of reach, how far from the base mounting the tool plate can extend, and the manipulator configuration. Referring back to Figure 2-6, the work volume of a cartesian manipulator is a cube whose height is the available travel along link 1, while the two horizontal dimensions are determined by the available travel along link 2 and link 3. A cylindrical coordinate robot, as illustrated in Figure 2-7, has a work volume centered on the elevation post, limited in height by the maximum elevation and extending from the center to the limit of the reach. It will be cylindrical in shape, though a wedge will be missing if the rotational joint cannot rotate a full 360 degrees. A spherical coordinate robot, as illustrated in Figure 2-8, will have a work volume that is a subset of a sphere of a radius equal to the reach. The exclusions result from limits on the elevation axis, and limits on the base rotation axis.

The work volume of a revolute arm is again a subset of a sphere, but the inner surface of the work volume is bounded by arcs determined by the interaction of the limits of rotation of the shoulder and elbow joints.

The concept of positioning in robotics is used to describe how well the robot is able to bring the end effector to a desired location, involving both mechanical and control aspects. There are typically three parameters that are often used to characterize positioning. "Repeatability" is the most commonly specified parameter which measures how closely the end effector can return to a position it was at previously. "Accuracy" specifies how closely the robot can position the end effector to an arbitrarily selected location. "Resolution" indicates the minimum displacement of target location that can be distinguished by the robot. Figure 2-11 illustrates these parameters. All three can be limited by the actuator's or the controller's ability to sense the joint position; it can also be limited by the software routines that perform calculations for joint position. Rotational joints tend to aggravate both of these situations. If joint position can only be controlled or sensed to a hundredth of a degree, this uncertainty is multiplied by the distance from the joint to the end effector. Furthermore, control of rotational joints involves more numeric manipulation than translational joints, making limits of computational accuracy a problem.

Sensing capability can enhance the effective repeatability and accuracy of a robot by allowing it to better define the target location. However, it cannot improve performance beyond the resolution limit of the robot. If the minimum resolution is greater than the distance to the target location, the robot will overshoot when correcting, and can oscillate between two points on either side of the target location.

Speed and load capacity at the end effector are largely determined by the mechanical components of the robot, but can be enhanced by control and sensing features. The mechancial limitations are based on how quickly the actuators can accelerate the arm plus payload and how quickly they can decelerate the load at the end of motion. The end effector can further limit these capabilities if it is unable to retain its grip through high accelerations. If the controller is capable of controlling acceleration and deceleration at the end effector, the manipulator can use maximum allowable accelerations to traverse a path in the minimum amount of time.

Load capacity also affects positioning parameters. Current robot manipulators tend to sag under load; while the joint positions may be upheld, the end effector is displaced downward. Sensing systems can detect this deviation from expected position, and modify the joint positions to accommodate for the sag, raising the load capacity while maintaining the specified positioning quality.

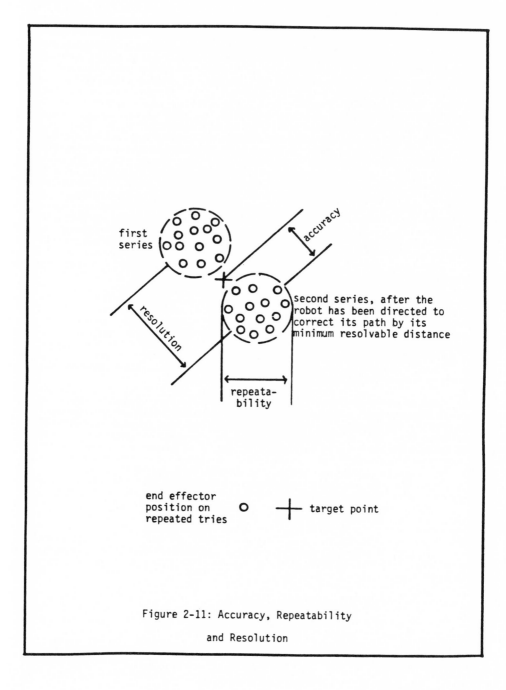

first
series

accuracy

second series, after the
robot has been directed to
correct its path by its
minimum resolvable distance

resolution

repeata-
bility

end effector
position on   O   +   target point
repeated tries

Figure 2-11: Accuracy, Repeatability

and Resolution

External Integration

This section has, thus far, looked at an industrial robot as an individual component of the manufacturing process. While some level of interfacing and coordination with surrounding equipment is included in most robots, e.g., sensing that a die is open or that a parts feeder is empty, the goal of much of the development of industrial robotic systems is more ambitious: a true Flexible Manufacturing System (FMS). In an ideal FMS, the starting point is raw material and the end point is the finished product. The FMS should also have the capability of producing a number of different parts and performing all operations automatically. The key to a successful FMS is supervisory control, for coordinating the individual robotic work cells and managing the transportation of work pieces from one cell to the next.

The basic type of robotic cell that has been integrated into Flexible Manufacturing Systems consists of one or more CNC machine tools that are loaded and unloaded by a robot. Coordination of this cell is made possible by converting status indicators on a console to electrical signals in a communications system, and by modifying control inputs normally operated by push buttons and switches to allow electrical actuation. As a result of these conversions, the entire cell can be monitored and operated by the supervisory computer.

The movement of workpieces from one cell to another is performed by a computer-controlled handling system. The supervisory computer detects the need for parts transport (i.e., a cell has finished working on a piece), provides a cart to transport the piece, routes the cart to the next cell and orders the cell to unload the work piece from the cart. Simultaneously, the supervisor has been performing the same chores for each of the other cell-to-cell transport steps and logging all of the information on the location and status of each work piece in process.

Beyond these coordination and service tasks, the supervisory computer provides a Management Information System (MIS) that formats and can present information on the status of all work in process. Sample capabilities include scheduling projected work up to thirty days in advance, setting machining time required for each part or family of parts, and even allowing for preventative maintenance schedules on each piece of equipment. In particular, the Printing Equipment Group of Harris Corporation has a supervisory computer that optimizes scheduled production by grouping batches of parts to be produced according to part sizes, thus minimizing the number of setup changes required.

This level of integration, while not common, has been achieved with today's technology, by careful planning and intelligent use of available hardware. It represents a major thrust, not of the development of robotic technology, but of ways to make use of current robotic technology.

2.3 State-of-practice by Application

In the following section, several prevalent robotic applications are described according to their state-of-practice. For the purpose of this section, state-of-practice is addressed with respect to the following four aspects: level of technical sophistication, degree of industrial usage, various driving forces (such as economics, productivity and safety), and technological barriers hindering further use. The set of applications chosen has been divided into two categories, fundamental applications and composite applications. The fundamental applications include welding, material handling, inspection, assembly, and painting/coating; and the composite applications are those of a less significant role.

Fundamental Applications

Welding

The technological sophistication of different robotic welding implementations varies from application to application. For most spot welding tasks, the most important characteristic of the robot function is repeatability. Smooth path control and external sensing are generally not necessary, as the robot needs only to repeatedly move to a given point, independent of the path it takes. Complicated arc welding, on the other hand, may require a much higher degree of sophistication. In general, the robot needs not only to go to a specific point or set of points, but to traverse a given path with controlled speed and acceleration if possible. In addition, it is desirable to have the capability of altering the preset path to respond to changing welding conditions. This, when possible, requires not only a high sophistication level of individual components such as sensor systems, but also a complex, integrated method of control.

Because the level of advanced technology necessary for spot welding is relatively low, robots penetrated spot welding applications early. Currently, automobile spot welding uses the largest number of robots of any manufacturing process in the U.S. In contrast, robots have not previously been as successful in penetrating arc welding applications. As the technology necessary for complex procedures has become available, however, the number of arc welding robots has risen significantly.

The driving forces for robotic implementation have been predominantly improved quality and reduced costs. While the robot is not necessarily always fast enough to justify its cost through increased throughput, the consistent quality of robotic welding is usually better than human welding. This is true both for spot welding, where fatigue due to heavy equipment and long shifts is sometimes a problem, and for arc welding, where consistency over a long weld path may lead to difficulties.

Continued and increased use of robots for spot welding is not, in general, dependent on further advances in new technology. One exception may be, however, price. As technology improves, it may not be possible to perform new functions with robotic spot welding, but it will be possible to perform established functions more economically. Increased use of robotic arc welding, though, is heavily dependent on advances in sensing and control technology. As sensing techniques improve, robotic arc welders will be able to autonomously adapt the weld path and parameters to meet varying weld conditions. This ability will reduce or eliminate the current need for expensive, precise fixturing mechanisms.

Material Handling

The technological issues involved in current material handling applications range from the more routine to the very complex. In the simplest
cases, the "pick-and-place" processes, the robot needs only to move to
a prescribed location, grasp an object, move to a second prescribed location,
and release the object. In the more advanced implementations, the robot
may use any combination of specially engineered grippers such as magnetic
or vacuum grippers, some method of smooth path control, or various sensors
to locate and verify acquisition of the workpiece. The level of sophistication, then, generally depends on the specific needs of each individual
implementation.

While the percentage of material handling processes performed by
robots is not very high, the number of robots involved in these processes
is large and, in fact, steadily rising. This is due largely to the vast
number of material handling types of applications performed in industry.
While not all material handling implementations are suitable for robotization,
there is still significant room for robot penetration into many material
handling operations, especially tool load/unload type operations.

The driving factor for robotization of material handling applications
depends heavily on the work volume. If the batch size is very large,
then hard automation is generally more economical than robots. Similarly,
if the batch size is very small, then human labor is usually more economical
than robots. There may, however, be overriding reasons for using robots
in applications where they would be less economical than other methods.
These reasons may include work in an unpleasant or hazardous environment,
such as the foundry environment, or highly repetitive or difficult work
which would cause fatigue in human laborers.

With the exception of high-precision material handling, most material
handling processes can now be robotized without further technological
advances, albeit some at great cost. There is a key trade-off in material
handling operations, namely precise fixturing versus the ability to locate
an object accurately and grasp it easily. For those applications where
positioning must be very precise, it is necessary to know exactly where
the workpiece is and where is the most efficient place to grasp it. This
can currently be done with fixturing techniques; however fixturing reduces
the flexibility of the robot and increases the system cost. As sensing
technologies and gripper designs improve and become cost-effective, applications requiring great precision will use sensing devices and multi-purpose
grippers rather than fixturing systems.

Inspection

Robotic inspection as a process generally uses the most technologically
advanced means available. As sensor technology improves, inspection applications become more varied. Sensing systems currently used for robotic
inspection include 2-D and lightstripe vision, as well as force sensing

and binary tactile sensing. However, other types of sensing are being implemented as advances in IR, ultrasonic, and eddy current sensing technologies have brought the price of these sensors down to a cost effective level. In addition, control technology is a key element of robotic inspection processes. To perform an inspection task, the robot needs an internal model of the ideal workpiece from which to make comparisons. In theory, this model could be as simple as a linear measurement, such as the part must be eight inches long, or as complex as a detailed 3-D model of the part. Robot controllers are becoming sufficiently sophisticated to hold, and in some cases even automatically generate, a complex internal model of the workpiece.

Until recently, sensing technology has been either unavailable or uneconomical. For this reason, robot penetration into inspection processes has been very minor. As the technologies improve and the prices drop, robotic inspection becomes more common. Additionally, because inspection processes are increasingly coupled to assembly tasks, robotic inspection will become more common with the rise of robotic assembly.

The primary reason for using robots in inspection tasks is quality control. The consistency and repeatability of the robot and the control algorithms that compare the workpiece to a model allow for not only greater, but more predictable levels of quality. Once a tolerance has been preset, the robot will reject any inferior part and accept any part that meets the tolerances, eliminating any subjectivity from the process. This consistency and predictability aid in manufacture and process planning. A Secondary reason for using robots for inspection is the capability of in-process inspection, which allows for inspection of workpieces in hazardous environments.

While the use of robots for inspection is increasing, further technological advances would speed the penetration of robotic technology into inspection processes. One of the most important factors that hold back the use of robotic inspection is not availability of new technologies but rather the need for decreasing the cost and increasing the speed of current technologies. Additionally, the areas of 3-D real time vision and precision tactile sensing arrays are very active research topics, and, when fully developed, will expand the scope of robotic inspection.

Assembly

Robotic assembly operations may be performed at a variety of sophistication levels. For easy-mating assemblies, low levels of sensor and path control sophistication are required, while for the more critical assemblies complex force sensing and vision may be necessary. In addition to advanced sensing requirements, critical path control may also be required. The geometry of assembling two closely fitted workpieces is not trivial; although a human can easily compensate for slight misalignment, a robot cannot always make the minute corrections in position and angle of attack to properly assemble two workpieces. While completely accurate and efficient assembly control methods are not yet available, partial solutions to this problem are available and are being used in production.

Because the technology necessary for close-fitting assembly has not previously been available, robots have not been used extensively in this area, and have in fact been used very little in easy-mating assembly. As the necessary technologies are improved and diffuse throughout industry, robotic assembly applications will become more prevalent.

The driving force for robotic assembly, as for robotic material handling, depends on throughput volume. For very large volumes, hard automation with fixturing systems is more economical than robotics, while for very small batches human labor can be more economical than robots. For those volumes of work where robots have the potential for being economical, robotic assembly has the advantage of increased consistency over human labor. Just as for inspection, the high repeatability of the robot affords a higher and more predictable level of quality control than human systems. A secondary incentive for using robots for assembly involves clean room and hazardous or unpleasant environments. Using a robot for an operation that must be performed in a clean room eliminates the complications of human preparation for the clean room.

The three most important technical barriers to extended use of robotics in assembly tasks are sensor technology for easier part acquisition, force feedback control, and advanced control technology for accurate assembly algorithms. Additionally, error recovery algorithms are currently not sufficiently sophisticated to do much more than simply abort an operation. Ideally, these algorithms should be able to isolate the problem and, if the problem is not critical, continue the assembly task.

Painting/Coating

In general, robotic painting and coating operations require a very low level of technological sophistication. For example, sensors are not widely used in painting applications. The most critical aspect of the robot technology necessary for painting tasks is smooth path control. In some of the more recent painting applications, however, the robot controller is called upon not only to direct the path of the robot and control the painting apparatus, but also to coordinate the painting with the movement of an assembly line and with other concurrent operations such as door opening.

Because robotic painting and coating operations require a minimum level of technology, in combination with the fact that this technology has been available for some time, robots have shown a heavy penetration into the painting industry, especially automobile paint spraying applications. In fact, several different robot manufacturers have built reputations solely on their paint spraying robots.

The majority of painting robots, as mentioned, are used by the automobile industry. Because the automobile industry deals with fairly high volumes of throughput, the robots are more cost effective than human workers in terms of increased throughput. In addition, there are several other important considerations for using robots in painting operations. One consideration is quality. If a satisfactory painting path is programmed into the robot, it will follow that path exactly, cycle after cycle, day after day. This will result in very consistent, high quality painting. In addition, the spray painting environment is potentially very hazardous to humans.   By replacing a human with a robot, the manufacturer not only removes a human from a hazardous environment but also eliminates the need for expensive ventilation systems and protective masks that are necessary when a human is performing the painting.

Work pieces or assemblies to be painted by robots still require accurate fixturing (expensive and not readily modified). The use of vision sensing would considerably enlarge the field of application for painting robots. What is required is the development of 3-D vision technology at acceptable prices.

Composite Applications

The final six robotic applications to be presented (sealing/bonding, finishing, investment casting, die casting, forging, and plastic molding), have been classified as composite applications for the following reason: while these applications have been significantly penetrated by robotics, they are technologically equivalent to one or more of the previously mentioned applications. For example, the robotic technology involved in robotized forging is similar to simple material handling, with the possible addition of of a specialized end effector or sensor. For this reason, these applications will be discussed as a group, with references to those applicable technologies that have been previously described.

In general, the composite applications are characterized by an adaptation of generic robotic technology to a specific task. Thus, there is usually a moderate to high level of technological sophistication among these applications.   For example, robotic sealing/bonding and investment casting are extensions of painting and material handling techniques, respectively. Added to these generic technologies in both of these applications, however, are the complex path control capabilities required for each application. Similarly, finishing operations can be accomplished with basic material handling techniques enhanced with advanced sensing capabilities such as force and torque sensing.

Each of these six applications enjoys a fair degree of robot penetration. With the exception of some of the very advanced sensing and path control capabilities, the generic technological capabilities have existed for some time.   This has given the technology a chance to penetrate and be refined by each application.   Die cast loading/unloading, for example, was among the very first types of robotic applications, with the first implementation appearing in the early 1960's.

Due to the varied nature of the composite applications, there may be many different factors affecting the considerations for using robotic technology.   Safety and environmental factors, for example, are major considerations in using robots for forging and plastic molding applications.   In contrast, increased consistency and quality of the workpiece is the primary driving force behind robotic implementation in investment casting applications.   For sealing/bonding applications, the increased speed of the robot results in a higher throughput and profitability compared to human labor.

For the most part, the composite applications are not dependent on further advances in technology to realize a greater penetration. The technology, as mentioned, is and has been generally available. The greatest barriers to further robotic implementation in these areas seem to be hesitancy on the part of end-users, not a deficit in technology.

2.4 Robotic Usage by Industry

Of all of the components of the American manufacturing industry, only a few are making significant use of industrial robots today. This section will briefly describe the robotics involvement of these industries, but two points should be kept in mind. First, some of these industries are more clearly focused than others. The aerospace industry is well-defined while non-metals light manufacturing is more of an organizational category than a coherent industry. Second, there is a significant amount of overlap between these industries. General Motors is clearly part of the automotive industry, but is also heavily involved in foundry activities. This type of cross-industry linkage can affect the level of technology implemented by a company as strongly as competition from other members of its own industry.

The industry descriptions that follow will present information on how long the industry has been involved with robots and factors that have encouraged and discouraged robotic implementation to provide a background for their current position. The current situation for each industry will be described, and illustrated by examples of typical or innovative implementations. Finally, a qualitative assessment of each industry's responsiveness to robotic developments will be given.

Automotive

The involvement of the American automotive industry with robots dates back to 1961, when General Motors installed a robot die casting unloader. While early industrial robots were limited in their capabilities, these capabilities were well matched to the demands of many tasks in automotive manufacturing. When the automotive industry began installing robotic spot welders, a pattern of robotic usage was established: simple robots performing simple tasks in high volume.

Many factors have encouraged the automotive industry to implement robots. The environment in which many assembly operations are performed is noisy and hazardous, while the jobs are monotonous and fatiguing. Escalating hourly costs for personnel and increasingly stringent OSHA requirements for work environment have steadily raised the cost of labor. Robots are seen as a method of holding costs down with the added benefit of improved quality, a matter of increasing concern among U.S. automobile manufacturers in the face of foreign competition.

A major barrier to robotic implementation in most industries is the high initial cost. This factor was less of a concern to automobile manufacturers because of the high volume of production; costs could be distributed over many production units. Furthermore, the automotive industry has, since the mid 1950's, accepted yearly retooling as a fact of life; thus the reluctance to invest in capital equipment has been less than in industries that retool on ten to fifteen year cycles.

Today, the automotive industry is the largest user of industrial robots in the U.S., with approximately 50% of America's installed robots. Spot welding is the most robotized application; at the end of 1983, almost 60% of General Motor's robots were spot welders. Machine loading is also heavily robotized, and spray painting robots are becoming common. The early pattern of simple robots performing simple tasks still holds true for the American automotive industry today.

However, this pattern in no way implies that this industry is complacent with respect to robots. In-house R&D efforts have kept the manufacturers abreast of new developments, and through-the-arc sensing robots for brazing body panels together and vision-equipped robots for assembly are being actively pursued. While the robots in American automotive factories may reflect a certain conservatism, this industry has demonstrated a willingness to implement new technology as soon as it considers the technology to be sufficiently mature.

Foundry

The foundry industry has been implementing robots since the early sixties. Early material handling robots were suitable for tasks like die unloading, the first foundry application for robots.

The major motivation for robotizing foundry work has been risk to human workers. Virtually every foundry process from pouring molten metal to the final cleaning of a casting exposes workers to heat, noise, fumes and dust. Robots have been used to reduce this exposure and also to relieve humans of the fatiguing tasks of manipulating hot, heavy metal parts.

The major barriers to increased robot utilization in foundry work have been limits of industrial robot flexibility and sensing. This is most clearly shown by the cleaning operations that until recently have remained a manual operation. The two major difficulties in automating the cleaning process have been the variability from casting to casting and the force or torque sensing required to control abrasive cut-off and grinding wheels.

The foundry industry today is one of the leading users of industrial robots in the U. S. Most of the robots in foundries are still performing material handling, with robotic unloading of cast aluminum transmission housings at Doehler-Jarvis being typical. Robots are also becoming common in investment casting where the quality of the cast part is largely determined by the consistency of the mold. Robots have demonstrated their ability to achieve greater consistency than humans, in addition to being able to handle mold trees several times heavier than humans.

The more demanding task of finishing castings is being performed by the Swedish firm of Kohlswa Jernverk using an ASEA model IRb-60 robot.

That installation uses torque and force sensing to control metal removal rate, and has demonstrated significant improvements in productivity over human performance due to the robot's ability to safely use higher powered grinding tools and to perform more consistently.

While the foundry industry may not be generally thought of as techno- logically innovative, with respect to robotics, they have established themselves as a major user of industrial robots. The implementations in this industry have overcome the problems associated with one of the harshest of the industrial environments, and, through sophisticated techniques like force controlled grinding, have demonstrated a willingness and ability to keep pace with developing technology.

Non-Metals Light Manufacturing

Non-metals light manufacturing shows its most conspicuous use of robots in the fabrication of plastic parts. The environment surrounding injection molding equipment is hot and fume-laden, and operator fatigue substantially reduces productivity. As with other industries, removal of personnel from a bad environment is a major incentive to introduce robots. Robotic implementation cost is the primary barrier in this industry; the small shops that make up a significant fraction of this industry cannot afford even a simple material handling robot. It is, therefore, not surprising that most of the robots in this industry are found in larger companies such as General Electric and Hoover.

Non-metals light manufacturing utilized about one sixth of American industrial robots by the end of 1982, and the bulk of them are used by larger companies in the injection molding process. A typical implementation is that used by Hoover in which a Prab-5800 robot unloads vacuum cleaner parts from the molding machine and presents them to a broaching machine for sprue removal. A more ambitious and better integrated project by General Electric involves automating their Louisville, KY dishwasher plant. In this plant, fully automatic injection molding machines are serviced by a computer-controlled conveyor system loaded and unloaded by robots produced by Cincinnati Milacron.

While it is not possible to categorize the technological responsiveness of this industry as a whole, it is clear that leaders like General Electric intend to keep up with new technology, if not to lead the way.

Electrical/Electronics

The electrical/electronics industry has long been taking advantage of automation in certain areas. Hard automation is firmly established for insertion of components into printed circuit boards in large batch electronics, while many of the processes involved in fabricating circuit boards (such as resist coating and etching) have been handled by automatic equipment.

The repetitive, labor-intensive nature of many tasks in this industry is considered an already solved problem. The large volume board stuffing is being done with hard automation, at a speed that robots cannot hope to match, while small batch board stuffing is commonly performed outside the U.S., in countries with low labor cost.

Nevertheless, robots have penetrated this industry, and robots involved in the electrical/electronics industry represented about one tenth of the American industrial robots at the end of 1982. This penetration has been based on either using simple robots to replace humans in low demand but particularly unpleasant jobs at a lower cost, or by having the robots combine tasks normally performed by several people.

A typical example of the first approach is used by Northern Telcom Canada Ltd. to assemble terminal blocks. This low technology component is made by loading binding posts and a support block into a hot press, with the press applying heat and pressure to seal the posts into the support block. A robot made by PUMA is used, and while not significantly faster than a human operator, the robot can perform the task more economically by being able to operate continuously and by not requiring the special protective equipment needed by humans when handling hot (500°F) parts.

An example of a sophisticated application in which several tasks are combined is being performed by the Digital Equipment Corporation. A robotic cell is used to insert keycaps into keyboard assemblies, and the first task performed by the cell is inspection of the keycaps prior to assembly. Using an Autovision II vision system, the robot examines all incoming keycaps, rejecting any with incorrect legends or flaws, and loading acceptable keycaps into magazines for use by the second robot that performs the actual insertion. This combination of consistent inspection with actual assembly results in better quality control and is likely to set a pattern for assembly applications in this industry.

The electronics industry has not been very swift in implementing robots, due in part to many potential high volume applications already being performed by hard automation. However, the level of interest in sophisticated robots, such as vision-equipped assemblers, is very high. As robots with enhanced capabilities become available, this industry is ready and willing to use them.

Heavy Equipment Manufacturing

The heavy equipment manufacturing industry began their major involvement with robotics for arc welding in the late 70's. Their interest in robotic welding has been motivated by the same reasons as other industries: the cost and limited supply of skilled welders, and the long term health risks associated with the welding environment. This industry, more than most industries that use arc welding, has frequent need to weld thick work pieces which are difficult to weld and generally require flux-cored welding wire, which is particularly unpleasant to work with.

However, the heavy equipment industry operates in relatively small batches. This tends to make cost justification of robots more difficult because of fewer production units over which to distribute costs. For this industry, robots must show major productivity gains to be cost effective.

Nevertheless, robots have made significant penetrations into heavy equipment manufacturing, with this industry having approximately 10% of American industrial robots at the end of 1982. Welding is the most common application, as typified by use of Cincinnati Milacron T3 robots by the Locomotive Products Division of General Electric to weld large structural elements for diesel-electric locomotives. While the volume of production of these units may not be large, these robots have justified their installation by performing all of the needed welds in as little as half of the time required by humans. International Harvester has invested heavily in robots for production of their series 50 tractors. Nine machining cells, each equipped with two CNC turning centers that are loaded and unloaded by Cincinnati Milacron T3 robots, are used to turn gear blanks for the transmissions. A material handling robot produced by Prab is then used to transport ring gears through heat treating operations, and a DeVilbiss three-robot system spray paints much of the tractor chassis.

The above examples, coupled with the maintenance of in-house robotics R&D groups by other companies in this industry, such as John Deere, indicate that heavy equipment manufacturers are interested in and willing to make use of robots as the technology becomes available.

Aerospace

The involvement of the aerospace industry with industrial robots is relatively recent, compared to the automotive and foundry industries. In 1975, General Dynamics demonstrated the feasibility of a robotic work station for drilling aircraft wings. However, it was not until four years later that their first production robotic work station, funded by the Air Force MANTECH program, went into operation drilling pilot holes in composite materials. Early industrial robots had little impact on this industry, largely due to the need for a higher level of precision than those early robots offered.

There have been many factors that have encouraged the introduction of robots into aerospace manufacturing. The Air Force, through the MANTECH and TECHMOD programs, has made plain its interest in seeing its contractors implement robotics. The competitiveness of this industry requires the use of the most cost-effective manufacturing techniques available. Beyond cost-effectiveness, sheer precision of fabrication is critical in this industry; each new generation of aircraft is more demanding in manufacture than the prior one. Human techniques, using purpose-designed tools and carefully worked out methodologies, have kept up with demands for increasing precision, but may have reached the limits of development. On the other hand, robotic techniques are still in the early stages of development and show much room for rapid improvement. Health hazards represent an area of major concern in aerospace, especially with respect to many of the coatings that are commonly spray deposited. Robots offer an obvious way to remove humans from these hazards.

The major impediments to aerospace use of robots has been the need for high precision, coupled to the small batch sizes typical of the industry. Drilling and routing to the required precision requires the use of templates by today's robots, and fabrication and maintenance of templates for each different part used is an expensive proposition. This, with the high initial cost of robots, results in a cost of implementation that can only be distributed over a limited number of units produced.

Today, the aerospace industry is only lightly penetrated by robots. Spray coating is the most common robotic application, with material handling and finishing (i.e. deburring and sanding) following. Some machining operations such as routing and drilling are being performed with robots, but templates are required. Typical spray coating applications are the use of Trallfa's robots by Fairchild Republic for painting parts of the A-10 and 747, while Martin Marietta is using a Cincinnati Milacron HT3 robot to spray ablator material on the external tanks of the space shuttle. In the field of material handling, Northrop is using a Cincinnati Milacron T3 to lay up plies for composite materials, an application in which robots are becoming prominent.

While the aerospace industry has been prominent in robotic R&D, it has been slow to implement robots in production. The reluctance to purchase expensive hardware for small batch production and limited lifetime contracts will probably continue to act as a deterrent to industrial robots in aerospace manufacturing.

2.5. A Composite Picture of Current Robotic Technology

The preceding sections plus Appendix A represent a fairly comprehensive assessment of the current robotic technology. What is intended to be achieved in this section is a balanced summary that can reasonably synthesize this substantial information base. The adopted approach to achieving this goal is to look at robotic applications in different industries across the board to highlight the major features and prominent trends. As a result of this assessment, current capabilities and technological barriers can be identified. They are, however, presented in Chapter 4 as an integral part of a technological forecast for ease of comparison instead of being included in this section.

One useful picture of the robotic technology can be obtained by analyzing the current applications according to their required level of capability sophistication. The objective in this exercise is just to arrive at a qualitative assessment of various applications as to where they are positioned in this "spectrum of technological sophistication". This picture will be helpful in understanding the present status and future potential developments of robotic technology along various application paths. It is extremely difficult to completely characterize the so-called "sophistication level". For the present limited purpose, several generic sensing and control capabilities are used as the key indicators to approximately define this spectrum. On the sensing axis, the sophistication level is envisioned to range from a single binary sensor to the sophisticated capability of real-time adaptive sensing. On the control axis, the sophistication level is characterized at the low end by a preprogrammed controller and at the high end by the capability of fully adaptive control and complete process planning.

In this spectrum, Figure 2-12 lists twelve application areas and shows where they are situated. In some cases, each application area is further divided into several generic categories that are distinguished

by their different capability requirements. This provides a more detailed
picture of what is being achieved in each application area. Figure 2-12
also separates those applications which are now quite established in terms
of industrial usage from those that are believed to only exist in isolated
cases, as a prototype, or only as lab-scale model. The former are included
in solid boxes while the latter characterized by boxes drawn with broken
lines.

Another picture of the current robotics technology can be illustrated
by studying the level of penetration of various robot applications in
different industries. Since reliable numbers of robots actually being
used in each industry are currently not available, it is more appropriate
to describe the robot penetration in a rather qualitative manner. In
Figure 2-13, the six industries which are presently known to use robots
are listed in one axis against the other axis containing the twelve current
robot applications. If an application has been well established in a
significant number of industrial installations, then it is indicated by
a solid circle. If an application is not reported in use anywhere and
is unlikely to be adapted by that industry in the near future, then it
is characterized by a hollow circle. Note that applications not relevant
to a particular industry are indicated by a dash. Those applications
that are marked by a half-filled circle belong to a group of applications
that have been practiced in isolated cases or are being demonstrated with
prototype units.

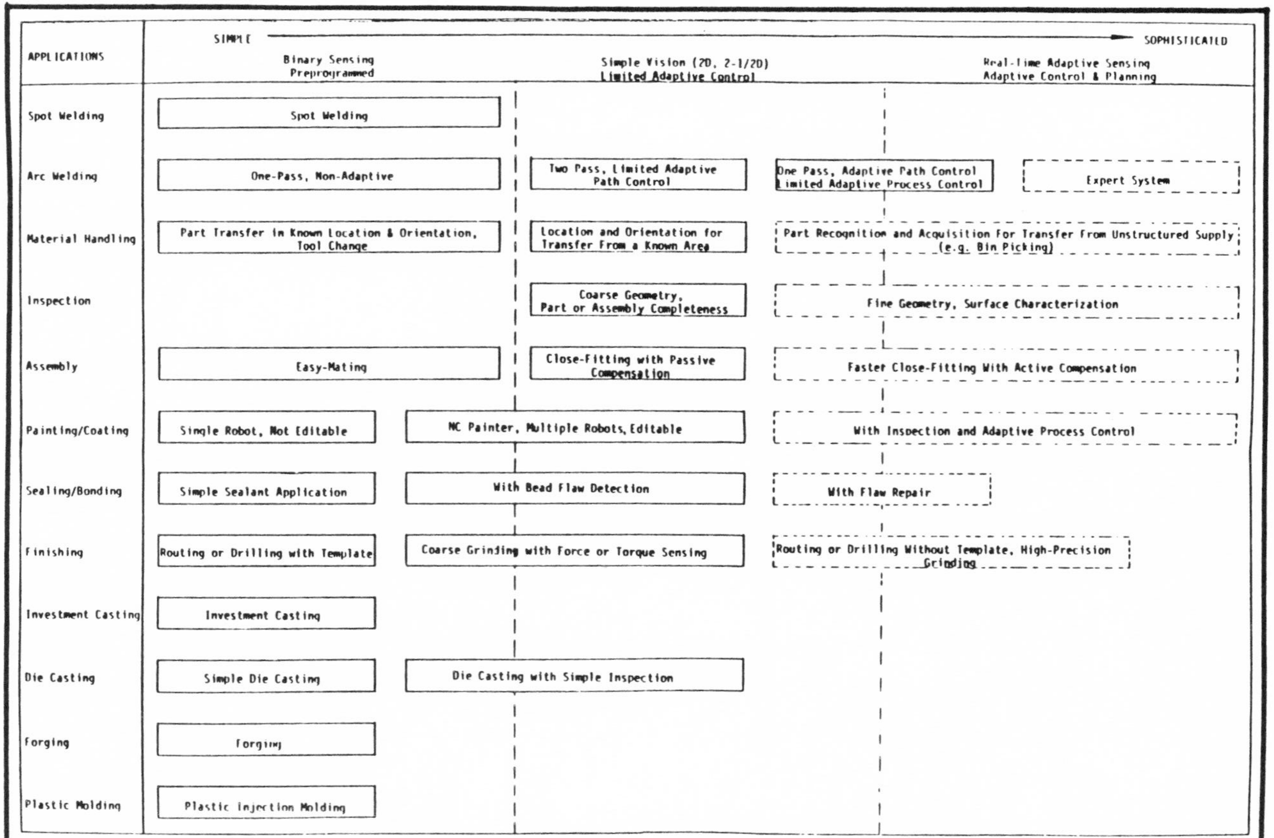

Figure 2-12: Established and Emergent Robotic Applications

APPLICATION / INDUSTRY	SPOT WELDING	ARC WELDING	MATERIAL HANDLING	INSPECTION	ASSEMBLY	PAINTING/ COATING	SEALING/ COATING	FINISHING	INVESTMENT CASTING	DIE CASTING	FORGING	PLASTIC MOLDING
AUTOMOTIVE	●	◐	●	◐	◐	●	◐	○	○	●	○	○
FOUNDRY	–	–	●	○	–	–	–	◐	●	●	●	–
NON-METAL LIGHT MANUFACTURING	–	–	●	○	◐	○	●	–	–	○	–	●
ELECTRICAL/ ELECTRONICS	○	◐	●	●	●	○	◐	–	–	○	–	◐
HEAVY EQUIPMENT MANUFACTURING	○	●	●	○	○	◐	○	◐	○	○	○	○
AEROSPACE	–	◐	◐	◐	○	●	◐	◐	◐	○	○	○

● = significant penetration;   ◐ = moderate penetration or being introduced;   ○ = no penetration;   – = not relevant

Figure 2-13: Penetration of Current Robotic Applications in Various Industries

3. R&D Activities in Robotics

The status and future developments of robotics technology are strongly influenced by current and planned efforts in robotics R&D. A major part of the present study is, therefore, to conduct a comprehensive survey of R&D activities in robotics both inside and outside the U. S. It is an ambitious task which is achievable only when its objectives are well defined and its scope is properly bounded. It is with this perspective that a practical approach of stressing different aspects and focusing on more accessible information for various R&D programs is adopted. For example, one must seek different types of information on industrial R&D from those obtainable from government agencies. Even among government agencies, one should take into account different practices in information dissemination because of their differences in missions and traditions. Another situation that should be addressed is the question of what type of information is available and how one should cover R&D activities in Western Europe, Japan and the Soviet Bloc. In view of publications readily available, only the general R&D structure and directions in Western Europe and Japan are discussed while a closer examination is performed for the countries in the Soviet Bloc. In general, the differences in emphasized aspects and depth of treatment in our coverage of different R&D efforts will be clear to the reader.

In this chapter, the discussion of R&D activities is divided into two major sections, U. S. and foreign countries, with a smaller concluding section to highlight the key trends and directions in robotics R&D. In the U. S., information is organized into groups of institutions of a similar nature. They include government agencies, robotics producers and end-users, and the academic/non-profit community. Because of the special focus on Air Force activities, the federal agencies are classified into three sub-groups: Air Force, other DOD, and non-DOD federal. Discussion on foreign countries are divided into three groups: Japan, Western Europe, and the Soviet Bloc. Countries having an influential presence in robotics will receive more attention than the minor ones. At the end, a separate section is devoted to an integrated synthesis which is intended to bring out the key features and observations drawn from the preceding discussion.

3.1. U.S. Activities

3.1.1. Air Force

The Air Force funding efforts in robotics research encompass a broad spectrum of projects, from very basic research to the development and implementation of applied technologies. In practice, there is a rough division of funding sources into three categories: AFOSR, concerning primarily basic research, AFSC, funding applied research and developmental applications, and AFLC, dealing primarily with direct Air Force applications in the form of application studies. The following discussion will consider the individual robotics R&D programs of each of these offices and commands in turn, followed by a table summarizing the overall directions of Air Force R&D.

The structure of the AFOSR funding effort is based on the concept of "centers of excellence". The majority of AFOSR's R&D funds are channeled into a small number of institutions, which then become the focal point for many different areas of basic research. The University of Michigan, one of two centers of excellence, receives approximately $1 million per year from AFOSR to conduct research on high performance manipulators, sensor subsystems, special-purpose computers and languages, knowledge systems, and sensor-based robot structures. All of these projects are integrated into a higher level study effort considering robot based manufacturing cells as building blocks for an integrated factory system. Stanford University, the other center of excellence, also receives about $1 million per year to conduct a similar program with slightly more emphasis on basic sensor and sensor-based control research. There are two other recipients of AFOSR funding that, although not centers of excellence, represent a sizeable research effort: SRI International, a non-profit research lab which receives approximately $200 thousand per year to conduct very broad-based research, including work on positioning accuracy and control; and Brigham Young University, which receives $100 thousand per year to study microcomputer control of robots.

The AFSC effort in robotics R&D consists primarily of providing and/or managing the funds for robotics research in the MANSCIENCE, MANTECH, and TECHMOD programs. Each of these programs, while concerned with manufacturing productivity as a whole, have individual projects that specifically examine the emerging role of robotics in manufacturing. There is some variation in the focus of these programs; they range from very basic to more applied emphases. The MANSCIENCE Intelligent Task Automation project, or ITA, is the most fundamental of the three projects. ITA consists of two parallel projects, each performed by a different project team. The first project team, Honeywell, Stanford University, SRI, and Adept Technology, Inc., have as their goal the formation of basic hardware and software tools to be used in automated assembly. To date they have completed the design of a micromanipulator, force sensing fingers, 2-1/2 D vision hardware, and are close to completion of parallel force control strategies. To complete these studies, the Honeywell team was funded with $3.35 million through the middle of FY85. The second team, Martin Marietta, Stanford University, ERIM, VPI, RPI, University of Massachusetts and McDonnell Douglas, are investigating the use of multi-arm systems, both in assembly and inspection operations. To date they have designed fiber optic and elastomeric tactile sensors, high speed feature detection algorithms, and a 3-level hierarchical planner. Almost complete are a laser scanner and several adaptive control schemes for servo controls. To complete this research, the Martin Marietta group was provided $3.24 million, through the middle of FY85.

The MANTECH "Advanced Robotic Systems for Aerospace Batch Manufacturing" project is much more sharply focused on the goal of advancing application technology than MANSCIENCE's ITA project. The Advanced Robotic Systems project is divided into three tasks, with different contractors responsible for each task. Task A, conducted by McDonnell Douglas, involves the enhancement of their Machine Control Language (MCL) to make it compatible with a variety of CAD systems for off-line programming. Task B, conducted by RVSI, involves the development of a vision system. Task C, conducted by Grumman Aircraft and Fairchild Aircraft, involves the control of drilling, trimming, and riveting procedures, aided in part by the off-line programming under development in Task A. Much of the robotic application technology developed through these and previous programs such as ICAM has been disseminated to industry both directly and through other Air Force and DoD programs such as TECHMOD. One TECHMOD program, the Rockwell International program to study the enhancement of mechanical tasks, has a sizable funding of its own, $900 thousand.

Another source of the Air Force's funding efforts in robotics research and development is the AFLC. The AFLC is active both in supporting ALC-sponsored application development studies and in funding AFLC REP TECH efforts. The ALC studies are generally very application-specific, such as a Georgia Tech study to examine the feasibility of using robotics in automated packaging and warehousing, and a General Electric Aircraft Engine Group study of automated turbine blade inspection.

Although there is some overlap of interests, the above three robotics R&D efforts supported by the Air Force generally reflect the R&D missions of their sponsoring agencies. The AFOSR generally supports basic, unfocused research, in the form of block grants to establish focal points of robotics research. The AFSC supports a mixture of basic and developmental research to advance the state of manufacturing technology and productivity through its manufacturing science and manufacturing technology programs. Finally, the AFLC supports application-specific studies and efforts that help to influence AFLC process planning and activities. The above information on robotics R&D activities of the Air Force is tabulated in Figure 3-1.

3.1.2. Other DoD

Navy:

The Navy's efforts in robotics research and development follow a similar pattern to that of the Air Force. The Office of Naval Research, in a role analogous to the AFOSR, is the arm of the Navy concerned primarily with funding basic research projects in robotics. In addition, individual commands such as NAVSEA and NAVAIR are responsible for supporting application-oriented research and development projects consistent with their overall goals. This section presents the directions of Navy robotics R&D, highlighting ONR, NAVSEA, and NAVAIR. Finally, summaries of individual R&D efforts sponsored by each agency are presented.

Although ONR has not created "centers of excellence" as has AFOSR, there are two universities, the Massachusetts Institute of Technology and Carnegie-Mellon University, that have been very heavily funded by ONR. At MIT the funding has been weighted towards sensor research, while at CMU the R&D has mostly involved control research. ONR funding at other research institutions covers all aspects of robotic research: software control algorithms at New York University, University of Massachusetts and SRI, manipulator design and control at the University of Utah, sensor research at Yale University, Case Western Reserve University and the University of Rochester, and system performance issues at North Carolina State University,

```
                    Air Force Robotics R&D
                       (Sheet 1 of 2)


AFOSR:  Unfocused sensor, manipulator and control studies

      Performer/Estimated Support:  University of Michigan  ($1M)

      Research Areas:               o  Center of excellence:  high
                                    performance manipulators, sensor
                                    subsystems, knowledge systems and
                                    problem solving, manufacturing cells.

      Performer/Estimated Support:  Stanford University  ($1M)

      Research Areas:               o  Center of excellence:  broad based
                                    research with emphasis on sensor
                                    hardware and sensor control.

      Performer/Estimated Support:  SRI, Intl.  ($200K)

      Research Areas:               o  Broad based, emphasis on location
                                    and control.

      Performer/Estimated Support:  BYU  ($100K)

      Research Areas:               o  Microcomputer control of robots

  ─────────────────────────────────────────────────────────────────

AFSC/MANSCIENCE (ITA):  Developing robotic components for manufacturing
      applications.

      Performer/Estimated Support:  Honeywell/Stanford/SRI/Unimation
                                    ($3.35M)

      Research Areas:               o  Formation of basic hardware and
                                    software tools especially applicable
                                    to automated assembly.

      Performer/Estimated Support:  Martin Marietta/Stanford/ERIM/VPI/
                                    RPI/McDonnell Douglas  ($3.24M)

      Research Areas:               o  Development of multi-arm systems
                                    for both assembly and inspection.
```

```
+---------------------------------------------------------------------+
|                                                                     |
|                      Air Force Robotics R&D                         |
|                        (Sheet 2 of 2)                               |
|                                                                     |
|   _____ |
|                                                                     |
|                                                                     |
|  AFSC/MANTECH (Advanced Robotic Systems):  Implementing robotic     |
|  technology                                                         |
|  to increase robotic manufacturing technology.                      |
|                                                                     |
|      Performer:                     Fairchild Aircraft              |
|                                                                     |
|      Research Areas:                o  Robotic drilling and trimming |
|                                                                     |
|      Performer:                     Grumman Aircraft                |
|                                                                     |
|      Research Areas:                o  Robotic drilling and riveting |
|                                                                     |
|      Performer:                     McDonnell Douglas Aircraft      |
|                                                                     |
|      Reserach Areas:                o  Enhancement of MCL to provide |
|                                        off-line programming with CAD |
|                                        links                         |
|      Performer:                     Robotic Vision Systems, Inc.    |
|                                                                     |
|      Research Areas:                o  advanced vision system       |
|                                                                     |
|   _____ |
|                                                                     |
|                                                                     |
|  AFLC:  Feasibility studies for specific applications.              |
|                                                                     |
|      Performer/Estimated Support:  Georgia Tech   ($80K)           |
|                                                                     |
|      Research Areas:                o  Examine the feasibility of    |
|                                        using robots in automated     |
|                                        packaging and warehousing.    |
|                                                                     |
|      Performer/Estimated Support:  General Electric   ($5.4M)      |
|                                                                     |
|      Research Areas:                o  Turbine blade inspection      |
|                                                                     |
|                                                                     |
|                                                                     |
|         Figure 3-1:  Summary of Air Force Robotics R&D Activities    |
|                                                                     |
+---------------------------------------------------------------------+
```

Purdue University, University of Maryland, and a Westinghouse research laboratory.

NAVSEA is currently supporting three robotics research projects, two involving issues of autonomous mobility, navigation, path planning and location control, and one involving welding techniques. The first project, located at the Naval Oceans Systems Center in San Diego, CA, is devoted to the design of a large scale autonomous mobile robot. The major research issues are obstacle avoidance and navigation by means of a hierarchical path planner. The second project, located at the Naval Surface Weapons Center in White Oak, MD, deals with a smaller scale mobile robot. The major research issue is navigation and decision making through use of coordinated sensory input. One projected application of this intelligent mobile robot is for use as a sentry. The third project is performed at the Philadelphia Naval Shipyard to develop and test adaptive seam welding techniques for ship hull fabrication. The focal point of the project is the design of a specialized welding end effector. Under a similar contract in 1983, the Philadelphia Naval Shipyard designed the Puma Arc Welding System (PAWS).

NAVAIR sponsors a very large effort to develop a robot to automate some aspects of refurbishing Navy planes. The Naval Air Rework Facility in San Diego, CA has funded the Southwest Research Institute (which should not be mistaken for the previously mentioned SRI, Intl.) with $2.3 million to help them develop a robot that will perform the inspection and de-riveting operations necessary in rebuilding airplane wings. The robot will use vision and eddy current sensing to inspect each rivet, through available decision algorithms will decide whether it needs replacing, choose the correct drill parameters to properly drill the rivet, and then change tools to punch out the rivet. The final plan requires the robot to be on a mobile cart, so accurate positioning techniques will be necessary.

The above R&D activities are organized and presented in greater detail in Figures 3-2 and 3-3.

Army:

The Army's efforts in performing and supporting robotics research are divided into two distinct categories. One thrust involves research and development of robotics for use in the battlefield. This effort, beginning approximately four years ago, stemmed directly from HQ Army, and has been a cohesive, directed project to answer the question of how applicable will robotics and artificial intelligence be to battlefield situations between now and 1990 and 2000. The work began with feasibility studies such as that conducted by SRI in 1982 and continues currently with basic research to study the long-term possibilities of intelligent battlefield robots.

The second major thrust of Army robotics research is in manufacturing technology. Unlike the battlefield effort, the manufacturing technology effort is not centered in one place in the Army, but is divided between individual commands. Each command is responsible for organizing and conducting

Navy Robotics R&D (ONR)
(Sheet 1 OF 2)

PERFORMER	RESEARCH TOPIC
Carnegie Mellon University	o Develop visual reasoning capabilities o Design high power/mass ratio manipulators through use of Lagrangian modelling o Integrate sensor and end effector capabilities
Massachusetts Institute of Technology	o Develop reasoning capabilities based on visual pattern recognition o Improve welding techniques through better sensory integration o Improve current tactile force sensing techniques and integration of information
New York University	o Continuing ONR grant to develop special purpose, process oriented robot language. Funding: $1.25M-$1.5M/Yr
North Carolina State University	o General study of measurement and interface technology, machine control and feedback control of machining processes
Purdue University	o Wide range of research problems pertaining to precise engineering issues involved in a flexible manufacturing system
SRI, International	o Development of process-oriented language
University of Rochester	o Vision-pattern recognition techniques
University of Utah	o Enhance control of three finger gripper through use of antagonistic tendons

Navy Robotics R&D (ONR)
(Sheet 2 OF 2)

PERFORMER	RESEARCH TOPIC
University of Maryland	o Combined effort with NBS to develop error compensation analysis, defect identification analysis, experimental identification of dynamic characteristics
Yale University	o Vision-scene understanding. Estimated funding: $50K
Case Western Reserve	o Various aspects of tactile sensing

Figure 3-2:   Summary of Navy R&D Activities Sponsored
by the Office of Naval Research

Navy Robotics R&D (NAVSEA and NAVAIR)

PERFORMER	RESEARCH TOPIC
Naval Ocean Systems Center/ NAVSEA	o Hierarchial path planner and obstacle avoidance control of large scale antonomous mobile robot. Estimated 1984 funding: $120K
Naval Surface Weapons Center/ NAVSEA	o Decision making capabilities through use of integrated sensory input in autonomous mobile robot
Philadelphia Naval Shipyard/ NAVSEA	o Specialized welding end effector to enhance welding techniques for ship hull fabrication
Naval Air Rework Facility/ NAVAIR with Southwest Research Institute	o Develop and test a robot wing-de-riveter, incorporating advanced sensing and decision making capabilities

Figure 3-3:   Summary of Navy R&D Activities Sponsored by NAVSEA and NAVAIR

research in robotics that may be applicable to its operations. The research and development conducted through these programs, then, tends to be very applied, application-specific work.

One exception to the individual nature of the second thrust is a department-wide interest in painting/coating operations. The Tank-Automotive Command (TACOM) is the lead command, and is coordinating efforts with Depot System Command and the Troop Support Command. Chemical agent resistant coatings, and camouflage pattern painting requirements are an area of emphasis.

The Army is constantly increasing its commitment to robotics support; as development projects are completed new projects are budgeted. The current commitment, in terms of on-going 1984 projects and 1985 projects through apportionment review amounts to approximately $3.1 million for manufacturing methods and technology efforts. The area of emphasis is primarily assembly and testing.

Presented in Figure 3-4 is a series of tables summarizing the individual R&D efforts of each of the active commands. The top section of each table identifies the command and gives a brief description of the thrust of the command's activities, the left-hand column describes each individual effort, and the right-hand column gives approximate funding levels of each project and outside performers, if appropriate.

DARPA:

DARPA's efforts in robotics research are divided into two thrusts: sensor-based control of robots for use in manufacturing, and the development of technologies necessary for an autonomous mobile vehicle. The mobile vehicle effort is in support of DARPA's long term goal of establishing a technology base for non-manufacturing military applications in maintenance logistics and weapons support. To this end, DARPA is concentrating its project funding in several areas: control of specialized manipulators, such as flexible or high-powered arms, and integration of advanced sensory input as a basis for both navigation and manipulator control. Specific topical areas supported by DARPA R&D funds are listed in Figure 3-5.

3.1.3. Non-DoD Federal

NASA:

The National Aeronautics and Space Administration has a significant funding program for robotics research as it applies to the problems of manipulating objects in space. The program is unique among federally-funded robotics programs as it is primarily concerned with integration, both within the robot in terms of integration of feedback control, and outside of the robot in terms of system input and output integration. Of a total funding effort of about $1.5 million in 1984, more than 60% is devoted to control issues. The program as a whole is broken down into four topical project areas.

```
Army Robitics
(Sheet 1 of 2)
```

MICOM

MICOM's R&D thrust has been in the assembly of electronic missile parts.

o Wire harness assembly. MICOM has   Hughes and Boing Aerospace
   been working to incorporate   have been working with
   assembly and testing of the   MICOM. Total funding
   harness.   has been $2.15M.

o Chip recognition. MICOM has been $700K for system build
   working for several years to in-   requirements.
   corporate material handling, optical
   pattern recognition, and assembly
   techniques into a single work cell.
   Prior efforts determined the system
   requirements.

DESCOM

DESCOM's efforts in robotics R&D have centered around operations
involved in the production and maintenance of tracked vehicles,
such as spraying and coating, blast cleaning, and assembly/dis-
assembly operations.

o Automated blast cleaning. This is $582K for cleaning,
   part of a three year effort to $299K for disassembly,
   automate processes involved in $325K for reclamation,
   reclaiming a double pin tracked $795K for welding,
   vehicle. Also included are
   robotic disassembly of double pin
   tracks, reclamation of hardware
   from the tracks, welding of
   suspensions, and camouflage
   painting.

AMCCOM

AMCCOM has concentrated its robotics R&D on the issues involved in the
manufacturing, inspection, and material handling of weapons, as well
as sensor-based robotic applications requiring high precision.

o Material handling for x-ray $709K in 1984
   techniques. A robot would
   increase the quality control
   during inspection of Howitzer
   carriages by increasing the
   consistency of placement.

Army Robotics
(Sheet 2 of 2)

o Robotic welding. Adaptive $291K for 1984
   control is being developed for $438K proposed for
   robotic welding of weapon   1985.
   components.

o Material handling and assembly $180K for 1984
   of smaller caliber weapons.
   This is a feasibility and
   application study attempted to
   increase the production quality
   and volume.

o Automated assembly and testing $1.946M proposed for
   of IR transducer. This is a   1985
   feasibility study to determine
   the requirements for such a
   system.

o Automated assembly of electronics $1.018M proposed for
   module and top sensor. This   1985
   feasibility study will determine
   the needs for automated assembly,
   highlighting optical and tactile
   sensors and control.

o Welding. There are two efforts $285K for RIA
   in robotic arc welding. One is $438K for the
   a development of general shop   continuation of
   welding techniques at Rock Island   ARRADCOM; both
   Arsenal and the other is a contin- for 1985.
   uation of ARRADCOM's welding project.

AVSCOM

This command has begun a small effort to incorporate robotics into forging
processes.

o Adaptive control forging. This $215K in 1984
   project will incorporate image $430K in 1985
   sensing and a thermal video
   subsystem to gather data which
   will be used to control form and
   heating of the workpiece.

Figure 3-4: Description of R&D Activities Funded
             by Various Army Command

DARPA Robotics R&D

<u>Performer</u>   <u>Topical Area</u>

System Integration and Demonstration,
Sensory control, and Advanced Mechanical Design

Honeywell,   o This is a collaboration with the
Martin Marietta   Air Force's MANSCIENCE ITA project
                                       which is concentrating on hardware
                                       and software tools for assembly
                                       operations, as well as control of
                                       multiple arms.

Case Western Reserve   o DARPA shares funding with the Navy's
   University   ONR to perform various tactile sensing
                                       research, including haptic sensing.

Carnegie Mellon University   o 3-D vision sensing for robot control.

Honeywell   o Vision-range sensors and control
                                       systems.

Stanford University   o End point control of flexible robots,
                                       path calculation and tracking hand
                                       control. Estimated 1984 funding:
                                       $300K.

University of Utah   o Electromagnetic machines with micro-
                                       actuators.

Duke University   o Cooperation with Lord Corporation
                                       to produce compliant, anthropomorphic
                                       structures and actuators

Robotic Support of Autonomous Mobile Vehicles

Carnegie Mellon University   o DARPA has been funding CMU for an
                                       extended period of time to do research
                                       in the field of spatial reasoning.

University of Maryland,   o DARPA is continuing a previous grant
Yale University   for work in spatial reasoning.

NBS,HEL   o Sensor and control integration for
                                       robust, ammunition-handling robot

Rockwell International   o Ultrasonic imaging sensors and
                                       algorithms for closed-loop control

Figure 3-5: R&D Areas Supported by DARPA

The first topic is vision processing. NASA funds the California Institute of Technology with approximately $550 thousand per year to study the software algorithms and parallel processing architectures necessary to process visual information. The objective is to speed the processing so that information can be used for real-time control of the manipulator.

The second research topic that NASA is concerned with is man-machine interaction. NASA's ultimate goal of a combined teleoperated/expert system robot requires a complex interface between the expert system of the robot controller and the human input system. NASA funds the Jet Propulsion Laboratory, a NASA captive laboratory, with approximately $250 thousand per year to study possible architectures for this man-machine interface.

The third research topic that interests NASA is supervisory control. This is essentially an extension of the first topic. Once the information from the vision sensor is processed, it must be incorporated into the robot control program to produce the desired response to the visual input. NASA funds JPL and indirectly both the University of Southern California and Stanford University with $175 thousand per year to study efficient methods of supervisory control based on visual feedback information. Included in this grant is a study of the precise control of non-rigid robot arms conducted by Stanford University.

The fourth research topic supported by NASA is that of systems integration. In a remote, teleoperated/expert system robot, there will be many varied forms of system input and output. Input may come from on-board vision sensors, on-board position sensors, and remote teleoperation signals. Output may be in the form of position control, manipulator control, and teleoperation feedback signals. The robot controller must be sufficiently robust to handle precise coordination of the system inputs and outputs. NASA is currently funding the Langley Research Laboratory with $500 thousand anually to study advanced system integration. There is a close coordination between Langley and the team developing the NBS system controller. In addition, there is a separate joint funding effort, about $100 thousand per year in total, between NASA and the NBS to study space station robotics.

NASA sees their robotics R&D efforts growing in the future. With the exception of the Stanford University project, which should remain stable, funding in the other research areas is expected to increase in support of the growing NASA space station project. For example, FY1985 funding for the JPL man-machine interface work will increase from $250 to $350 thousand. These four R&D thrusts are summarized in Figure 3-6.

NBS:

The robotics program at the National Bureau of Standards is unique among federal robotics research programs in that the majority of the research is performed in-house with a large portion of the funding support received from other federal agencies. While divided into four distinct efforts, all of the robotics research conducted at the NBS has an underlying objective of formulating standards for the robotics industry. Work is concentrated not only on developing a robot subsystem but developing the subsystem to be compatible with other subsystems in a predictable manner. These

NASA Robotics R&D

PERFORMER	RESEARCH TOPIC	EST. SUPPORT
California Institute of Technology	o Vision information processing, Software algorithms and hardware parallel processing techniques	$550K
Jet Propulsion Lab	o Man-machine interface: control methods and architectures of a complex interactive man-machine interface ($350K budgeted for (1985)	$250K
JPL-Stanford-USC	o Supervisory control: incorporating visual feedback response into robot control program	$175K
	o Control of non-rigid robot arms	
Langley Research Lab	o Systems Integration: Complex integration of various interactive inputs	$500K
	TOTAL	$1,475k

Figure 3-6: Summary of Robotics R&D Programs Supported by NASA

four efforts include software control hierarchies, vision sensing, tactile-sensored and quick-change grippers, and the establishment and support of the Automated Manufacturing Research Facility (AMRF).

The control system under development at NBS is designed to be both flexible and versatile. The system, based on the use of discrete "state tables" to define a world model, was originally designed to be used as a software development tool. Its emulation, simulation, single-step, and reverse-time capabilities will allow the programmer to write, test and debug robot controller software and make it ready for the shop floor with a single system. Other research institutes such as Westinghouse have used the system for their in-house non-proprietary research.

The emphasis of the vision research at NBS is to find a solution to the bin-picking problem, i.e. real-time acquisition of randomly-ordered parts in a factory environment. The novel technical aspect of this project is the use of two planes of structured light to illuminate the object. The use of the second plane of light gives information concerning the pitch and yaw of the object, in addition to the usual distance information obtained from one plane of light. The connection of this work to other work at NBS is the fact that the visual information is processed through the use of a world model. This world model is used as a means of standardizing information transmission protocols.

The gripper research at the NBS is divided into two projects: sensored grippers and quick-change grippers. Work in sensored grippers includes a two-finger gripper that is both force- and position-servoed, and the incorporation of tactile sensors and wrist position sensors into a complex gripper. Standardization of the mechanical and information interfaces is the focal point of the quick-change gripper research.

The largest robotics effort at the NBS is represented by the establishment of the AMRF. The aim of the facility is to provide a working factory environment for use as a developmental testbed. Many R&D projects have been conducted in this testbed: development of a universal calibration scheme, modification of a control system to include shop floor control, and robot-to-robot, robot-to-control, and robot-to-NC tool standardization techniques. Funding for the AMRF is not from a single source; the NBS solicits project funding from potential users of the technology under development. Members of industry have loaned or donated $800 thousand worth of equipment for use in the AMRF. Contributions to this facility also come from DARPA through the Air Force ITA project, the Navy's ManTech program, and the Army's Aberdeen Proving Ground. In addition, universities have occasionally been invited to use the AMRF for their robotics research projects.

NSF:

The National Science Foundation has as one of its missions the support of basic and applied research at a fundamental level. In the field of robotics research, NSF has followed this principle and funded broad-based basic research in robotics. All aspects of robotic technology have been represented in the NSF program, from sensor and control research to issues of robotic system performance. When individual project information is classified according to the control, sensing, manipulation and system performance taxonomy, however, overall qualitative R&D trends emphasized by this program become apparent.

The research area most heavily funded by the NSF is that of visual imaging. Program managers at the NSF believe this to be the fastest growing part of robotic technology, and plan to continue this policy. The current funding level of visual imaging projects is just over $1 million per year, compared to a total annual budget of just over $4 million in robotics. Other major areas of funding include tactile sensing and speech understanding with budgets of about $300 thousand each, control research with an annual budget of about $800 thousand, manipulator, actuator and end-effector research with a budget of approximately $1 million per year, and system performance research with a funding level of about $500 thousand per year. NSF is actually funding more basic research in robotics than the $4 million total implies, because projects supported by other NSF programs are also relevant to the robotics field. For example, multiprocessing and VLSI research supported by the electrical engineering program and control theory research conducted through the systems group program are also useful in solving robot controller problems.

The short-term future of robotics funding at NSF is expected to remain steady with some increases in both current research areas and new areas. For example, a small effort of several hundred thousand dollars per year has just begun with the aim of studying possible applications of robots in the construction environment. Figure 3-7 presents a detailed breakdown of the NSF robotics programs by university and associated research issues in FY83.

3.1.4. Industry-Wide R&D Directions

So far, the U.S. R&D activities associated with the funding agencies have been reviewed. However, the R&D community of robotics also includes industrial laboratories, university research programs and several not-for-profit laboratories. With the exception of the industrial laboratories, they are primarily R&D performers and therefore play a relatively passive role in influencing the current emphasis and future directions of robotics R&D. The industrial laboratories may play a more active role since spending of their in-house research dollars is principally dictated by the corporate policy. It should be pointed out, however, that these industrial laboratories occaisionally compete for government R&D funds. As a result, it is useful to assess R&D activities from the perspective of a performing group. Due to the limited scope of this chapter, R&D activities associated with the performing groups are described in Appendices B and C. Here the overall trends of industrial R&D are summarized, because the robot producers and end-users represent an independent force driving the general R&D directions.

The work performed by these industrial participants ranges from basic research to application development. There is, however, an approximate division among the industrial participants on the basis of their research

NSF Funding Analysis for FY83
(Sheet 1 of 5)

Research Topic	Performer	Support Level (S K)
Sensing:		
Vision:		
1) Picture algebra and picture data structures	Illinois Institute of Technology	42.1
2) Distance sensor for robotics	Kazuko Enterprises	113.9
3) Cost effective sensor systems for robots	Draper Labs	98.9
4) Structural and syntactic pattern recognition	Purdue University	105.1
5) Complex surface recognition for robot vision	University of Tenn	99.8
6) Fast pyramid algorithms for motion analysis and image	RPI	66.9
7) Incoherent optical processing using grating imaging	University of Michigan	20
8) Incoherent optical processing using grating imaging	Oakland University	25
9) Image processing for machine vision research	University of Wisconsin	36
10) Low-level functions of machine vision	Northeastern University	52
11) 3-D digitizer for creation of hierarchical models for robotic vision	VPI	7.2
12) Automatic visual inspection of printed circuit boards	SRI, Intl.	150

NSF Funding Analysis for FY83
(Sheet 2 of 5)

Research Topic	Performer	Support Level
13) Recognition of parts and their orientation	L.N.K.	110
14) Computational & geometric aspects of pattern recognition & vision	Johns Hopkins Univ.	40
15) Dynamic scene analysis	Univ. of Michigan	39.5
16) Structural matching and geometric reasoning for object classification	VPI	72.9
17) Theory and techniques for low-level vision	Univ. of Rochester	51.2
18) Integrated Architecture for industrial 3-D vision	Machine Intelligence Corporation	35
19) Multi resolution image analysis	Univ. of Maryland	157.4
	Subtotal	1,322.9
Tactile:		
1) Automated tactile sensing	Case Western Reserve University	130.9
2) Integrated PVF2 transducer arrays	Stanford University	100
3) Thin film touch sensors	University of Texas	48
4) Robotics force sensor arrays	Bonneville Scientific	35
	Subtotal	313.9
Speech:		
1) Parallelism in speech processing	Purdue University	88.9
2) Speech synthesis & recognition by computer	Louisiana State University	16.5

NSF Funding Analysis for FY83
(Sheet 3 of 5)

	Research Topic	Performer	Support Level
3)	Robust natural language processing	New York Univ.	60
4)	Robust natural language processing	Burroughs Corporation	24.9
5)	Knowledge acquisition in speech understanding	Carnegie Mellon Univ.	157
6)	Automatic speech understanding	MIT	87
7)	Natural language utterances	SRI, Intl.	94
8)	Natural language processing	Duke University	1
9)	Natural language information with database systems	Univ. of Pennsylvania	34.9
10)	Research in natural language processing	Univ. of Pennsylvania	145.5
		Subtotal	709.7

Control:

1)	Dynamics and control of kinematically redundant systems	Ohio State Univ.	82.8
2)	Advanced intelligence control for trainable manipulators	RPI	14.5
3)	Intelligent bubble storgage for robots	CMU	14.5
4)	Research on geometric modelling	Univ. of Rochester	164.2
5)	Robust control of mechanical motion	Cornell University	50.0
6)	De-coupled motion of robot manipulator	Tennessee Technical University	48

NSF Funding Analysis for FY83
(Sheet 4 of 5)

	Research Topic	Performer	Support Level
7)	Development of evolutionary programming techniques	Wayne State Univ.	58
8)	Strategies for data acquisition and utilization	Univ. of Pennsylvania	72.1
9)	Decision making in advanced robotic systems	Polysystems Analysis	35
10)	Advanced control of flexible manipulators	Scientific Systems	35
11)	Computer graphics & design for robotics	Univ. of Alabama	37.1
12)	Visual-tactile coordination for robot control	Univ. of Massachusetts	196.2
		Subtotal	937.9

Manipulation:

1)	Mobile robots for manufacturing	Univ. of Virginia	99.9
2)	Shape and dexterity of workspaces of manipulators	UCLA	48.0
3)	Design conditions for robot manipulator and end-effector orientation	Arizona State Univ.	75.3
4)	Local and global kinematics of multi-degree of freedom arm	Stanford University	75.0
5)	Instantaneous kinematics and geometry of robot manipulators	Univ. of Florida	290.3
6)	Bracing approach to lightweight robot arms	Georgia Tech	79.7
7)	Investigation of novel robot arm	Oregon State Univ.	69.7

NSF Funding Analysis for FY83
(Sheet 5 of 5)

Research Topic	Performer	Support Level
8) Robotic material handling vehicle	University of Utah	48.0
9) Synthesis of spatial mechanism	Univ. of Florida	80
10) Equipment for precision machining systems	Univ. of Florida	49.3
11) Computer aided analysis of mechanical system	University of Iowa	17.6
12) Instantaneous space kinematics	Oklahoma State Univ.	56.4
13) Materials handling	Georgia Institute of Technology	25
	Subtotal	1,014.2
System Performance:		
1) University/Industry Cooperative Research Centers		
1a) University of Rhode Island		220
1b) Georgia Tech		200
2) Optimization of robot mech. operation	Adv. Tech. & Research	35
3) Systems Design	Carnegie Mellon Univ.	73.1
	Subtotal	528.1
	TOTAL	4,826.7

Figure 3-7: Detailed Analysis of NSF Funding in Robotics

emphasis. Those companies that use robots but are not directly involved
in the production of robots tend to concentrate their efforts on short-term
application development, while those involved in the production of robots
or robot components tend to perform more basic, long term research. A
number of companies that produce robots for both internal use and external
marketing, however, tend to perform a broader range of R&D. In the following,
the individual research issues of each of the programs are identified
and grouped into five generic research categories. From this grouping
one can see overall industry-wide trends emerging in such research areas.

Mechanical:

In general, industrial research on mechanical robotic components
is concentrated on improving existing components, such as the high-precision
drives being developed by Allen Bradley, rather than on breaking new ground.
Several exceptions are the three-wheeled mobile cart for the PUMA developed
by Adept Technology, and the design of quick-change end effectors by GMF
Robotics.

Control:

The area of robot control is one of the most active in industrial
robotics research. There emerge three primary directions in this area
of research, which are focused on process planning, integrated control
of entire manufacturing processes, and sensor-based control. The development
of robot programming languages is a unifying thread among these projects.
For example, quite a few industrial laboratories are studying the language
requirements for integrated control systems, sensor-based control systems,
and process planning languages. Another common element among control
research programs is the development of geometric models. These models
are used as a basis for vision algorithms, trajectory planning and integration
with CAD systems. There are several dominant industrial forces in the
control field, which are represented among others by McDonnell Douglas,
General Electric, Automatix, IBM, Cincinnati Milacron and GCA.

Sensing:

Sensing research, another very active field of industrial robotics
research, is divided into two disciplines, vision sensing and tactile
sensing. Tactile research and development generally takes the form of
either short-term adaptation of binary sensors to application needs or
long-term development of more advanced tactile capabilities. Leaders
in the development of advanced tactile capabilities, such as increasing
spatial resolution and shear force sensing, are the Lord Corporation,
AT&T, and General Electric. Industrial vision sensor research, much like
tactile research, currently has two directions, short-term application-
specific and long-term developmental. The short-term vision research
tends to concentrate on inspection processes, while the more general research
is aimed at in-process control. Key industrial research laboratories
working on inspection include Fairchild, Westinghouse, Digital Equipment

Corporation, and Northrop, while General Electric, General Motors, and Honeywell are studying more complex uses of vision in manufacturing processes. McDonnell Douglas, Fairchild, and IBM are all working on 3-D vision. In addition to tactile and vision sensing, there are other smaller sensing research efforts, such as the true volume sensor research conducted by RVSI.

System Performance:

Positioning and speed control are two important issues in industrial R&D. Both Northrop and General Dynamics are developing improved positioning capabilities for aerospace applications such as drilling and fabric lay-up, while Allen Bradley is developing increased speed capability for sealing/ bonding applications.

Applications:

There is a great deal of application-specific research and development conducted in various industries. This research, due to its nature, covers a wide range of processes. However, there is currently a significant trend towards directing efforts into two processes, seam welding and assembly. Automatix and General Electric have been strong forces in the development of advanced seam welding techniques; while General Motors, IBM, Westinghouse, Adept Technology, Digital Equipment Corporation and General Electric concentrate more on robot assembly systems.

3.2. Foreign Activities

3.2.1. Japan

The robotics R&D efforts in Japan differ from those in the U.S. and Western Europe in that the Japanese government plays a more active role in influencing the general directions of the robotics technology. The structure of research institutions in Japan is similar to that of the U.S., consisting of four groups: national research institutions, public universities, private universities, and industrial laboratories. In general, research performed at the government-sponsored institutions has a fundamental orientation, while work at the private companies is of a more applied nature. Additionally, the quantity and scope of robotics research in Japan is quite extensive. For this reason, it would not be practical to present here a detailed list of each research project. For such an in-depth view of individual research projects underway in Japan, the reader is referred to any one of numerous reports written on the subject, (cf. "Trip Report; a Visit to Japan" by Thomas Binford of Stanford University). Rather, it is the intention of this report to provide a structured summary of the Japanese robotics R&D effort. As will be shown, the character of the robotics effort in Japan largely reflects its research climate as influenced by the government. The government funding strategies, as well as national policies, have a direct effect on the specific research topics studied in Japan.

Funding Structure:

The general methods of research funding in Japan consist of government support of universities and national research institutions, government incentives to industry research, industry support of in-house research, and industry support of public and private research institutions. While these funding channels are qualitatively similar to those found in other countries such as the U.S., the relative levels of funding in each of these channels are different. Until April of 1982, the vast majority of funding for robotics research came from industry. In contrast to practice elsewhere, however, most of these funds tended to remain in-house, rather than be used to support university programs. In addition, government support, although on the rise, was on a much smaller scale. In 1980, for example, the government budget for robotics research was under $2 million. Reversing this trend, the Japanese government initiated in 1982 a seven year, $130 million program to advance the available robotic technology. A second national seven year program, the "Jupiter" project, began in 1983 with an estimated funding level of between $55 and $80 million. In addition to this increased funding, Japanese robotics R&D is also steered by the government policy of targeting its funding to specific applications to increase the effectiveness of the associated programs.

The Japanese industry's role in funding robotics R&D has been very similar to that played by industry in other countries, namely to drive research and development efforts in the direction of application-oriented issues. Unlike the industrial efforts elsewhere, however, industry in Japan as a rule does not cooperate closely with research institutions. For example, of a total 1980 industrial R&D budget of almost $16 billion, only about $1.3 billion, or approximately 8%, went to support universities and national research institutions, while the remaining 92% was used for in-house research and development.

Research Directions:

The current government policy regarding robotics research is to target those projects with a potential for private sector commercialization or for removing humans from dangerous environments. Under this guideline, the government keys their activities to the development of certain critical technologies, and funds those research projects heavily, even to the exclusion of other basic research issues. In this manner, the government hopes to realize the greatest gains in a specific technology with the least possible resource input. This is the same strategy that the government applied to the development of digital technology, which formed the basis of the great boom in the Japanese electronic industry. This targeting strategy today is manifest in the form of the two national programs. The overall objective of the first program is to improve the available generic robot technology so that individual industrial companies can modernize and automate their facilities. The goal of the second program, the Jupiter project, is to improve those robot technologies necessary to remove human workers from critical or hazardous environments, such as nuclear, undersea, and rescue situations.

The divergence of the industrial programs from national institution and university programs is evident not only in the funding trends but also in the research areas addressed. For instance, one of the targeted areas of research in the 1982 government project is actuator technology, or "mechatronics". In contrast, analysis of patent information for the past several years reveals that development of actuator technology in private industry is relatively inactive. This segregation of research projects, combined with at least a minimal level of communication between the two types of research groups, leads to a very well rounded and complete research and developmental technology base.

Specific Research Topics:

As mentioned earlier, there are currently two national robotics programs running in parallel in Japan: one began in 1982, and the Jupiter project began in 1983. While the goals of these projects are different, the technologies studied in the projects are similar. The highlighted issues of the first project include the sophistication and miniaturization of vision sensing systems, sophistication of touch sensor technology, and various aspects of actuator technology, or mechatronics. The scope of the Jupiter project is more extensive, incorporating those issues previously targeted and including many more. The following table (see Figure 3-8) is a list of those technological issues targeted by The Japanese government as key barriers to the widespread use of robotics for critical or hazardous work. Over the course of the Jupiter project research in each of these areas will by funded by the government. Numbers in parentheses indicate the number of projects already underway in those areas as of the beginning of the program in 1983. A topic with no number indicates that although the program has targeted that area as important, no projects had been initiated as of the start of the project.

In summary, the Japanese robotics effort can be characterized by the institutional and funding structures within the research community, the overall research directions, and the specific research topics studied. The institutional structure, although similar to that of the U.S., is funded in a segregated manner, with industry supporting in-house applied research almost exclusively, and government targeting funds for specific areas of basic research. These research areas are directed towards the key issues necessary for the advancement of specific goals, such as improving generic robotic technology for manufacturing and industrial needs, or for improving those technologies that will enable robotic applications for hazardous and critical work, as seen in the individual topic areas of research studied under the Jupiter project.

3.2.2. Western Europe

United Kingdom:

The outstanding feature of robotics R&D in the U.K. is that it is driven to a large extent by engineering and application thrusts, rather than by the scientific issues as in the U.S. and Japan. This comes about

Japanese Robotics R&D
(Sheet 1 of 3)

Mechanical:

- Actuator (27 Actuator/Manipulator)

    o Compact AC servo and microservo motors

    o Weight/Output ratio approximately equal to muscles

    o 3 axis actuator

    o High capacity batteries for mobile robots

- Manipulator (27 Actuator/Manipulator)

    o Light, multi-articulated arms incorporating advanced materials

    o Improvement of 3-roll wrist

    o Master/slave manipulator system

- End Effector (2)

    o 3-fingered dextrous hand

    o Force sensored gripper

- Locomotion (30)

    o Mobile robot capable of navigating in complex environment

    o Multi-ped robot capable of climbing stairs, walls, pipes and trenches

- Others (5)

Control:

- Hardware

    o High-speed dedicated processors

    o Parallel processing architectures

```
                        Japanese Robotics R&D
                           (Sheet 2 of 3)


-  Software (59 Language, 9 Path Control)

      o   Hierarchical control algorithms, both within the robot and
          for system integration and task organization

      o   Task specific, skilled algorithms, such as control for
          assembly using gripper force

      o   Acquiring and using a knowledge base

      o   Standard programming language and operating system for 16-bit
          processor environment

      o   Language capable of voice command recognition

      o   Autonomous control for mobile robots, navigation

      o   In-process fault diagnosis


Sensing:

   -  Vision  (15)

      o   Miniaturization of camera system

      o   Fast pattern recognition, goal of <0.1 Sec.

      o   Increased spatial resolution, goal of 4k x 4k element
          semi-conductor

      o   High speed processing system for tracking motion

      o   3-D vision

   -  Tactile

      o   Flexible matrix touch sensor with high spatial resolution

      o   Shear sensor

      o   Force-displacement sensor

   -  Hearing  (5)

      o   Continuous voice recognition of unspecified speaker

      o   Direction of abnormal in-process sounds
```

Japanese Robotics R&D
(Sheet 3 of 3)

- Proximity/Ranging

    o Laser ranging system

- Others (30)

    o Light, compact gyroscope for robot

Application:

- Assembly

- Finishing

Adapted from:

- JIRA, "Report on Research and Development trends by Universities,
  National and Public Institute, etc, regarding Industrial Robots"
  (March 1983); and

- JIRA, "Report of Long-Term Forecast on Technology of Industrial
  Robots" (March 1983)

Figure 3-8:   Robotics R&D Areas of Concentration
in Japan

mainly due to the different character of government funding programs in the U.K. An important characteristic of the U.K. government policy is the push for immediate industrial modernization. There is a willingness on the part of the government to fund industrial modernization efforts that have a short time frame. Specifically, one program currently underway provides up to 33% of the cost of feasability studies, application, and manufacture of robots to interested companies. This attitude extends to university research as well. One of the major university R&D programs in the robotics field was created in 1980 through the Science and Engineering Research Council (SERC). The important aspect of this program is that it is a cooperative effort between government and industry; half of the resources come from the government and the other half from industry. This funding structure, compounded by the fact that there is very little military support to the universities for basic research, creates a heavy dependence on industry. This dependence on industry pulls robotics research in the direction of solving short-term, application-oriented problems, rather than building a base of scientific knowledge from fundamental research. A detailed examination of the focused areas of research for some of the major institutions reveals this application-oriented characteristic.

Of the universities that are active in fundamental robotics research, three have programs that have been sizable and successful in creating a groundwork for robotics research: Edinburgh, Warwick and Oxford.   The work at Edinburgh has concentrated on studying the kinematics and geometry of assembly operations, as well as the development of the robot language RAPT.   This language has since been refined by GEC, Britain's leading robot manufacturer. Although in the last several years some key researchers have left Edinburgh to work in the U.S., leaving the program with few people and a low funding scale, there is still significant theoretical work being done there. At Warwick University the work has concentrated on mobile robots. Warwick hopes to consolidate some SERC funding and establish a nationwide center for research in mobile robots and particularly automated warehousing. The research at Oxford university is more application-oriented than work at Edinburgh and Warwick, concentrating on automating factory processes such as arc welding. Structured light and adaptive control are the highlighted issues there. The structured light system, similar to that designed by Automatix of the U.S. has been successfully demonstrated and is rapidly penetrating industry.   There are a number of universities with smaller, usually single-focus programs, such as Imperial College, working on adaptive and logic control, Liverpool, working on control issues and the University of London, working on vision.

The industrial R&D effort in the U.K., as mentioned earlier, is directed toward solving application-oriented problems. The largest industrial R&D effort comes from GEC. GEC has established two separate in-house research laboratories, the Hirst Research Laboratory, conducting research in 3-D stereo-vision and tactile sensing, and the Great Baddow Laboratory, which is working on adapting RAPT for process planning, as well as a very stiff, accurate robot called GADFLY.   There are several smaller efforts centered in the aerospace industry. Some of these companies are funded heavily with military money to study robotic vision. On a smaller scale, there are many industrial participants performing in-house application specific R&D in robotics. Most of these efforts are at least partially connected with, if not subsidized by, government programs to advance the implementation of robotic technology. As an example of this cooperation between industry and government, one should also include the Production Engineering Research Association (PERA). In addition to being a key performer of industrial robotics R&D, PERA is the most important U.K. supplier of the government-funded Robot Advisory Service. Finally, the National Engineering Laboratory (NEL) in Scotland is a key factor both in performing robotic R&D and disseminating the emerging technology to industry.

In summary, it can be seen that the overall thrust of robotics research in the U.K. is directed toward short-term modernization of manufacturing processes, rather than long-term fundamental research. This is due in general to the overall industrial climate of the U.K. and specifically to the types of funding programs that the government and industry support.

France:

France has recently embarked upon a major R&D effort to upgrade the general level of manufacturing technology. Included in this effort is a three year, $350 million program from 1983 to 1985 to fund robotics research, train robotics specialists and promote the implementation of robotics in industry. Robotics R&D in France then is characterized by a heavy support program from the government, with the aim of building a solid scientific knowledge base. For this reason, government-sponsored research programs tend towards basic research, while industrial support drives the more applied, developmental research. The end result is a very well rounded robotics program.

The robotics research institutions in France consist of three types: government in-house laboratories, industrial in-house laboratories, and university laboratories. Government funding generally flows from government agencies to all three types of laboratories, while industrial funding remains for the most part in-house, with some channeled to university programs. Although this structure appears similar to that of the U.S., there are two notable differences. The first is that there is much less cooperation between industry and universities, in the sense that industry generally expects to get finished products from university testbeds, as opposed to merely ideas and concepts. The second difference is that there is a much stronger emphasis on in-house government research. For example, CNRS (the National Scientific Research Council) operates several major laboratories with an emphasis on robotics, including INRIA (National Research Institute on Computer Science and Control) in Paris, LAAS (Computer Science and Systems Analysis Laboratory) in Toulouse, and IMAG (Computer Science and Applied Mathematics Laboratory of Grenoble) in Grenoble. In addition to CNRS laboratories, there are other in-house government robotics laboratories, such as the one operated by the CEA (Atomic Energy Commission) and the recently established National Robotics Laboratory in Marseilles. CNRS is responsible not only for funding but also for initiating research programs. The thrust of one such program is to advance automated assembly techniques.

The industrial R&D effort in France has centered for the most part around the automotive industry. The largest effort in this field has

```
                          French Robotics R&D

           INSTITUTION                        FOCUSED AREA OF RESEARCH

Universities

    Technical University of        o  Real time vision processing, geometric
       Compiegne                      data-base design, systems integration

    University of Lille            o  Basic research issues


Government In-House Labs

    LIMSI                          o  Manipulation for assembly, visual
                                      inspection, work cell integration

    INRIA                          o  Perception: Laser illumination for
                                      assembly & inspection, 3-D vision,
                                      obstacle avoidance

    LAAS                           o  Sensors and sensor data processing,
                                      systems integration:  control for
                                      assembly, perception, planning ARA
                                      Project, mobile robot

    IMAG                           o  Robot programming tools, (developed
                                      language LM) automatic assembly,
                                      expert manufacturing planning,
                                      vision: gray scale & 3-D using laser

    CGA                            o  Inspection of nuclear facilities

    LAM                            o  Modelling and control of manipulators,
                                      simple vision, coordination of
                                      multiple robots

    DERA                           o  Control systems; flexible automation
                                      and robots for space systems

    National Robotics Lab          o  Various applied research issues


Industry Labs

    Renault                        o  Industrial robot research vision
                                      for inspection, controllers, process
                                      planning

    MATRA                          o  fast vision module
                                   o  assembly robots




            Figure 3-9:  Robotics R&D Directions in France
```

been on the part of Renault. Renault began building robots in-house for its automotive assembly line, and continues to design robots and controllers for other users in industry. One characteristic of Renault's efforts has been a close collaboration with the government-run INRIA laboratory. With regard to universities, the Technical University of Compiegne and the University of Lille represent the two largest robotics programs in French universities.

Figure 3-9 summarizes the key research areas sponsored by the French R&D community on robotics. From the detailed picture of research topical areas, it can be seen that the thrust of robotics research in France has been to build a solid base of fundamental research. This research is performed within a climate of heavy government support in an overall effort to upgrade the manufacturing technology base of the country.

West Germany:

Unlike robotics efforts in other countries, robotics research in West Germany is much more centralized. At the center of the German robotics effort, performing the majority of the research and development, are the Fraunhofer Gesellschaft Institutes. The Fraunhofer Institutes are actually a series of twenty-six individual not-for-profit research institutions, funded one third by general government block grants, one third from industries, and one third from specific government contracts. Three of the institutes which are very active in robotics work are the IITB in Karlsruhe, the IPA in Stuttgart, and the IPA in Berlin. Several universities worth noting here because of the large size of their robotics programs are Karlsruhe University, the University of Aachen and the University of Stuttgart. In the industrial sector, there are a large number of robot producers, such as KUKA and Volkswagen, that have substantial programs of application- and production-oriented robotics research.

The IITB in Karlsruhe is currently working on three robotics projects. The first project involves using structured light to guide robotic arc welding. The research carried out has been a forerunner to the vision work done at Oxford University. The second project at the IITB is called the "very advanced industrial robot". Although the name is reminiscent of the Japanese "fifth generation computer" project, the IITB project involves merely a multi-sensored robot. The emphasis of the project is modularity and sensor-based control through the use of a high-level language. The third project is focused on machine vision. The novel aspect of this project is that its thrust is not to improve software but to produce a hardware-intensive, fast, marketable vision module.

There are approximately 30 people working in various robotics projects at the IPA in Stuttgart. One of the larger projects at the IPA has been the development of a robot measuring station. The station is designed to assess various functional capabilities of new robots. Another large robotics project at the IPA involves the coordination of a flexible manufacturing cell. While in the past the cell has had few sensors, current work involves adapting various kinds of vision and tactile sensors for use in the cell.

The robotics program at the IPA in Berlin is also large, consisting of many different robotics projects. Three of the major efforts are devoted to controllers, modeling, and adaptive sensor control. The controller project involves designing a controller for the German-produced KUKA robot. This controller is also used by the Daimler-Benz automobile company to do highly accurate and difficult to reach spot welding tasks. The modeling project consists of the development of COMPAC, a package for 3-D surface modeling. The sensor-based control project emphasizes arc welding. The research issue there is how to use magnetic sensing of the arc parameters to guide the robot arc welding gun.

The three largest university robotics efforts in West Germany reside at Karlsruhe University, the University of Aachen and the University of Stuttgart. At Karlsruhe University, the robotics program has employed as many as 25 people. The research issues under study include several different types of vision systems, a portable robot programming system, and a highly instrumented robot gripper. Robotics research at the University of Aachen is divided into two directions. One direction is in robot language development, similar to the RAPT development effort at Edinburgh and GEC. The second direction has been the development of a modular robot. The aim of the project is to develop a robot constructed of standard, modular parts, i.e. interchangable actuators and linkages, as well as the software to control it. They have in fact marketed a working version of this modular robot. The third sizable university robotics effort is at the University of Stuttgart. The Stuttgart program is centered on sensor-based control of manufacturing processes, specifically in welding and grinding operations.

West German industrial R&D is driven by in-house application problems. The best example of this is found at Volkswagen. Volkswagen started by using imported robots in their automobile assembly plant. As the need for application-specific developmental R&D rose, Volkswagen stopped modifying foreign robots and began producing their own robots. This effort has grown significantly, and now Volkswagen is marketing several different lines of robots worldwide.

One additional point should be mentioned concerning the institutional structure of robotics R&D in West Germany. Each of the three Fraunhofer institutes mentioned is closely associated with a university in its respective city. In fact, the university professors who are responsible for the robotics programs at the universities are also directors of the institutes. This powerful link provides for rapid and effective diffusion of emerging technologies into industry.

In summary, the overall thrust of West German robotics R&D is very application-oriented, much more so than, for example, the French effort. This is partially due to the funding structure of the German robotics R&D. In a situation similar to that of the U.K., the research institutions in West Germany are heavily dependent on industry and short-term government contracts for their support. This has pulled the research more towards the development of robotic applications.

Other Western European Countries:

Robotics R&D efforts throughout the remainder of Western Europe take the form of scattered programs, with no cohesive structure or overall research goals. In these countries there are one or two isolated research efforts, but little or no evidence of country wide policies or programs.

After France, West Germany and the U.K., the next largest robotics R&D program is in Sweden. Robotics research in Sweden is focused on industrial development, with the largest program conducted by ASEA, a manufacturer of one of the most accurate robots built. The majority of ASEA's research is dedicated to application oriented programming. In addition, there are several trade institutions, such as a welding institute, that study the application of robotics relevant to their field of interest. The above-mentioned welding institute is currently studying the use of vision to monitor the bead to control arc welding processes.

The character of the robotics R&D effort in Norway is similar to that in Sweden. Trallfa, the leading manufacturer of spray painting robots worldwide, conducts in-house research and development, as well as cooperates closely with research institutions such as the Central Institute for Industrial Research, the National Institute of Technology, the Roaglund Research Foundation, and the Christian Michelsen Institute. The robotics research issues highlighted in these programs are mainly in control applications.

The robotic research in Italy follows a similar pattern to that in Sweden and Norway. Several medium-sized robot manufacturers and end-users, Olivetti, DEA, and Fiat specifically, fund their own in-house developmental research and cooperate with university research institutions, such as the Milan Polytechnic Institute. Work at the Milan Polytechnic, the oldest and largest robotics research program in Italy, includes natural language understanding, automated problem solving, and sensor-based control. Sharing the resources of the Milan Polytechnic has been the Laboratorio per Ricerche de Dinamica dei Sistemi e di Bioingegneria (LADSEB), a national institute for bioengineering and systems dynamics research. LADSEB has been working on robot programming languages, geometric modeling and robot actuator control.

The Belgian robotics R&D effort is typical of the smaller robotics programs in Western Europe. There is very little industrial involvement, as the robotics industry is struggling to overcome pressures from imported technology, and has little capital to support university programs. The government supports robotic research and development, but at a low and uncertain level. The one very active research institution is the Katholieke Universiteit Leuven, where current robotics projects include force sensing, active compliance, sensor-based control and programming language development.

Other Western European countries in which there is scattered work in robotics research include The Netherlands, Switzerland and Finland. The Dutch government supports the robotic industry as a whole with approximately $5.5 million in the form of incentives for industry to cooperate closely with research institutions, robotics education programs, and subsidies and loans to stimulate pilot demonstrations of robotics and flexible manufacturing. Robotics R&D in Switzerland is for the most part performed at two technical institutes, the Zurich Federal Institute of Technology and the Lausanne Ecole Polytechnique Federale. Funding for these programs, which are directed largely towards basic research issues, comes from a close cooperation with industry, as well as from limited government funds. It should be noted that research and development in Finland is not funded by the government, nor does the government engage in administrative practices for the industry's protection.

3.2.3. Soviet Bloc

The Eastern European nations have been involved in robotics research for over 20 years, with the first industrial robot being produced in 1971. The reasons for their interest in robots are similar to those of the West, namely the problems of labor shortages, training requirements, dangerous and monotonous jobs and the need for higher quality products at reduced costs.

This section presents a sample of the institutions involved in robotics R&D, with selected highlights of the research conducted at each institution. Additionally, when possible the areas of future R&D efforts that individual countries can be expected to follow will be included. Before describing the efforts undertaken by specific countries, it should be pointed out that, with the exception of Yugoslavia, all of the countries to be mentioned belong to the Council for Economic Mutual Assistance (CEMA), which coordinates joint R&D efforts to unify the design of robots. These efforts are carried out through the Experimental Machine Tool Research Institute of CEMA.

Although the Eastern European countries have been involved in robotics for some time, at present the CEMA members are approximately a decade behind the West in robot technology. This is due in a large part to lack of digital microprocessors, limiting the capabilities of commercialized robots largely to pick and place operations. The shortage of computers also affects research in such a way that the efforts are confined to theoretical research. This problem of lack of computers is not as acute in Yugoslavia, which has a closer relationship with the West. The Yugoslav robots are, with one exception, controlled by microprocessors including the Intel 8080 microprocessor. Due to their advantage in computing power they are considered a leader in Eastern European robotics.

The deficiency in computing power of the CEMA nations is expected to be significantly alleviated in 5 to 7 years. By that time the Soviet electronics industry would be capable of manufacturing precision digital electronics. This improvement will allow the production of more complex adaptive and artificial intelligence control systems, thus greatly increasing robotic capabilities. It will also presumably drive their previously theoretical research into a more practical direction.

Soviet Union:

Institutions:

Research in Soviet robotics is coordinated by the Council on the Theory and Design of Robots and Manipulator Devices. The Leningrad Polytechnical Institute's Special Design Bureau of Technical Cybernetics has been designated as the leading institution for research and devlopment of robots. The Bureau oversees over 50 research institutions and manufacturers in the Leningrad area in robotics R&D. Members include the Pozitron Production Corporation, the Optical Design and Precision Mechanics Bureau, the Leningrad Institute of Aviation Instrument Building, the Electrotechnical Institute and the Refrigeration Industry.

There are several leading academic institutions in Soviet robotics research other than the Leningrad Polytechnical Institute (LPI). These include the University of Kiev's Institute of Cybernetics, the University of Moscow's Institute of Applied Mathematics, and the Moscow State University's Institute of Mechanics.

The education of robotics specialists was made the responsibility of the Ministry of Higher Education and the State Professional Educational Authority. The first universities to teach a robotics engineering curriculum were the Bauman Technical Institute in Moscow and the Leningrad Mechanical Institute. Presently, most major engineering schools offer courses in robotic technology.

R&D Focuses:

Soviet robotics R&D currently involves a wide spectrum of research areas. The general topics under consideration are robot control, sensing, mechanical structures and applications. R&D is on-going in all of these areas, but it is limited by a lack of computers and integration capacity. At present the Soviet electronics industry is not capable of producing precision electronics. There are a limited number of 8-bit microprocessors, mostly copies of American chips, but the Soviets lack the ability to program them effectively.

Because the robot controller depends so heavily on electronic computing power, this area has been most affected by the lag in Soviet digital technology. As a result of this, research in the area of control has been for the most part on a very theoretical level. Examples of such work are in the development of mathematical systems to aid in programming control systems, and in the verification of various mathematical theorems in AI research at the University of Kiev. Another topic of theoretical research influenced by the lack of computers is a proposed algorithm which tracks and approximates the contours of objects without computers but rather with logic conditions.

Although most control system research is conducted on a theoretical basis, some experimental R&D is performed. One such example is a robot at LPI which uses digital control. It is operated by two computers, an ASVT M600 and a Minsk 32, with a five-level hierarchical control system.

The robot has demonstrated the ability to grasp irregularly-shaped objects, negotiate obstacles and assemble parts. Other robots with advanced control systems include a three-legged walking robot at the Computer Center of the Academy of Sciences and the OSU Hexapod, a six-legged walking robot developed at the Moscow State University. This robot is being developed for the timber industry with the help of Dr. McGhee of the Ohio State University.

The institute most active in robotic sensor research has been LPI. A variety of sensors have been developed at LPI which include TV vision, laser vision, ultrasonic sensing, tactile sensing, force sensing, and hearing. An example of a robot at LPI employing sensory capabilities is LPI-2, which has TV vision and an ultrasonic locator for finding objects, as well as two grippers with force and tactile sensors. Another LPI robot is a TSIKLON-3B robot with hearing capabilities. It can respond to 200 spoken commands such as, "open gripper" and "rotate waist". This capacity to respond to spoken commands represents a big step for the Soviets in overcoming their lack of programming ability.

In addition to LPI, vision research is conducted at the Leningrad Institute of Aviation Instrument Building in visual identification for sorting of objects on a moving conveyor belt. Researchers at the University of Moscow are conducting research in the theory of image identification. Additionally, the University of Kiev has been a leader in Soviet vision sensing research.

The Soviet research in mechanical systems is primarily focused in the areas of actuators, grippers and modularization. Because electric drives are the most common type of actuator in the Soviet Union, research has concentrated on their refinement and improvement. Harmonic drives are under study as well, presumably because of minimal weight and size and a self-locking characteristic preventing unwanted joint movements.

Gripper development is an active field of research at LPI. An example of current work is the development of soft grippers, which are capable of handling sensitive objects such as light bulbs. Another gripper effort involves an electromagnetic gripper. This gripper has reduced search and pick-up time for a particular experiment from 702 sec. to 40 sec. by "grasping" an object in a closely packed container with an electromagnet instead of a conventional gripper.

Development efforts are being made in robot modularization. The modular approach is being pursued by manufacturers to allow several robots to be constructed from standard module parts. This approach allows, for example, over 100 arms to be made from 16 different modules.

Research in industrial applications is concerned primarily with the development of standardized flexible manufacturing facilities. The goal is to achieve high flexibility and automation in manufacturing processes. Again, the institution which appears to be involved most heavily in application R&D is LPI.

Funding information on Soviet robotics R&D was not available; however, an estimated measure of the level of effort can be determined from the number of research institutions and employed researchers. From an estimate by Kent Schlussel of the U.S. Army Foreign Science and Technology Center, there are approximately 60 research institutes active in robotics research. The largest of these is LPI, with about 400 researchers. Following LPI are the University of Kiev and Moscow State University, each with approximately 200 researchers.

Future Directions:

Perhaps the best prediction for future Soviet robotics R&D comes from the Deputy Chief Designer of the State Committee on Science and Technology, P.N. Belyanin. He indicates that due to the increase in the availability of microprocessors Soviet robotics will utilize adaptive control to a greater extent and will develop single computers for control of several robots. This advance can be expected some time after the Soviet electronics industry develops the capability to produce reliable microprocessors, which is expected in 5 to 7 years. It can be assumed that with greater adaptive control capabilities robots will be assigned more complex tasks, such as assembly. Further, with improved computer availability robots like the LPI-2 with artificial intelligence may reach commercial production. Other expected improvements include increased modularization of robots, increased speed, and durability and reliability with reduced size and costs.

Bulgaria:

Institutions:

Bulgaria is the research coordinator for robotics and applied aspects of automatic machine theory within the Experimental Machine Tool Research Institute (ENIMS) of CEMA.

The producers of Bulgarian robots are the Beroe and Gidrazlika combines, the Sofia Machine Tool Institute, the Plovdiv Technical Design Institute, the Bulgarian Academy of Sciences and the Robotics Research Center of the Sofia Higher Engineering Institute. Additionally, the American firm Versatran collaborates with Bulgarian domestic producers in the production of several robots.

R&D Focuses:

As with other Soviet Bloc countries information about Bulgarian robotics research areas and trends is difficult to obtain. The information available, however, indicates that Bulgarian R&D is mainly concentrated in the area of industrial applications, such as painting and loading/unloading. Additionally, there is some control and AI research carried out at the Robotics Research Center of the Sofia Higher Engineering Institute, while the Beroe combine is the center for the development and commercial production of robots.

Czechoslovakia:

One of the best gauges of robotics research conducted in Czechoslovakia comes from an examination of the Czech robots displayed at the Third International Exhibition of Industrial Robots in Brno, Czechoslovakia. At this exhibition, five Czech robots were displayed, three of which were joint Soviet-Czech developments. All had hydraulic drives, NC control and modular design, hence it can be assumed that Czech R&D, with the apparent help of the Soviets, is at this stage.

Future efforts in Czechoslovakia will be devoted to adaptively controlled robots for assembly, finishing and other applications requiring high speeds, accuracy and reliability. Also, parts transfer robots will be developed for in-process handling of parts up to 160 kg.

East Germany:

Robotics research and development in East Germany is conducted for the most part at the Dresden Technical University's Production Engineering Department, the Cybernetics and Information Institute of the Academy of Sciences and the Fritz Heckert Machine Tool Combine in Karl-Marx-Stradt.

The information available on East German robotics R&D indicates that present research involves the areas of flexible assembly processes for small to medium volume production, tactile sensors and control systems. The research in assembly processes is conducted at the Dresden Technical University and at the Fritz Heckert Machine Tool Combine. Each center has constructed assembly cells for process simulation. The Cybernetics and Information Institute conducts research in tactile sensing and control systems.

Hungary:

Institutions:

Although Hungary does not produce any robot at this time, and conducts little research in the field, an infrastructure exists for Hungarian R&D. Specifically, the Professional Council on Robotics and the Ministry of Industry coordinate development projects, while coordination of robotics applications is the responsibility of Technical Institute of the Machine Industry. There are presently two research centers in Hungary active in robotics research, the Computer and Automation Institute of the Hungarian Academy of Sciences and the Czepel Machine Tool Factory.

R&D Focuses:

The two institutions currently performing robotics research are involved primarily in industrial applications. The work done at the Czepel Machine Tool Factory is conducted with Bulgarian robots coupled with servicing high-precision NC lathes. The development of domestic robots is the responsibility of Microelectronics Enterprise, which is preparing to produce simple measuring and testing robots. For this goal, $1.8 million in government aid and an equal amount from Microelectronics Enterprise have been dedicated. Additionally, the Computer and Automation Institute plays a key role not only in Hungary but also throughout the CEMA nations. For example, most of the software for Bulgarian robots was and still is developed at this institute, which is the focal point for coordination of software compatability throughout CEMA. Finally, the Institute is conducting significant R&D on the application of artificial intelligence in robotics.

Future Directions:

The principle goal of Hungarian robotics R&D is to begin production of simple robots starting in 1985. However, Andras Gabar, Deputy Minister of Industry, indicated that the long term direction of robotic development in Hungary will not be in domestic production of complete robots but rather in joint production with other CEMA nations, in particular, Bulgaria and East Germany.

Poland:

Institutions:

In addition to participating in robotics R&D with other CEMA nations, Poland has its own robotics program involving adademic and industrial research centers. Presently, there are two research centers in Poland for robotics. These are the Institute for Biocybernetics and Biomedical Engineering of the Polish Academy of Sciences and the Technical University of Warsaw. At the University of Warsaw there are three separate institutes conducting robotics research.

Industrial research centers have been under the supervision of the Ministry of the Machine Industry since 1970. Those industrial research facilities active in robotics R&D are the Institute of Precision Mechanics, the Machine Tool Research and Construction Center, the Machine Technology and Construction Basic Research and Development Center and the Industrial Institute of Automation and Measurements.

R&D Focuses and Future Directions:

In the near future, specialized robots with few degrees of freedom will be developed; however, in the longer time frame, more versatile modular robots are expected in Poland. One of the areas of future Polish R&D, besides modularization, will be in control systems utilizing microprocessors and simplified programs. Additional efforts are expected to improve accuracy, reliability, arm speed and load capacity.

Yugoslavia:

Institutions:

Yugoslavia, because it is not a member of CEMA, conducts its own independent robotics R&D programs. The predominant robotics research center in Yugoslavia is the Robotics Department of the Mihailo Pupin Institute. Other institutions active in robotics research include the Joszef Stefan Institute, the Factory of Hydraulics and Pneumatics, and the Factory of Domestic Equipment.

R&D Focuses:

In the past, Yugoslavian R&D has concentrated on the development of multi-degree of freedom, articulated manipulator systems. These studies have proven fruitful, as evidenced by the number of complex manipulators produced and in use in Yugoslavia. These manipulators are of varying design, using electric, pneumatic, and hydraulic actuators. Presently, research efforts are focused on the integration of microprocessor control with these manipulator systems. Through the import of foreign hardware, such as the Intel 8080 microprocessor, Yugoslavian R&D efforts have been able to concentrate on efficient programming, teaching and control methods for the robots.

3.3. Trends in Robotics R&D

In looking at the vast amount of information concerning research in the field of robotics, one sees that the world-wide R&D effort is not merely a collection of individual, undirected researchers studying various aspects of robotic technology, but that there are several key, unifying aspects of the research effort. With respect to the overall picture, it is apparent that, with essentially only one exception, every country active in robotics research has, to one degree or another, a national direction or program. These programs range from the more coherent, such as Japan's series of national projects or France's country-wide effort to build a manufacturing technology base, to the less structured but still prominent directions, such as the U.K.'s national emphasis on the solution of short-term manufacturing problems. The importance of these programs is that they are the force which directs the thrust of the research efforts. The notable exception to this trend is the United States. Although the amount of robotics research conducted in the U.S. probably exceeds that in any other country, there is no coherent national climate or program directing this work.

A closer study of robotics R&D in the U.S reveals that, although there is no unified national direction, the individual research funding sources such as the Air Force, the Army, the Navy and other government agencies have their own robotics R&D program goals. These funding agencies, together with private industry, are the forces which determine the direction in which robotics R&D will go. The Air Force, for example, has as the thrust of its robotics R&D program the advancement of aerospace manufacturing technology. This goal is manifest in several different R&D efforts. The first of these efforts, the Advanced Robotic Systems for Aerospace Batch Manufacturing project conducted under the MANTECH program, focuses on off-line programming with CAD/CAM links, drilling/trimming control, and drilling/riveting control, while the second effort, the Intelligent Task Automation (ITA) project conducted under the MANSCIENCE program, focuses on the study of various aspects of robotic assembly, including sensor and control issues.

In contrast to the Air Force, both Navy and Army R&D programs focus on improving the current capabilities and meeting the short-term needs of each of the services. Maintenance and support are key factors in these programs. Both the Navy's NAVAIR-sponsored wing de-riveter and the Army's DESCOM-sponsored vehicle maintenance project are examples of the emphasis on maintenance. In the area of support, the Army is conducting several studies on battlefield robotics, such as weapons loading/unloading, while the Navy is concentrating on autonomous robots and navigation for undersea work and robotic sentries.

Three other government agencies active in funding robotics research, NASA, NBS, and NSF, each have their own motivations and particular directions. NASA, pursuant to its program goals, is active in funding research in teleoperation and remote sensing in an effort to develop robots for use on a space station. Similarly, NBS is active in performing robotics research in the areas of standardization and system integration. The Automated Manufacturing Research Facility of NBS serves as a testbed for new developments in interface capabilities and system standardization. Additionally, NSF funds, in accordance with its mission, research projects in fundamental issues of robotic technology.

As a final comment on the federal driving forces in U.S. robotics research, it should be noted that, due to the funding policies of the NSF, there is a fair amount of very basic, undirected research conducted at the university level. Occasionally, this research will produce technological progress that in turn will push further research. One of the first examples of this in the robotics field is that of robot vision. Two-dimensional vision capabilities adequate for such applications as simple part identification and inspection were developed in laboratories before there was a significant need for them in industry. As industry slowly began to take advantage of the capabilities, a new push for the refinement and enhancement of these capabilities began.

Industrial robotics R&D has, in general, taken a different direction than that of federally funded R&D. This is due largely to the structure of the robotics industry. Because the robotics industry is in its adolescence, it includes many smaller, very competitive companies. Robot producing companies such as Automatix and Adept Technology have been very active in targeting a particular market, such as vision or arc welding, and focusing their R&D program in a direction to secure that market. Additionally, robot end-users have targeted their R&D programs on application developments that will increase their productivity.

In summary, one can see that there is a vast amount of robotics research being conducted in many different aspects of the field, both within the U.S. and abroad. It is apparent, however, that there are several key research topics that are receiving the greatest effort, both in terms of number of projects and in funding amounts. These topics are presented here as a brief overview of the direction of robotics R&D efforts.

Mechanical:

    o standardized and quick-change grippers
    o sensored grippers

Control:

    o hardware architecture
    o sensory integration
    o hierarchical control
    o modeling/simulation/emulation
    o high level programming languages

Sensory:

    o processing and interpretation of visual images
    o tactile sensing arrays
    o speech understanding

    Although this list shows where the bulk of the current R&D efforts
are directed, it is not necessarily comprehensive in the sense that it
excludes the category of application-specific development. With the exception
of the NSF funded research, almost all of the robotics research performed
in the U.S. is driven to some extent by application need. When considering
the driving forces behind robotics research, one sees that there are several
research topics that are not currently stressed but that might yield substantial pay-offs in terms of increased effectiveness and productivity.
These areas include control and structure of compliant manipulators, as
well as robust fault tolerance and error recovery algorithms. Additionally,
one can see that there is a large amount of effort directed in the field
of speech recognition. It is currently under debate whether this capability
will be of significant use to industry in the foreseeable future.

4. A Technological Forecast of Robotics

This report has, so far, concentrated on defining the current status of industrial robotics. On the basis of this status, the present chapter describes the development paths that robotics will take in the future. After a brief description of the methodology used to develop the forecasts in this chapter, section 4.2 discusses each component of a robotic system in terms of its current status, developmental needs, approaches to future development, and expected short and long-term results. The last part of section 4.2 takes a broader view and discuss integration of robotic components and integration of the robot as a whole with surrounding equipment. Section 4.3 separates robotic applications into three categories, Low Growth, High Growth, and Blue Sky, according to the effects of emerging technology on each appliation. The chapter closes, in section 4.4, with the general trends in robotics, with respect to both technology and the robotics industry.

4.1 Methodology

Forecasting technological developments in the field of industrial robotics is difficult because of the rapid change and growth that typifies an adolescent technology. Informed forecasting requires a thorough knowledge of the robotics R&D community and an understanding of the point of view of robot users in industry.

Examination of the robotics R&D community began with an assessment of the technology being worked on in the laboratory. Extensive reading of the recognized journals in the field provided a starting point by indicating the major areas of research activity and identifying many of the key research groups. Attendance at meetings and conferences provided more opportunities to assess current research work and informal discussions with researchers in attendance helped to develop a sense of how the robotics R&D structure operates. Additionally, extensive discussion with the research oriented members of our expert panel provided background information on some of the important research groups, critical assessments of major projects and indications of the directions in which the research groups would like to go.

However, there is another key element involved in making predictions on research: funding. Examination of robotics research included a major effort to identify the funding structure that supports the research. There is very little undirected research funding today; the organizations paying the bills generally have specific problems or at least areas that they want addressed. The funding structure was analyzed not only to locate the major sources and their goals, but also to project future funding levels to allow tentative prediction of the R&D situation in the future.

Information on the robot user point of view was acquired largely from trade journals and interviews with users of industrial robots. These sources provided concrete information on the status of industrial robots in use today and the developmental needs of end-users. This information was supplemented by discussions with the expert panel which included robot users. In addition, the panel discussions illuminated the role of in-house R&D performed by robot end-users in enhancing current implementation and solving detail problems not considered by other researchers.

4.2. Capability Projection

4.2.1. Mechanical

Manipulator

Current:

Most of today's robot arms are clumsy and slow and generally achieve rigidity (as required for current methods of control) by means of brute strength, i.e., massive components.

Needs:

Greater speed
Better flexibility
Better absolute accuracy
Better agility
Better efficiency

New Directions and Approaches:

Composite materials and more use of tubular cross-section components can achieve rigid but lightweight structures.

Parallel linkages rather than serial can improve load capacity, and use antagonistic drive to improve precision.

New types of bearings such as air bearings and ion implanted surfaces can improve joint performance.

Small, lightweight and precise robots (like the SCARA) can be more suitable for many tasks.

Short Term:

Rigid but lightweight arm structures will become available, with better payload/robot weight ratios than conventional arm structures. Improved joint and bearing designs will result in reduced friction and stiction, improving precision. Small precise robots for tasks such as assembly will become an increasingly larger part of the robotic population.

Snakelike manipulators with many degrees of freedom as have been demonstrated for nuclear plant inspection, will come into use primarily for applications where

agility is more important than speed or load handling
capacity.

Long Term:

As control problems associated with non-rigidness
are solved, robots with light flexible arms will become
common.

Non-discrete joints will become available on some
industrial robots where agility is important.

Arm development may diverge into two major families:
    o flexible arms
    o parallel linkage arms
with flexible arms dominating light load applications and parallel
linkage arms dominating heavy load and high precision applications.

Actuator

Current:

Three types of actuators are currently used in industrial
robots, each with some shortcomings:

    o pneumatic - soft operation, difficult to control.
    o hydraulic - messy; precision mechanical
       components subject to disruption by impurities
       in fluid supply.
    o conventional electric - low power to weight
       ratio, adds a lot of weight at the joint,
       backlash prone, not stiff under load due to
       reduction gearing.

No currently available actuators incorporate intelligence
or control at the actuator to modify actuator response,
one promising approach to equalizing arm kinematics
over its range of motion.

Needs:

Future actuators will need better efficiency; current
arms can typically lift about one tenth of their own
weight, with actuator power being one of the major
limitations. Less backlash and better stiffness under
load will be necessary.

New Directions and Approaches:

Development of direct drive electric actuators is
an active area, and addresses many of the problems
with conventional electric actuators.

Development of improved pneumatic actuators can result
in better control, allowing advantage to be taken

of the desirable aspect of pneumatics, e.g., easy availability of compressed air, high strength to weight ratio, clean operation. Miniaturization while retaining efficiency will be important as interest in micro-manipulation increases.

Modularity of actuators can improve speed of maintenance and speed up design considerably. Moving the actuator off the arm and just transmitting power to the joint can improve arm efficiency.

Incorporation of a dedicated processor for each joint/actuator can allow equalization of the joint kinematics over the entire range of motion.

Short Term:

Improved electric actuators, e.g., rare earth magnets and direct drive, will improve power/weight ratios and precision by eliminating reduction gearing backlash and play.

Some initial versions of tendon drive will appear, although the difficulty of transmitting high torque is likely to restrict this to low power joints such as dexterous hands.

Long Term:

Eventually, reliable tendon drives with high torque transmission capability will become available, used not just for hand drive, but for some of the arm joints.

Distributed actuators, i.e., muscle type with power developed over a volume instead of a line or plane, will be available. The actuator as an integral part of the arm with motion achieved by flexing along the arm length rather than pivoting about a fixed joint will appear for special applications.

End Effector

Current:

Today's common end effectors are crude and inflexible with respect to tasks.
Very few of those in use have any sensing at all, and those that do are limited largely to simple binary tactile sensing.

Because of their lack of versatility, end effectors

generally have to be custom designed and fabricated by the user for each application. Due to the lack of standardization, there is practically no interchangeability among today's end effectors.

Needs:

Versatility is essential, i.e., either end effectors that can handle a wide variety of shapes and sizes or quick change capability allowing the appropriate end effector to be mounted simply and quickly.

Miniaturization, or at least reduced bulk, is needed to reduce interference with surrounding objects during task performance.

Real time sensing at the end effector for adaptive control is needed for many delicate or critical tasks.

New Directions and Approaches:

Standardization of end effectors by performance and interfacing is being actively pursued. Distributed processing for end effector mounted sensors has appeared in the laboratory, but needs much more development work. Varying approaches to dexterity are under development, including but not limited to articulated hands, which can provide a flexibility in task performance well beyond that of any simple gripper.

Short Term:

Quick change capability based on some level of standardization of interface can be reasonably expected in the near term.
Some local sensing at the hand, such as use of an ultrasonic proximity sensor, will be available as a standard part of many off-the-shelf grippers.

Small, coarse tactile arrays will be commercially available but with limited sophistication of processing. Mounting of video cameras on the end effector for part location and identification will become common.

Long Term:

A true general purpose hand, with high resolution force sensing "skin", will become commercially available, providing in a single end effector the capabilities needed to perform the vast majority of applications.

Mobility Mechanisms

This section discusses only the mechanical aspects of locomotion.

Current:

Rail and gantry systems currently provide some mobility for industrial robots, but they are not flexible. They provide only extended reach along one or two fixed axes.

Wheeled systems, while useful in some applications, are restricted to highly structured environments, i.e., smooth floors and a known smooth path. Even then, their ability to precisely locate a tool (or even themselves) is poor.

Needs:

Even operating in an indoor environment, improvements are needed. More precise positioning and repeatability is necessary and whatever drive mechanism is used should be able to traverse a factory floor with moderate level of litter. Once at a destination, if the robot is to be used to perform manipulation, the drive mechanism must be stable enough to act as a fixed base for the robot's manipulator. In addition, mobile robots that mount a manipulator require much better energy efficiency in the manipulator and in their power source.

New Directions and Approaches:

Establishing robot location by sensing fixed beacons rather than using wheel rotation sensors can improve precision.

Legged locomotion systems are under active investigation, but the mechanical and control complexities need a great deal of work.

Much of the development work on mobility systems is aimed at producing teleoperated devices, but this work is directly applicable to mobile robots also.

A number of novel approaches to locomotion are being examined, including hybrid systems that use wheels when suitable and arms to lift or pull the robot over obstacles when needed.

Short Term:

In the near term, mobile robots riding on wheels with some type of supervision will be able to surmount minor irregularities in floors and modest amounts of rubbish without losing orientation or position information. Mechanical registration techniques will be used to allow a mobile robot to position itself at a work location with precision comparable to that of a fixed robot making a mobile manipulator-equipped robot usable for precision work.

Inspection and simple maintenance required in high hazard areas, such as nuclear power plants, will be performed by robots, with teleoperation capability allowing human supervision.

Long Term:

For highly unstructured environments such as construction sites, active tracked suspensions and legged systems will become available. Robots that can climb by gripping and pulling themselves will be available for work on scaffolds and in outer space.

4.2.2 Control

Current:

Most controllers in use in industrial robots today are primitive by current technological standards. Most operate in an open loop mode, and those that include sensing generally do it in a crude way. Early controllers, due to their limited capacity and versatility were suited to non-sophisticated applications; these are the applications that today show the greatest robotic penetration. In turn, the successful use of simple controllers in these applications has tended to de-emphasize the need for more sophisticated controllers in the minds of many robot users.

Needs:

Controllers need to incorporate more of the currently available computer technology in order to:

o better integrate sensory data, at much higher speed;

o have greater capacity to handle complex control problems, such as 6 degree-of-freedom arms with optional path planning and adaptive motion; access data bases;

o utilize off-line programming techniques; communicate with other machinery and computers.

New Directions and Approaches:

Much of the development work on controllers is aimed towards treating the controller as a computer, and using many of the methods developed to improve the efficiency of computer systems.

o Distributed processing
    The advantages of this are more than simply
    load sharing; a satellite processor can
    be designed for optimal performance for
    its specific tasks, without requiring general
    purpose capabilities. Furthermore, the
    bandwidth required can be greatly reduced
    because the information being conveyed to
    the controller can be a sensing result,
    not all of the sensory data. The development
    of dedicated special purpose processors
    is a very active area of R&D.

o Networking
    The way in which various processing modules
    are logically connected can have an effect
    on efficiency of the coordinated effort.
    Additionally, the controller should be
    able to accept information and commands
    from above, i.e., a supervisory computer
    that could incorporate expert systems, AI,
    etc.

o Software hierarchy
    Software is expected to perform a wide range
    of tasks, and the level at which tasks are
    ideally performed is not constant. The
    development of operating systems for robot
    controllers will allow task-appropriate
    access to the computer, from machine language
    I/O routines and housekeeping modules, canned
    and ready to run, up to high level language
    compilers and interpreters that allow the
    robot to be programmed in an easy-to-use
    language, with the results automatically
    converted to efficient execution code prior
    to use.

Near Term:
    Distributed processing, downward from the controller,
    using dedicated processors connected to sensing
    systems is likely to become available with vision
    systems containing outboard processing for image
    processing and pattern recognition.

    When suitable tactile sensor arrays are developed,
    much of the distributed processing technology
    for vision may be transferrable to tactile sensing.

    Controller processing will become faster and
    more tolerant of errors as a result of improved

hardware and more sophisticated software. More complex path control will become available, including limited dynamic accommodations, and some optimization of arm trajectory.

Controllers will have software operating systems to handle the housekeeping of distributed processing and to support compilers for high level language programming. Offline programming will be common on sophisticated industrial robots.

Long Term:

Controllers will tie into local area networks to communicate with surrounding machinery such as parts presenters.

Controllers will cease to be directly programmed by humans. AI systems, working from CAM produced information, will use graphics and expert systems, with human supervision, to develop the programs needed by the robot on the plant floor. These programs will be downloaded directly to the controller via an integrated communication network.

As a result of increasing integration, the controller will lose much of its identity as it becomes simply one link in a processing hierarchy, extending from a VLSI chip on the back of a tactile sensor to the top level supervisory computer that oversees the operations of the entire plant.

4.2.3 Sensing:

Vision

Current:

Today's systems are too slow in processing visual data to produce results from vision sensing in real-time and too expensive for many users. Resolution is poor, requiring very prominent features for recognition. Software is neither well developed nor efficient, while depth mapping for 3D is very slow. Lack of standardization makes it difficult to interface vision systems, and standardization is hampered by a lack of consensus on what type of information should be communicated.

Needs:

Vision systems need to become much faster in order to be used effectively, and less expensive to be used more widely.

New Directions and Approaches:

VLSI chips are being developed for dedicated high
speed processors, optimized for this use and separated
from the main robot controller.   Development of
edge imaging and pattern recognition methods to achieve
higher speed and better object recognition is very
active.

Short Term:
(All in benign environment)

Use of dedicated VLSI processors will speed up 2D
and 2 1/2D vision to real-time capability for use
in adaptive control.

Range mapping will become faster and provide richer
range maps, but real-time 3D vision will take longer
to implement than 2D or 2 1/2D.

Better resolution without excessive processing time
will allow richer feature sets for better recognition
of objects.

Long Term:
Sufficient speed will become available for real-time
3D vision, including shape extraction and comparison
with CAD/CAM models.

Increasing sophistication of abstraction and recognition
methodologies, applying signal processing techniques,
will allow vision to be used in non-controlled environments, with noisy data.

Standards for signal interfacing, using symbolic rather
than numeric communication, will finally be adopted,
allowing vision systems to be utilized as plug-in
modules.

Tactile

Current:
Today's tactile sensors as used in production industrial
robots are limited to either simple contact sensing,
or force and torque sensing on a single axis. Sensors
lack dynamic range and are not very robust.

Needs:
High reliability and long lifetime are essential for
industrial applications.

Better resolution in arrays is needed for more precise part location and shape mapping.

Tangential force sensing is needed to detect imminent slippage of gripped parts.

New Directions and Approaches:
VLSI technology incorporating processing and the sensing array itself is being examined to produce monolithic sensor/processing chips.

Exotic sensor materials, such as PVF2, are being sandwiched with wear resistant rubber to produce robust epidermal sensing arrays.

Many of the imaging and pattern recognition methods developed for vision systems are being studied for application to tactile data.

Short Term:
Modest size, modest resolution tactile arrays with dedicated processors, already demonstrated in the lab, will become available on commercial robots, but will not be very common due to performance limitations.

Force sensing along a single axis will become common and multi-axis force sensing will appear on the shop floor.

Long Term:
High resolution wide range force sensing arrays will become available commercially with a sophistication level comparable to that of vision systems.

Sensor arrays with their associated processors in a single package will become available, and standards for interfacing will allow plug-in installation.

Processing of tactile data will become sufficiently improved to allow real-time acquisition of 3D shapes by touching.

Proximity/Ranging
There is very little use of proximity sensing or ranging by industrial robots today. IR sources and detectors are occasionally used to detect obstacles or locate objects, but these are low sophistication implementations.
Ultrasound has been used for coarse location of objects and to detect intrusions into the robot's work volume, but range information is not always reliable and the beam is unfocused, making precise object location difficult.
Eddy current sensing is being explored to locate and

characterize rivets in aircraft but this is a highly specialized application. The complexity of eddy current sensing equipment makes it rather unpromising for large scale general industrial usage.

Laser rangefinding is well developed and is being used for generation of range maps for 3D; it is clearly a workable means of getting range to a point. However, like eddy current sensing, it requires a lot of expensive technology to acquire a simple range, and the possible eye hazards to personnel in the vicinity can be a serious barrier to use in an industrial environment.

A major effect on proximity/ranging development may result from the introduction of the development kit containing the Polaroid ultrasonic ranging system. At a very modest cost, this kit supplies a complete system from transducer to electronics to give range results in an electronically readable form. This system shares the shortcomings of other ultrasonic systems but the easy availability of the kit has triggered a great deal of interest.

Using the kit as a test bed, researchers have demonstrated its utility by mounting it on a variety of grippers to detect gross object location and to aid the gripper in homing in on the object. The near field limitation on ranging has been greatly improved by adding an active damping system to the transducer, and other refinements are likely to appear soon, due in part to the number of researchers now working with the system.

Methods developed for using the Polaroid system and improving it are likely to spawn a new generation of compact inexpensive ultrasonic sensor systems, purpose-built and marketed as easy to use plug in modules.

Sonic

Currently, there is very little use being made of sound sensing for industrial robots. While acoustic signatures have been used to monitor processes such as the seating of snap fit parts, the majority of interest in this area centers on speech recognition for command purposes. This capability is available, but there are three major limitations:

- Vocabulary is limited

- Commands are only recognized when spoken by a single person.

- Recognition is tone sensitive, and becomes unreliable in stressful situations.

It is arguable whether or not speech recognition at this level is a meaningful capability; in the near term it does not seem likely to come close to the flexibility and reliability of keyboard communication.

In the long term, speech recognition will come into its own through the development of artificial intelligence and natural language capability. While someday humans may communicate verbally with high level supervisory computers in the factory, we are not going to see a pick-and-place robot on the shop floor conversing with people.

Smell   Any method of sensing that can produce results in the form of electrical signals can, in principle, be used by robots. While not in use yet, olfactory sensing has the potential for subtle process monitoring. A great deal of preliminary work is needed to identify the chemical emissions of industrial processes and what they imply about the status before any real use can be made of a robotic sense of smell.

4.2.4 Integration

There are two types of integration involved with industrial robots:

o Internal - coordinating the robot components, especially sensing systems.

o External - coordinating the robot as a whole with surrounding equipment.

Current:

Internal integration is very difficult unless all components are acquired from the same vendor who also assembles the system. Incorporation of other suppliers' units, such as vision systems, is difficult due largely to the lack of standardization of interfaces and protocols.

External integration is crude. Most current industrial robots operate as an "island of robotics", and connect with surrounding equipment via parts feeders and fixturing. While CAD/CAM databases often coexist with industrial robots, no one at this time has implemented direct communication and coordination.

Short Term:

Internal integration will improve as communication and interfacing standards develop. Sensing systems will be the first well modularized components, allowing selection of any one of a variety of vision systems for a particular robot.

External integration will reduce the robot's dependence on expensive and inflexible fixturing and feeders. They will be replaced by simpler mechanical systems as robots become more flexible and less demanding on peripheral equipment. Some of the burden of parts presentation will be taken over by simple robots, such as mobile carts and pick-and-place robots. Coordination with external computer systems such as graphic modelling systems will become common for sophisticated installations.

Long Term:

Internal integration will be greatly improved by industry-wide standards for interconnection. A buyer will be able to add to or upgrade his robot's capabilities by plugging in modules for control or sensing functions. This will also result in reduced down time for robots by allowing rapid replacement of failed modules to bring the robot back into production.

External integration will connect and coordinate entire production lines, including many robots. CAD/CAM Systems will connect with graphics aided robot programming systems which will then down load the resulting programs to the robot production line. Each of these systems will keep a high level supervisory system informed of progress of all projects. This supervisory system will perform the necessary planning, stock and machine allocation, maintain inventory and maintenance schedules, and support a sophisticated Management Information System that provides any requested information about any section or level of the entire system.

4.3 Application Projections

With respect to the effect of future developments in robotics, there are three major categories of robotic applications:

Low Growth Applications - Developments in robotic technology will not produce sweeping increases in robotic penetration.

High Growth Applications - As developments in the laboratory and development stage become commercially available in the near term, these applications will show very rapid increases in robotic penetration.

Blue Sky Applications - These applications require capabilities that are still in early developmental stages. Robotic penetration will be very slow starting, and will not become significant in the near term future.

Low Growth Applications

Robots in these applications are generally characterized by:

- binary sensing

- programming by lead-through or walk-through

- preprogrammed unvarying path

- operation in very structured environment

- no use of knowledge or internal models.

These applications are currently well penetrated by robotics, and today's robots perform well. Primary barrier to further robotic penetration is cost of the robot and cost of set-up for a task, i.e., large batches are needed.

Spot Welding -

o requires that a tool be moved to a point and squeezed

o is being performed successfully open loop, with no adaptive control

o no great interest in more technical sophistication

o shows high penetration (mostly in automotive), but has reached a plateau

o requires large batches to be economical due to the high cost of the robot and clumsy programming

Spray Painting and Coating -

o requires that a spray gun be moved along a smooth path while triggering the spray

o is being done successfully open loop

o could be enhanced by sensing, but there seems to be little interest in this among users

o shows good penetration (mostly in the automotive industry)

o requires large batches for economical use due to the very high cost of painting robots

Forging -

- o requires that a hot work piece be placed in a die, and removed after forging

- o is commonly done by pick-and-place robots

- o shows some use of IR sensing for verifying that the work piece is gripped

- o uses the robot as a peripheral device; robots won't bring about major changes in forging

Investment Casting -

- o requires that a form be dipped in slurry and dried (repeatedly) to build up a mold

- o is being done open loop

- o produces consistent molds - the key to successful investment casting

- o shows modest penetration that will increase steadily but not going to be overwhelming

- o much investment casting done by small jobbers in small shops, limiting penetration due to cost

Sealant/Adhesive Application -

- o requires that an applicator gun be moved along a preprogrammed path

- o can be performed with no sensing but can leave voids undetected

- o simple vision allows void detection and repair

- o is time critical since work time for hot melt or epoxy adhesives is short

- o reduces waste; more consistent bead gives higher quality bond or seal

- o is an increasingly popular way of joining parts because it is cheaper than mechanical fasteners

- o robotic implementation is growing because of growth of process and robots work well in unpleasant, low paid job

Die Casting -

o robots are used to eject casting and to clean and lubricate
   dies

o with minimal sensing, the robot detects remnants in
   the die, and cleans the die only when necessary

o major effect on casting process is prolonged die life
   due to consistent lubrication

High Growth Applications

Robots performing in these applications will be generally characterized by:

- vision up to 3D

- force and torque sensing

- force sensing arrays

- adaptive control of path and process

- off-line programming capability

- enough adaptability to operate in less structured environment
   than low sophistication robots

- use of knowledge and simple internal models.

These applications are lightly penetrated by robots today.
There are major technical barriers to further penetration such
as limited sensing capability, insufficient speed, inadequate
precision or dexterity and inadequate adaptability to variations
in task.

(In this section, open bullets indicate the use of currently
in place technology while solid bullets indicate technology
and capabilities expected to be available in the near future.)

Material Handling -

o simple pick-and-place

- is being performed with minimal or no sensing

- requires parts presentation with precise location and
   orientation; this can result in large fixturing costs

- functions very efficiently as interfaces between robot
   cells, and as integration of production lines improves,

demand will increase

o parts acquisition with 2D vision

  - has been used to acquire and orient parts from a parts
    table

  - is beginning to be able to handle overlapping parts,
    but requires a controlled environment e.g., backlighted
    table

  - reduces the demands of parts presentation but is still
    short of bin-picking

  - this level of implementation will remain in use even
    when real time 3D vision becomes available for reasons
    of simplicity and cost

● parts acquisition from jumbled pile

  - this is an active research area

  - solution to bin picking in a commercially available
    form seems to be imminent

  - eliminates most of the cost of external parts presentation
    equipment

  - allows use of robots in industries that normally store
    parts in bins

  - will enhance robotic assembly by reducing peripheral
    equipment requirements

  - may see major application as first robot in integrated
    manufacturing lines, feeding new materials from unstructured
    storage.

Arc Welding -

o open loop, non-adaptive

  - simply moves torch along smooth but fixed path

  - requires fit-up and fixturing at a quality level that
    is unrealistic for most industries

  - this level of implementation is fading as adaptive path
    control (seam tracking) becomes common

o   adaptive path control

- once started on a seam, the torch follows the actual
  seam, compensating automatically for small errors

- most of today's systems perform in two passes, first
  to locate and memorize the seam with the actual welding
  performed on the second pass

- one pass systems, sensing during welding, are becoming
  available, and are faster than two pass systems

- through-the-arc sensing not only allows seam tracking,
  but also limited process monitoring

- these systems cannot do a good job of accommodating
  wide gaps along the seam either by adding more filler
  material or by rejecting the assembly as unweldable

● adaptive process control

- weld parameters are monitored during welding with either
  through-the-arc sensing or vision systems that watch
  the size and shape of the weld puddle

- this sensing allows process parameters to be controlled
  to compensate for poor fit-up and variability of materials

- improves the success rate on thin materials and difficult
  welds, e.g., thin pieces to thick pieces

- reduces the amount of effort needed to prepare the assembly
  for welding

Routing, Drilling and Grinding -

All three of these processes are force critical rather than
position or speed critical, i.e., the parameter to be controlled
is force or torque.

o   open loop, no force or torque sensing

- In order to prevent excessive forces or torques from
  developing, feed rates must be kept very low, resulting
  in very slow processing.

o   adaptive feed control with force or torque sensing

- Each of these processes involves bringing a rotating
  cutter against the work piece while moving the entire
  tool along or into the work piece.

- The process can be monitored by sensing torque required
  to drive the rotating cutter, or by sensing the force

required to move the tool along its path; both approaches have been used.

- While the cutting process can be controlled, where the cutting is done remains a problem in terms of positioning accuracy, and allowance (especially in grinding) for tool wear.

- To achieve high precision in drilling and routing today, precise templates are used to bring the tool to the work at the precise location desired.

- Templates are expensive to make and use, and generally a different template is required for each type of workpiece to be handled.

- Even with cost of templates, robotic drilling is becoming increasingly popular in aerospace for drilling rivet holes in skin sections because of the requirement for a large number of precisely located and drilled holes.

- Robotic grinding today is generally used for non-precision applications such as grinding flash off of castings.

• real time sensing

- Tactile or proximity sensing can be used to locate regi- stration features on jigs, improving the precision of tool location without templates.

- Real time sensing of arm deviation from expected path when under load will make precision drilling and routing possible without templates, giving a large boost to aircraft skin applications.

- Eliminating the cost of templates will reduce the batch size required to justify a robot, making it attractive to a wider range of users.

Inspection -

o Inspection is being done with simple vision systems and proximity sensing.

- Limited resolution and speed restrict inspection to, in general, verifying that a part is in place.

- Similarly, inspection robots can easily verify that the workpiece has a hole in it, but verification of the diameter or location of the hole with high precision requires sophisticated and complex computation.

- Surface texture inspection is not usable today; today's

inspection systems cannot distinguish between a scratch
and a crack in a surface.

Improved robotic inspection is going to result from two approaches:
improved robot performance and transplantation of existing non-
robotic inspection technology.

● improved robotic performance

  - High precision positioning over all of the robot's work
    volume will allow tactile inspection by mapping surfaces
    through feel; this has already been suggested as a method
    of inspecting aircraft structural components containing
    fillets, webs and cut-outs.

  - The resolution of vision systems determines how small
    a detail the inspection can detect; as vision resolution
    improves, texture inspection will become more feasible.

● transplanted technology

  - Currently available Non-destructive Evaluation (NDE)
    techniques, using the robot as a positioner and interpreter,
    are under development now.

  - X-ray examination can make use of robots to speed up
    the process by positioning heavy x-ray equipment more
    rapidly than humans. Removal of the human operator
    also allows the use of higher radiation levels, improving
    resolution and penetration.

  - Eddy-current measurements currently being investigated
    for locating rivets in airframe components can be refined
    and used for detection of deep flaws.

  - High sensitivity magnetic sensors or sensor arrays will
    allow an inspection robot to map the magnetic field
    at a surface with an applied external field. Surface
    defects distort the magnetic field and appear as anamolies
    in the surface field map.

Assembly -

  o Easy mating assembly is being done with minimal sensing as
     a special case of material handling, where the part is simply
     moved to the final assembled position. This approach is
     used for high clearance, compliant parts.

  o closely fitted assembly

     Today's implementations are seriously restricted by technological
     limitations:

- Sufficient precision in part positioning prior to insertion is not currently available; compensation for this is either by use of Remote Centers of Compliance (RCC's) or search algorithms that are slow to execute.

- Chamfering, which requires a change in parts fabrication, is generally needed.

- Most successful implementations involve radially symmetric parts assembled on a shaft, such as washers, sleeves and bearings slipped onto a motor shaft.

- Jamming is far too common due to lack of adaptive control for final positioning; as a result, parts are frequently damaged and the assembly process disrupted.

- Insufficient error detection prevents flawed assemblies from being detected prior to subsequent operations.

In spite of these difficulties, assembly is showing some robotic penetration in suitable labor intensive, high volume applications, and interest in improvements is very high.

● Developments in the near future, such as:

- improved mechanical precision

- vision to guide robot to the hole more precisely

- bin picking to make parts presentation more economical

- improved insertion algorithms to reduce jamming and speed the process

- error recovery routines to keep robot working will make robots much more attractive for assembly applications, and by replacing skilled (i.e., expensive) labor, will improve the economics of robotic assembly. Due to the combination of large volume production and pre-existing familiarity with robots, the automotive industry is likely to be an early user of robotic assembly.

● micro-assembly

- requires high precision while operating at a very small scale

- necessary miniaturization is not going to be available in the near future

- interest is high in the electronics industry; electronics will remain a high growth industry, and by the time miniaturization is available, the industry will be well

acquainted with the use of assembly robots at the printed
circuit board level

- as soon as the technology is available, this application
will grow very rapidly.

Blue Sky Applications

Robot that will perform these tasks will require combinations
of the following characteristics:

- autonomy

- mobility

- expert systems

- artificial intelligence

These applications have no current robotic penetration. The demands
that they make on adaptability, decision making and mobility are
not available in today's technology. Furthermore, unlike the barriers
in the previous section, work on the capabilities required for these
applications is still in the preliminary stages of development.

In the near future, the work that will be done that relates to these
applications will be based on teleoperated machines. This allows
a human operator to fill the need for intelligence and decision making,
the robot attributes that are the farthest from commercial availability.
As a result, much of the development and engineering required for
intelligent autonomous robots will be available as soon as AI capability
becomes available.

Nuclear Plant Maintenance

Inspection and maintenence in the radioactive areas of nuclear
power plants is generating a great deal of R&D effort, especially
in France and Japan. Most current work is focused on the development
of teleoperated inspection devices, but the ultimate goal is the
development of mobile equipment capable of performing any needed
repairs in the high radiation areas of the plant. Such equipment
will require a high level of mobility to navigate in a cluttered
environment and the ability to reach almost inaccessable areas.
Performance of repairs will require the appropriate tooling for replacing
tubing segments and re-welding seams. Inspection and certification
of repairs will also be required.

An autonomous robotic system to perform these tasks may eventually
be developed, but teleoperation capability will be retained to provide
a back-up control system. Indeed, once the difficulties involved
in performing these tasks via teleoperated devices have been solved,
the need for the next step, development of an autonomous robot for
this role, may not be very pressing. For regulatory reasons, the

level of human monitoring in these plants is likely to remain high. In addition, the detail differences from plant to plant, and the complexity of diagnosing and repairing unpredictable failures, are likely to result in a teleoperated solution being accepted as the final result.

Housekeeping

Cultural biases notwithstanding, housekeeping encompasses a great variety of complex tasks involving judgmental decisions at almost every step. These tasks must be performed in a very unstructured environment, subject to frequent rearrangement. In a home, the proximity of children and pets requires careful attention to safety factors.

A truly sophisticated housekeeping robot, able to accomplish the entire range of household chores, will tax AI capabilities to the limit, and will not be commercially for a long time. However, a robot with a limited task repertoire, such as carpet vacuuming, could be marketed in the foreseeable future. While it might not be cost-effective in the average home, this does not mean it would not sell; Americans have historically had a love of gadgetry. Additionally, the increasing number of elderly in our population can provide a ready market for robots to perform simple but tiring chores, provided that the human/machine interface can be simple and non-threatening. Such a robot could open up the home market, and generate strong demand for more sophisticated home robots in a shorter time frame than might otherwise be expected.

Construction Labor

This is personnel intensive work that will be very difficult to robotize because of demands of the construction environment. Mobility problems, both mechanical and control, are a major barrier for this application. A robot in this application may have to transport itself and its load over uneven, some times muddy terrain, climb ramps and scaffolding.

In addition to the difficult terrain, construction sites are inherently changeable environments, both in terms of progress on the structure and in in terms of large amounts of material temporarily stored in arbitrary locations. Mechanical reliability of complex locomotion systems in an environment loaded with dust, grit and dirt is a problem requiring a great deal of work.

Interest in construction robots will be driven by economic considerations. Not only are serious accidents expensive, but development of chronic and disabling conditions as a result of the physical strain of the work must be considered in the cost of labor. Current practice of paying modest hourly rates with no fringe benefits for laborers will have an increasingly difficult time in attracting sufficient manpower; the result is likely to be major increases in the cost of keeping personnel on site.

What is likely to appear before robotic construction site laborers, is the use of robots for off-site construction tasks. Much of the construction performed today consists of joining prefabricated units. This moves much of the labor of construction back from the building site to a factory, an environment that is much easier to robotize.

Maintenance by Expert Systems

The maintenance and repair of mechanical systems is very labor intensive and requires a high level of expertise when performed by humans. There are indications that the automotive industry may lead the way in this type of application. Several years ago, Volkswagen introduced a diagnostic system for some of their models, in which an automatic electronic diagnostician was plugged into a specially designed connector built into the car. This idea was apparently premature, and faded from sight. However, the increased use of electronic controllers for automobile engines, and the rising use of electronics in automotive instrumentation indicate that an automated diagnostic systems may not be far in the future. With expert system capability, the automated diagnostician could guide a human helper through the needed repairs and adjustments step by step, reducing the need for skilled human labor.

One can foresee more ambitious robotic applications in automotive maintenance, considering the number of automobile parts that are commonly rebuilt. These range from carburetors and alternators up to automatic transmissions. In the future, robotic systems will combine the precision and dexterity developed for assembly operations with inspection capability. With the addition of an expert system that can recognize parts needing replacement, a robot could perform the entire task, requiring only the old unit and a kit of replacement parts. This application may be nearer in the future than many others in this section since the rebuilding business is already somewhat concentrated, and trading in worn-out units to replace them with rebuilt units is a well established practice.

Hazardous Environment Rescue

Fire fighting frequently involves entry to burning structures to locate trapped people and transport fire fighting and support equipment. The number of firemen killed each year in the line of duty is a clear indication of the level of risk involved. Due to the direct reduction of high risk to humans, the economics of this application will have less effect than the high demands on mobility and autonomy. For rescue purposes, the robot must be able to select a path through a cluttered and rapidly changing environment, recognizing and assessing severity of thermal hazards. For delivery of equipment, (hoses, air bottles, etc.), the robot must also be able to find the destination where the equipment is needed. AI will be essential for risk assessment: a path that is likely to result in the destruction of the robot would not normally be selected, but a path with a high probability of destroying the robot may be acceptable when human life is at stake.

Orbital Construction

The high cost of labor has been a strong incentive to use robots in many applications, and the cost per man-hour of work performed in earth orbit may be the highest of any human endeavor. A major part of this cost is for transportation of the astronaut and life support equipment, and, due to the limited time that humans spend in orbit, this is a frequently recurring cost. Unlike a human, a robot needs to be transported only once, and only one way, while the required support equipment is only a fraction of that needed for a human. As a result, the cost of orbital construction may be significantly reduced by the use of robots.

If the development of space continues and grows as expected, orbital construction will become increasingly important. Space structures will be assembled from pre-fabricated components and can be designed for robotic assembly.

Some of the problem areas for this application are:

o establishing location - Sensing of active beacons seems feasible, though the work site is likely to be cluttered, requiring redundancy.

o motion control - Zero-g maneuvering is a complex problem, though much has been done in developing maneuvering systems for untethered operations from the Space shuttle.

o environmental protection - Methods to protect sensitive electronic and mechanical components from the orbital environment (vacuum, radiation) are not simple, but have been developed for the space program.

o system integration - Coordination of a robotic assembly crew is much more complex than coordination of several industrial robots. Mobility in three dimensions and the level of autonomy in each robot are two major sources of the additional complications.

o compensation for reaction forces - In a zero-g environment, application of force to a tool or workpiece requires that the worker be solidly anchored to avoid being moved out of position by the reaction force. This problem has become apparent for shuttle personnel working in orbit, and solutions to the problem are being developed. A robot is going to need an anchor point at each work station (which can be incorporated in the component design stage) and a way to attach itself. An alternative to mechanical anchoring to deal with rotational reactions is the use of counter rotating flywheels. Translational reaction forces would still require compensation with gas jets, using up fuel, but the elimination of the

complexities of mechanical anchoring might make this the
preferred way to go; it is too early to tell.

Based on the above discussion, we can summarize these three groups
of applications in terms of their current levels of implementation and
the effects of emerging technology on their future direction.

Low growth applications are those in which current robotic capabilities
are well matched to the application requirements. The robots in
these applications perform satisfactorily with little or no sensing
capability. Advances in sensing systems that will be appearing in
the near future can extend their capabilities somewhat, but will
produce refinement rather than qualitative change. Programming methods
used are generally walk-through or lead-through today, but incorporation
of off-line programming, if the needed accuracy can be incorporated,
will make their implementation easier but will not change their fundamental capabilities. Today these robots usually operate as "islands
of automation"; in the future they will be more closely integrated
with surrounding machinery. They are demanding of their environment,
requiring rigidly structured and undeviating surroundings.

The net effect of developments in robotics on these machines will
be to make them more flexible, easier to program and less demanding
of their environment, all of which reduce costs. This will result
in lowering the batch size that is economical to process robotically,
but will not fundamentally change the current situation. While penetration with respect to the application will rise steadily, robots
in these applications as a percentage of all robots will fall as
robots in medium sophistication applications become more widespread.

The category of high growth applications will show the strongest
growth over the next 2 to 6 years. Todays limited implementations
have demonstrated feasibility, and limitations are clearly identifiable.
Furthermore, possible solutions to today's technical barriers are
already in the laboratory. The barriers in these applications are
predominantly in sensing and sensory integration, and as has been
described above, major advances and improvements are expected in
these areas in the near term.

Two key aspects of these applications that indicate the likelihood
of rapid growth are that the barriers are susceptible to solution
by development of current laboratory technology and that they are
personnel intensive processes. The first will make robotization
possible, the second will make robotization attractive.

Blue Sky applications are those that are not likely to utilize robots
for quite some time. The key element in each of them is the ability
to perform tasks autonomously. While current sensing techniques
may not be adequate for these applications, the major barrier to
implementation is the lack of Artificial Intelligence in a robot.
Since work on AI is still in an early stage, predictions about   i t s

development are difficult. Furthermore, in the time frame in which
AI can be expected to develop sufficiently for these applications,
the strength of motivation for implementing robots for these tasks
may change considerably. As an example, for construction work, today's
primary motivation would be economy since human labor is expensive
and likely to become more so. Economic changes that would drive
the cost of labor up sharply would accelerate interest in this robotic
application, while a depression that drove the cost of labor down
sharply would reduce interest.

4.4. General Trends

4.4.1. Trends in Robotic Technology

The general trends we expect to see are:

o Separation of high sophistication robots from simple
   robots:

   Simple specialized robots for suitable applications,
   such as pick and place and assisting sophisticated
   robots, will become smaller, less expensive and
   better integrated with adjacent equipment. They
   will be recognized as having an important role to
   play, not thought of as just a second rate version
   of a sophisticated robot.

   Sophisticated robots will be much faster and more
   flexible, able to handle a greater variety of tasks
   and capable of dealing with more deviation from
   expectation in performing a task.

o Sensing will become both faster and better, and integration of sensory information will be much more efficient.

o   Mobility will be easily available due to improved
   mechanical mobility systems, more energy efficient
   design, and better self contained power sources.
   Improved sensing will provide precise location information
   and navigation methods will allow flexible path selection
   and optimization.

o Future robots will take advantage of lighter materials
   and more efficient design, and the lightweight, composite
   material arm may replace today's massive metal castings
   in heavy duty industrial arms.

o   Perhaps the greatest change will be the extent of
   robot integration. Sophisticated robots will communicate
   downwards to dedicated satellite processors, sideways
   to adjacent robots, and upwards to supervisory control

systems.

o Robotic work stations in which sensor mediated manipulation and inspection are being developed for specific
   tasks in a fully automated production line. While
   very expensive to develop, the ability to market the
   basic system to a variety of customers with only minor
   changes will make this a cost-effective route for
   development, and tend to produce, indirectly, a certain
   level of standardization.

o Lower cost is going to be a major trend in both sophisticated and simple robots. Some of the cost reduction
   will result from improved technology, either providing
   capabilities at lower cost, or providing more capabilities
   at the same cost. Another major factor in cost reduction
   will be economics of scale as the increasing number
   of robots being manufactured moves robot manufacturing
   more into mass production.

o Modularization will become well developed, as interface
   standards become accepted, allowing robots to be purchased
   in much the way a car is purchased: a basic model
   with a buyer specified list of options. Additionally,
   this modularization will improve the situation for
   third party component integration.

o Higher volume production and improved standardization
   will bring robots closer to being off-the-shelf items.
   This will reduce lead time when ordering, and increase
   confidence levels that subsequent robots will match
   the performance and specifications of the first ordered.

o Hybrid robotic/teleoperated devices will become common,
   leading the way in applications that will eventually
   be handled by fully autonomous robots, with teleoperation
   serving only as a fall-back method of control.

4.4.2 Trends in the Robotics Industry

At present, the number of companies in the U.S. producing and
marketing robots or robot components is quite high, more than
today's market can support. A shakeout is occurring, and many
of these companies are likely to disappear. Small, undercapitalized
companies are at greatest risk, but larger companies are not
immune from the pinch. Note that Copperweld has recently dropped
their robotics product line. Small companies are severely hampered
by the industry-wide lack of standardization; an excellent vision
system that is difficult or impossible to integrate with commercially
available arms is not going to be easy to market. Conversely,
controllers that cannot easily integrate external inputs will
become increasingly more difficult to market.

Indications of the state of flux in the industry can be seen in the Westinghouse/Unimation situation. In a relatively short period, Unimation split into Unimation-West and Unimation-East, Unimation-West became Adept Technology while Unimation-East was bought by Westinghouse.

During the next several years, the robotics industry is likely to be in a state of flux; however, some trends seem likely to appear:

Many of the larger firms will be marketing Flexible Manufacturing Systems, and robots as components of FMS.

More companies will market robots to Original Equipment Manufacturers (OEM's), who will use the robot as the basis of a retail product, adding features and customization.

Suppliers of complete turnkey systems will become more prominent, offering completely integrated and supported systems, minimizing the hidden costs of a robot.

Greater product differentiation and market segmentation will develop as vendors try to carve out either application specific or component specific markets. The first of these can be seen today in the way that DeVilbiss has established itself in the field of spray painting robots; the second is currently hampered by lack of interfacing standards.

Robot leasing and rental companies may become more prominent, in a way similar to current practice for main-frame computer systems. This will allow companies to begin using robotics with a much smaller initial investment, and will, to some extent, protect the user from being burdened with obsolete equipment.

The introduction of large numbers of inexpensive hobby robots will create a large pool of people working in robotics. This will bring a great deal of ingenuity to bear on some of the problems in robotics, and may become a major force in technological innovation, much like the effect that amateur radio operators have had on radio technology.

5. Conclusion

In consideration of the scope of this report, from a detailed analysis of the current status of robotic technology to a forecast of the future of robotics, this report is best concluded by giving the reader an understanding of the key interplays that connect R&D, the level of available technology and robotic applications. This chapter is divided into two parts, Dynamics of Technological Innovation, which describes the way in which research affects the level of available technology, and Evolution of Robotic Application, which illustrates the way in which available technology determines robotic applications.

Dynamics of Technological Innovation

The capabilities of industrial robots have progressed from the early industrial robots that could do little more than pick-and-place aplications, but this progression has consisted of surge and consolidation stages, rather than steady development. From the introduction of industrial robots on the shop floor in the early 60's through the late 70's, industrial robots acquired continuous path control, better repeatability and improved reliability. However, the result of these improvements was to enhance robot suitability for the early applications such as material handling and spot welding, rather than to extend robot utilization to new and more demanding applications.

Meanwhile, a university oriented robotics R&D community was beginning to take shape as robotics began to be seen as an interesting and potentially high growth area of funded research. While the capabilities required in a robot for the early applications were considered a solved problem, it was clear that many other industrial processes would require sensory and control capabilities well beyond those available from contemporary robots. Furthermore, many of the new entrants into robotics R&D came from areas such as computer science, prosthetics and control theory, and brought with them more of a basic, long term research orientation, in addition to infusing robotics with technology from the other fields. The results of these R&D efforts have been appearing since the late 70's as a second surge in the capabilities of robots on the plant floor.

The robots of this second surge differ from the earlier robots in two main areas: sensing and control. Vision systems are common on these second surge robots, typically 2D or 2-1/2 D. Robots without vision are likely to have force or torque sensing. Both types of sensing are used to provide the robot with information about its task, and with sensory integration provided by the controller, allows the robot to modify its actions to suit the situation, i.e., adaptive control.

With these capabilities, industrial robots have started to penetrate more demanding applications, with arc welding being the most prominent today. However, the application that many believe will dominate industrial robotics by the late 80's is assembly. Penetration of robots into this field is just beginning, but the potential market is tremendous, and robots designed explicitly for assemby tasks are becoming available.

Another important characteristic that will be seen as the robots of the second surge are installed is extended integration. The use of industrial robots as "islands of automation" is going to be supplanted by Flexible Manufacturing Systems, which integrate and coordinate an entire production line of robotic and automated work cells. This approach to implementing robots reflects a fundamental change in philosophy. The "island of automation" idea was based on the view that a robot is a replacement for a human worker, and this was the basis for most implementations of first surge industrial robots. Implementations of the second surge industrial robots will reflect the more subtle view of a robot as a production tool, one part of a production system.

While there is still a great deal of work to be done on robotic sensing and control, it appears that the basic elements that will drive a third surge in industrial robots are now in the early developmental stages in the laboratory. In a way similar to that in which second surge robots moved into applications characterized by the need for sensing, the third surge will carry robots into applications characterized by the need for autonomous action.

Evolution of Robotic Application

The early robots were typically capable of moving an end effector to specific, repeatable locations, and with the advent of continuous path control, could perform the intervening motion over a smooth, controlled path. Unloading die casting machines (a specific type of material handling), spot welding and paint spraying were all tasks within these capabilities and were early applications of robots. These applications set the pattern for the first wave of industrial robots: simple robots performing simple tasks. With an early start and a history of successful implementations, these applications showed a rapid growth of robotic penetration through the end of the 70's. Today, these are the most heavily penetrated applications, but the growth in penetration has levelled off. Technological developments expected in the near term will not open up significantly larger markets for robots in these applications, and the percentage of robots in these applications versus all industrial robots in use will decline as more demanding applications such as arc welding and assembly become robotized.

The beginning of the second wave of robot penetration can be seen today, with arc welding being the prime example. This second wave will be characterized by robots with greatly enhanced sensory capabilities, as compared with the minimal sensing typical of the first wave robots; this increased sensing is required by the applications in which these robots will be used. Arc welding is showing a significant amount of penetration today, using sensing and control technology that have only recently become available. The quality of sensing available today is also sufficient for initial implementations of robotic assembly, but developments that are now on the way from the laboratory to the shop floor are going to dramatically enlarge the potential market for robots in these applications. As a result, the penetration of second wave robots is going to grow rapidly in the near future and these robots, especially in assembly applications, will eclipse the first wave robots, and become the dominant portion of the robotic population. Along with improved sensory capability and control, the robots of the second wave are going to be better integrated with surrounding equipment, including other robots. The logical extension of this integration is the Flexible Manufacturing System (FMS) in which an entire production line of robots and automated machine tools are integrated and coordinated by a supervisory computer system. Such systems have been assembled today; the major change expected in the near future is greater ease of integration due to industrial robots being designed with integration capabilities from the start.

How the third wave of robotic penetration will come about is not clear, but areas in which long term research is now being performed give some indications of what can be expected. The key to the third wave will be the incorporation of Artificial Intelligence into robotic systems. In the same way that the second wave robots opened up entirely new applications for robots, the third wave robots will extend robotic capabilities into yet another level of applications. Maintenance and repair performed by expert systems is being examined by the Army as a means of reducing personnel requirements, especially under battlefield conditions. Autonomous robots for construction in outer space is likely to be another third wave application. The thrust of third wave robotics will, when it appears, be the replacement of highly skilled humans with robots.

Appendix A

Current Applications

Section 2.3 of this part summarized state-of-practice applications of robots in industry. This appendix is supplied for the reader who is interested in a greater level of detail than presented in the body of the report. The successful performance of many manufacturing tasks requires process considerations that may not be obvious to a reader without considerable exposure to the specific application. Similarly, the implementation of robotic systems in the industrial arena can give rise to difficulties with and limitations of current robotic products that are unfamiliar to a person whose background is not in robotics.

This appendix highlights aspects of the intersection of manufacturing technology and robotic technology. It is not intended to be definitive, and readers interested in greater detail than provided by this appendix are referred to the Bibliography (Appendix D) included in this report.

This appendix is divided into sections by specific application areas, starting with what are termed, for the purposes of this appendix, major applications, followed by minor applications. Any application in which robotic technology is very prominent is considered a major application, regardless of the level of robotic sophistication. In the major application area of welding, resistance (or spot) welding is an example of a relatively undemanding robot application that is prominent due to the high level of penetration of robots. The very simplicity of the application resulted in early and successful robotization of the process and predictable equipment investment paybacks. On the other hand, a sophisticated and demanding application in which there is a great deal of industrial interest, such as robotic assembly, is considered major even though robot penetration is not large at this time.

Application areas that are primarily extensions, special cases, or combinations of major applications are considered in this appendix as minor applications. They are presented in less detail, with referrals to the appropriate major application sections.

Each application section is organized in six parts: (1) Process Description, (2) Process Considerations, (3) Basic Elements, (4) Justifications, (5) Current Technological Constraints and (6) Application Examples. The first two parts deal with the generic process, the next three parts deal with robotic implications for the process, and the last part illustrates approaches that have been used to robotize the application.

Process Description: This section briefly describes the fundamental steps required for the specific industrial process. The steps are presented in sequential order reflecting manufacturing practice.

Process Considerations: This section points out aspects of the process that are either crucial to satisfactory performance of the task or that make performance of the task particularly difficult.

Basic Elements: This section describes the generic robotic components that are in use in this particular application. It is often seen that the same process can be implemented at various levels of sophistication, especially with respect to robot sensing systems.

Justification: This section points out the aspects of the application that tend to favor a robot over a human worker.

Current Technological Constraints: While all of the applications described in this appendix have been implemented, robotic penetration remains limited. This section identifies, for each application, some of the limitations in current robotic technology that prevent deeper penetration by robots into the application.

Application Examples: For each application, this section briefly describes specific implementations of robots, noting which Basic Elements are used, and illustrating various approaches to the points raised in the Process Considerations sections.

1. Welding

There are two major processes in use in industrial welding: resistance (or spot) welding and arc welding. Although the two processes are very different, we have followed standard practice by including them under the major category of Welding. The emphasis in the following analysis is on arc welding because, of the two processes, it is the more demanding, and has shown less penetration by robots than spot welding.

Process Description

Arc Welding
- o align parts to be welded
- o heat parts at seam by generating an arc between welding electrode and work pieces
- o apply filler material as needed
- o monitor weld for bead width, penetration depth, seam filling

Spot Welding
- o align pieces to be joined
- o clamp pieces between welding electrodes
- o heat pieces at weldpoint by passing a high current between welding electrodes.

Process Considerations

Parts alignment is vital to satisfactory performance in both types of welding. The two aspects of parts alignment can be characterized as set-up (how are the parts to be joined positioned relative to each other), and seam alignment (how well are the surfaces or edges to be joined aligned with each other). Both of these aspects are established by the fixturing used to hold the parts and the dimensional correctness of these parts. Set-up determines if the unit as a whole will be acceptable and is not affected by the actual welding operation. Seam alignment affects the welding operation by dictating the amount of filler material required. If the seam alignment is very poor, an acceptable weld may be impossible. Figure A-1 illustrates the two aspects of parts alignment. Example b in this figure illustrates poor set-up due to improper fixturing; while the seam could be welded, the finished unit would be unacceptable. Example c shows good set-up but poor seam alignment as a result of a bad edge on the horizontal piece. The poor seam alignment illustrated in example c is a common occurance when welding heat treated parts due to dimensional changes and warpage caused by heat treating.

In addition to positioning the work pieces correctly with respect to each other, positioning of the welding tool with respect to the work pieces is also critical for successful welding. For spot welding, the electrodes must be brought together from each side of the work pieces, aligned with each other and perpendicular to the surfaces of the workpieces. If the work pieces are deeply contoured, access to the inner side of the weld can be difficult, while large work pieces require a long, precise reach to bring the welding electrodes together at a point far from the perimeter of the workpieces.

Arc welding, as a line (and, in some cases, volume) process involves additional geometric and kinematic complexities. Motion of the welding torch along the seam must be a smoothly controlled path to maintain a uniform weld seam. Since the arc is affected by the geometric relation of the electrode to the work pieces, motion control must not only move the torch along the proper path, but also control the torch orientation with respect to the work pieces. In order to maintain the proper heating rate of the work pieces, the speed that must be controlled is that of the electrode tip with respect to the work pieces, taking into account any rotation of the torch to track a contour.

Another critical factor is temperature control of the parts at the point of welding. This control is exerted through control of the electrical parameters of the welding operation for spot welding: for specific thicknesses of specific materials, a controlled amount of current is passed through the work pieces at the weld point for a sufficient length of time to melt the work piece surfaces together. For arc welding, an additional parameter that affects heating is speed along the seam. Inadequate control of temperature of the seam boundaries produces bad welds: if the temperature is not raised sufficiently high, the weld penetration will be inadequate, while temperatures that are too high can produce burn through and seam gaps. (See Figure A-2.)

Basic Elements

Mechanical - Robot arms used for welding require a great deal of dexterity in order to properly locate and orient the welding tool, as described in the Process Considerations section. For arc welding, six

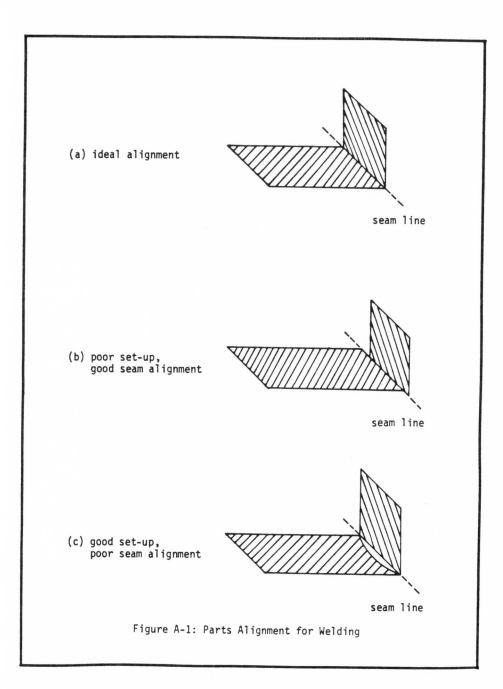

(a) ideal alignment

seam line

(b) poor set-up,
good seam alignment

seam line

(c) good set-up,
poor seam alignment

seam line

Figure A-1: Parts Alignment for Welding

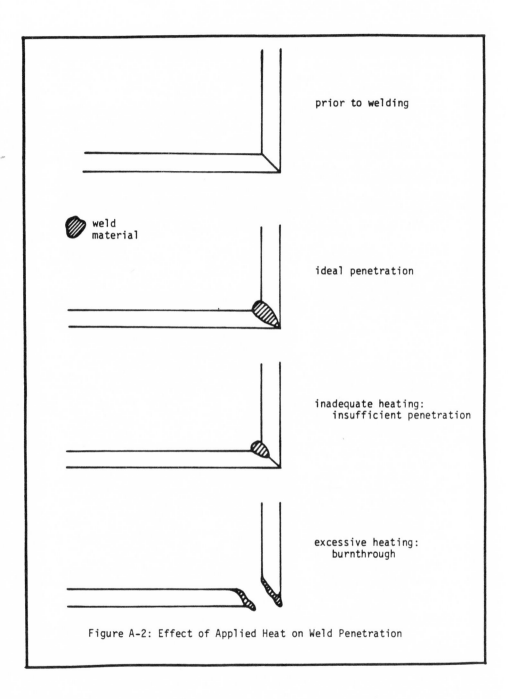

prior to welding

weld material

ideal penetration

inadequate heating:
  insufficient penetration

excessive heating:
  burnthrough

Figure A-2: Effect of Applied Heat on Weld Penetration

degrees of freedom are usually required, three to smoothly control torch location as the seam path is followed, and three to maintain the correct orientation of the electrode with respect to the work pieces. Load capacity is another important aspect of mechanical performance for welding; not only can the welding tools be heavy, but the power leads are thick and rather stiff. Additionally, inert gas arc welding requires a gas supply hose that adds to mechanical load. These supply lines add a component of resistance to flexing at each joint, and require additional force to overcome.

Controller - Spot welding, due to its relative simplicity, can be performed by simple controllers operating in an open loop mode; arc welding requires more sophistication from the controller. Sophisticated path control algorithms are required to move the electrode tip along a smooth path while controlling the orientation and speed of the electrode tip (a much more complicated problem than that of controlling the speed of the end effector). Seam tracking for adaptive path control to accommodate discrepancies between actual and expected seam location requires a controller that can integrate sensory information. Interfacing with the environment for purposes of control of welding parameters (such as arc current or rate of feed of welding wire) can be used to enhance the adaptive capability of a welding robot, but adds to the required sophistication of the controller.

Sensing - The first sensors used for robotic welding were simple tactile probes that rode along the weld seam to guide the welding torch. More recent applications have used through-the-arc sensing, i.e., monitoring welding current and voltage. The principle behind this method is that the position of the welding tip with respect to the surface of the work pieces determines the effective length of the arc which in turn affects the voltage required to maintain a constant current. (See Figure A-3.) Using an explicitly programmed back and forth motion perpendicular to the seam, a robot can constantly verify the location of the center of the joint, and this information can be used to keep the weld centered on the seam. This same technique has also been used for applications requiring large amounts of filler material to be deposited to reinforce the seam.

Vision sensing is beginning to appear for welding applications in industrial environments, reflecting improvements in flexibility, reliability and cost of robotic vision systems attained in the last several years. Two major problems are being handled with vision systems: seam tracking and weld characteristic monitoring. Visual seam tracking detects the center of the seam by recognizing the discontinuity in reflected light from the two work pieces or by interpreting the image of a strip of light projected onto the seam at an angle. (See Figure A-4.) For weld monitoring, visual systems have been developed that examine the shape and size of the weld puddle. This information can be used to indicate the penetration depth of the weld, whether the weld seam is forming symmetrically and whether the welding speed is appropriate.

Justifications

The consistency of robots in welding is a major advantage over humans.

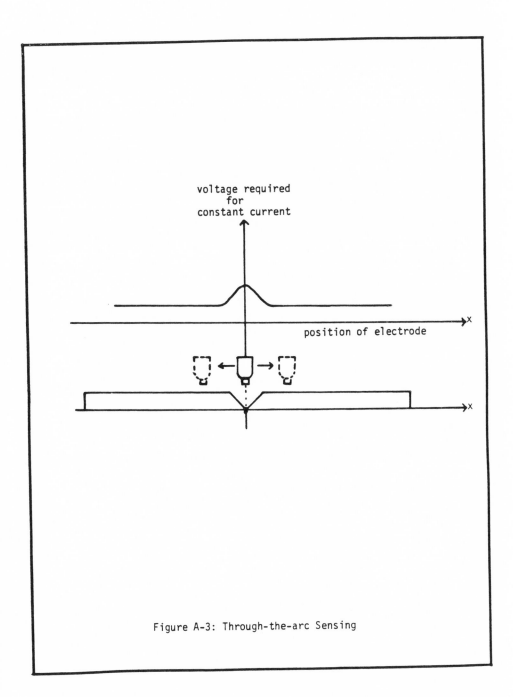

voltage required
for
constant current

position of electrode

Figure A-3: Through-the-arc Sensing

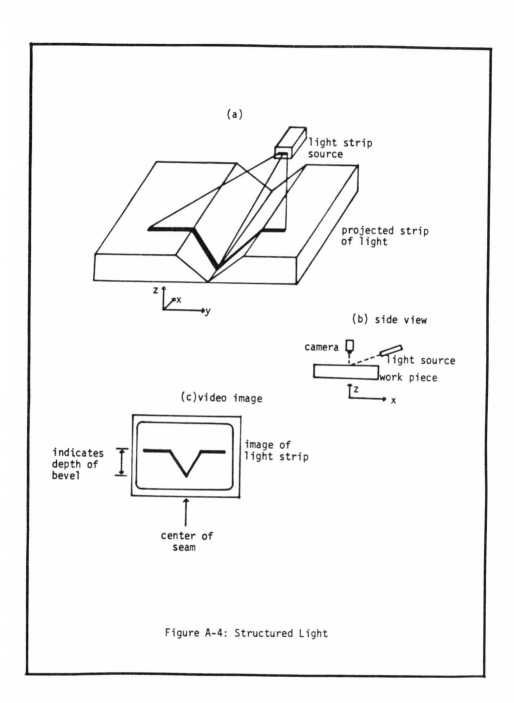

Figure A-4: Structured Light

In spot welding, if an assembly requires twenty spot welds, the robot will always make twenty spot welds (something that apparently cannot be assumed for human welders). If the robot system is properly set up, each weld will be executed properly, even those that are difficult to reach. Consistency with robots is also a major advantage in arc welding: properly set up and supplied, a robotic arc welder will produce a weld, each time, as good as an expert welder will produce at his best.

Environmental factors in welding have an adverse effect on the productivity of human welders. The heat in the vicinity of welding operations can become oppressive, while the fumes, especially when using flux cored welding wire, are unpleasant and can be hazardous. Protective gear, including gloves and especially a welding mask, are heavy and uncomfortable, causing fatigue. Since the arc produces significant amounts of ultraviolet light, exposed areas of skin rapidly develop sunburn, uncomfortable in the short term and potentially hazardous in the long term.

Another major advantage for robots in arc welding is the limited pool of available skilled human welders. Becoming an expert welder requires training and years of experience. This, coupled with the unpleasant aspects of the work, limits the number of people entering the field, while the negative aspects of the work encourage personnel to leave the field. As a result, the supply of expert welders is limited and the cost of using expert welders has risen steadily.

Current Technological Constraints

Current implementations of robotic welding require elaborate and costly fixturing in order to keep deviations in the parts alignment within the relatively narrow range of accommodation and compensation available to today's robots. Improvements in sensing systems are steadily expanding this range of accomodation, but a human expert welder can successfully weld a seam whose mis-alignment is beyond the capability of even a sophisticated robotic welder.

Selection of a sensing system for robotic welders requires serious trade-offs between flexibility and speed. While vision-based systems provide very good adaptive control, they tend to be relatively slow due to the processing requirements of the image interpretation. Visual sensing is further complicated by the light level at the work piece. Two-pass vision systems first scan along the seam to be welded without striking the arc in order to memorize the exact path needed for the weld; this minimizes the vision difficulties but increases the time required for the process and does not allow the vision system to monitor the weld parameters during welding.

There are many applications of welding, especially in shipbuilding, that require very long welded seams, well beyond the reach limitations of current fixed industrial robot mechanisms. Even the work now in progress to develop longer flexible arms will only push this limit back a bit farther. The alternative approach of developing mobile welding robots has run into major difficulties because of the precision needed in locating the welding torch. Inaccuracies in robot location due to shortcomings in the robot's navigation system and path distortion as a result of drag from trailing power cables have prevented the achievement of sufficiently precise torch location for successful application.

Application Examples

Since the end of 1981, the Locomotive Products Division of General Electric has been using robots to weld bolsters, the structural elements of a locomotive frame on which the power trucks are mounted. Robots are used in conjunction with 6,000 lb capacity positioners to weld these assemblies, consisting of steel plates up to 1 1/4" thick. The introduction of the robots has reduced the time required to perform all of the needed welds to one half of that previously required.

Robots are also used by the Aircraft Engine Business Group of GE to weld stainless steel components of fan frame hubs for jet engines. Cycle time, including part loading and unloading, has been reduced from the four hours required for manual welding to one hour. The actual arc on time with the robot has been reduced to 24 minutes, reducing the heat build-up in the assembly, and the greater precision in control of arc current, torch speed and orientation have improved the quality of final assembly.

2. Material Handling

Material handling, in one form or another, is the basis for virtually all robotic applications. The primary function of a robot is to move an object, be it a tool, inspection device, or work piece, from one point in space to another. In a stricter sense, however, material handling refers simply to moving work pieces. This could include re-orienting, palletizing, or simple pick-and-place operations.

Process Description

Broken down to its basic components, material handling can consist of:

- o locating the object to be moved
- o grasping the object
- o moving the object through a prescribed path
- o orienting the object
- o depositing the object in a prescribed location

Process Considerations

Material handling, though composed of a series of simple tasks, involves some subtle considerations. Locating the work piece, for example, is not a trivial task. Depending on what manufacturing process preceeded the handling step, the work piece may or may not be presented with a known location and orientation. Considerations involved in actually moving the object include the weight, momentum, and inertia of the object and the desired path, acceleration and deceleration speeds. A third process consideration in material handling is the geometry of the object to be manipulated. Small, delicate objects cannot be handled with the same methodology as large, solid objects. How and where objects are grasped can be critically important. A final consideration in material handling is the terminal position of the object. Work pieces which are palletized must be manipulated with much higher precision than those being placed randomly on a conveyor.

Basic Elements

The basic hardware and software elements present in robotic material handling have evolved directly from the process under consideration, for example, the question of locating the object to be moved. Classically, this has been done with the use of elaborate fixturing techniques. If the work piece is always "fed" to the robot in a very precise location, the robot need only to go to that location and grasp the object, unaffected by variations in part location. Fixturing, however, must be specially designed for each application, and is hence expensive. With the advances in sensing techniques such as vision pattern recognition, prices for sensing systems have become competitive with prices for some fixturing systems. Because sensing sytems are inherently more flexible than fixturing systems, there has been an increase in the percentage of sensing systems associated with material handing processes, from simple binary verification of part acquisition to complex determination of part location and orientation.

The question of part orientation also has an influence on the elements involved in robotic material handling. Robot manipulator arms are available with different numbers of axes, or degrees of freedom. The more axes a robot has, the more dextrous it is, but also the more expensive it is. For orientation applications, a robot with many degrees of freedom is required.   Conversely, simple pick-and-place operations require robot arms with few degrees of freedom.

End effectors have also evolved under the influence of process considerations. Gripper geometries are largely determined by the work piece or pieces involved, and have been generally custom engineered for each application. Current designs include: vacuum, two-finger, and jaw grippers, magnetic pick-ups, and combinations of these.

The robot controllers used in material handling applications have varying degrees of complexity, depending on the other elements involved. The advances in controller capability have been driven by the increasing demands of the specialized robotic hardware developed for individual applications. Robots with six degrees of freedom require a more complex controller than robots with only three degrees of freedom. Current controller technology available for material handling applications include: fine path control, algorithms to calculate kinematic and dynamic properties needed for varying arm speeds and payload weights, obstacle avoidance, complex grasping algorithms, and sensory integration capabilities.

Justifications

The highly repetitive nature of most material handling applications makes material handling an ideal candidate for robotic automation. Any operation as monotonous or tedious as a pick-and-place type of movement, especially with heavy loads, produces worker fatigue. Robot automation removes this from the process. The accuracy of robots is another advantage of robotic automation, especially in a palletizing or de-palletizing operation.

Current Technological Constraints

Even though sensing is becoming more widely used in robotic material handling, there are some advances in sensing technology that would allow robotic automation to penetrate a wider variety of applications. For example, more accurate slip sensing would enable real-time recovery techniques to be more effective; faster pattern recognition algorithms would allow more effective, real- time location, as in the bin picking example.

Application Example

A large Japanese manufacturer has an application example which illustrates several of the beneficial aspects of robotic material handling. In this application, a robot is used to palletize and de-palletize different types of bricks. The robot uses a gripper specially engineered to handle the sometimes brittle bricks with a minimum of breakage. The following benefits were realized with the robotic system:

        1)   Labor savings - With the addition of the robot, one less worker per shift was necessary.
        2)   Increase of productivity - Even with one less worker per shift, productivity doubled.
        3)   Quality improvement - With the specially engineered hand and accuracy of the robot, the defective part rate dropped significantly.
        4)   Safety/Environment - The heavy loads, dust, high temperature and safety hazards of working with the heavy load previously caused a high labor turnover rate. With the implementation of the robot, the manufacturer eliminated its dependence on an unreliable work force.

It should be noted that many of the application examples in subsequent sections are in fact expanded, specialized versions of material handling techniques.

3. **Inspection**

Process Description

Inspection, as it is performed in the industrial environment, usually consists of examining a work piece either during or just after manufacturing process. This complicates the inspection process, adding the necessity of determining where and in what orientation the workpiece to be inspected is. For manufacturing applications, a general inspection scheme consists of:

- o getting the part from its previous position
- o establishing a known orientation for the part
- o matching the object with an appropriate reference model or models
- o determining if the work piece is within acceptable tolerances of the reference model
- o sorting the object by part type or quality control

Process Considerations

Selection of the inspection point is the first process consideration encountered in industrial inspection. Ideally, it would be desirable to inspect a work piece through an entire manufacturing process. Economically, however, this is not usually practical. It is necessary, then, to choose the most logistically beneficial point or points in the production process to inspect the work piece.

A second consideration in industrial inspection is that of comparison method and thresholding. In general, a work piece can be inspected for many different qualities; it is important for both quality and economic factors to inspect only those properties of a work piece that can distinguish between desirable and undesirable pieces. In addition, it is necessary to determine exactly how close a measured property must be to the reference model to be considered acceptable.

An increasingly important consideration in industrial inspection is that of flexibility. It is often desirable to have the capability of inspecting several different types of parts, either simultaneously or in different batch runs. This requires the ability to accurately choose from among several reference models, depending on which part was to be inspected.

Basic Elements

Robotic inspection is usually performed in one of two modes: either by having the robot move the work piece in front of a fixed sensor, such as a camera, or by having the robot move the sensor around the work piece. In general, it is more efficient to have the robot carry the lighter of the two objects. In either case, it may be necessary for the manipulator to have a high degree of dexterity and accuracy, depending on the geometry of the object to be inspected.

There are currently three main types of sensing hardware available for use: tactile, vision and specially-engineered. Tactile sensors used for inspection can be either point sensing (including simple touch probes or contact switches), or tactile arrays. Although current tactile arrays used in manufacturing consist of binary elements, tactile arrays with force sensing elements have been demonstrated. Vision sensing is also in one of two modes: imaging, in which an object is noted as either being there or not, with possibly some image enhancement techniques, or pattern recognition, which can include scene interpretation. In addition, there are many specially-engineered sensors such as infrared sensors to detect heat given off by a work piece and magnetically induced eddy-current sensors used in metal-crack detection.

Justification

There are many justifications for using a robot in an industrial inspection scheme. The robot's immunity to fatigue and use of high-precision criteria allows for more consistent quality control and sorting results. Using a robot for inspection may allow in-process inspection to be performed in a hazardous environment. Robots can use sensory properties not available to humans, such a IR and eddy-currents. In addition, the use of robots for inspection allows for the electronic integration of inspection into the manufacturing process, providing an enhanced degree of flexibility.

Current Technological Constraints

The current constraints in robotic inspection concern both software and hardware issues. Pattern recognition algorithms, both for tactile contour maps and visual scene understanding, are too slow to allow real-time processing of the sensory information. Tactile sensing arrays are not sensitive enough to give real-time texture information and the size and cost of most sensing hardware makes it inappropriate or unfeasible in a large number of applications.

Application Examples

A very sophisticated inspection process is used by a major computer manufacturer to orient and inspect keycaps prior to loading them into magines for use in an automated assembly system. Key caps are shipped in bulk by the supplier, separated by keycap shape, color and legend. Keycaps are dumped into a bowl feeder that orients the caps and feeds them into a track leading to a visual inspection station. This inspection system rejects keycaps with defects, incorrect legends, flawed legends or surface defects, loading acceptable keycaps into magazines that are used subsequently by the keyboard assembly system.

In order for the vision system to "learn" the characteristics of a specific key cap, the operator steps through a menu driven procedure that inputs characteristics of the key (such as light legend on dark background), establishes the inspection window (i.e., what part of the visual field to process) and specifies the legend expected for the key. The system then prompts the user to feed a small number of keycaps known to be good through the inspection system to fine tune the inspection criterion. The result of this learning process is saved on a database and used to provide specifications anytime a batch of that specific keycap needs to be inspected. The entire inspection requires about two seconds per key.

4. Assembly

There are two major categories of assembly applications: closely fitted and easy-mating. The first of these deals with tight tolerance components that are generally fragile and require precise assembly motions. Easy-mating applications generally deal with larger components that are somewhat compliant.

Process Description

- o Acquire parts
- o Orient and Set-Up Parts
- o Perform Assembly - slide, insert, snap, press, stake
- o Inspect
- o Deposit finished assembly (Palletize)

Process Considerations

The range in weight and size of parts to be handled can vary widely, from a small spring up to a cast assembly housing. Furthermore, press fitting or staking as part of the assembly operation may require load capacity (i.e., strength) well beyond the weight of the parts involved. Any tooling that grips the parts must be able to accommodate the variety of shapes and sizes of parts involved in the operation, and handle them gently enough to avoid marring or deformation of sometimes delicate components.

Closely fitted components require precise assembly motions due to their tight tolerances, and may not incorporate any aids in positioning such as beveling or chamfering. Attempting to assemble close tolerance parts that are not properly positioned is likely to damage the parts, ruining the entire assembly. While press fitting is intended to require force for insertion, misalignment of the parts can raise the force required and ruin the parts.

Detection of errors during assembly operations is critical in terms of the unit being worked on and in terms of the assembly process itself. A flawed assembly is not only defective by itself, if it is not detected, it can be incorporated into a larger system that will then also be defective. If a particular step in the assembly sequence begins producing a high error rate it may indicate a problem with the assembly technique or with a batch of components.

Quality control can be implemented as a part of the assembly process and may include inspection of incoming components, inspection during asembly and inspection of the finished assembly.

Basic Elements

Mechanical - Manipulators used in closely fitted assembly are generally small and precise to match the requirements of a specific assembly task. The required load capacity need not be great, but should be adequate for press fitting, staking or application to assembly of units other than that originally implemented. End effectors used in assembly operations are usually specially designed for the specific parts to be handled. Remote Centers of Compliance (RCC's)[*] are becoming popular for closely fitted assembly due to their ability to compensate for small amounts of misalignment of parts.

[*] An ideal RCC is perfectly compliant perpendicular to axis of insertion and completely non-compliant along the axis of insertion.

Controller - Robot controllers for assembly applications can be set up for varying levels of sophistication. If parts are presented in an unstructured way, the controller must have the ability to search for and recognize the parts needed. Low clearance mating operations, since they frequently require positioning accuracy better than the manipulator accuracy, require that the controller be able to use some type of adaptive part mating algorithm for final alignment.

Sensing -

   o   Tactile - Binary sensing is useful as a simple test of whether or not a part is in place,i.e., to sense that a part has been dropped. Force sensing allows monitoring of parts alignment during insertion since misalignment causes excessive resistance. The abrupt change in applied force when mounting snap-on parts can be used to determine that the part is completely seated.

   o   Vision - Vision is becoming popular in assembly applications due to its flexibility. It is being used to locate parts for grasping, checking orientation and inspection of parts prior to assembly.

   o   Proximity - Light Emitting Diodes (LED's) and phototransistor detectors have been placed in the end effectors in some applications as an alternative to binary tactile sensing to verify that a part is in the gripper.

   o   Hearing - Hearing has been used in assembly operations both to verify that a snap-on part has seated and to detect the sound of a dropped part so that corrective action can be taken.

Justifications

Consistency is a major advantage of robots in assembly work. If the parts supplied to the robot are within specifications and the robot programming is set up properly, the robot will assemble each unit in precisely the same way. In contrast, human performance on monotonous tasks varies, making quality control difficult. Furthermore, human assemblers sometimes apply excessive force to poorly fitted parts in order to finish a unit.

This is a source of marginal or defective assemblies that can be eliminated by programming an assembly robot to limit insertion force.

In a clean room environment, a great deal of personnel time is spent preparing to enter, and the number of times an employee may leave and reenter the clean room per shift can be quite high. Once set up to operate in the clean room, the robot loses no production time leaving and reentering the clean room.

Current Technological Constraints

Parts acquisition is a difficult problem in robotic assembly at this time. Structured parts presentation (e.g., palletized supply, indexed presenter, etc.) solve the problem but can be prohibitively expensive, especially for small batches. Using the robot's sensing capability to locate and orient parts requires sophisticated (by today's standards) and costly sensing and control components. Furthermore, today's technology has not yet completely solved the "bin-picking" problem (acquisition of parts from a jumbled and overlapping pile), although there is a great deal of promising research addressing this problem, and a solution seems to be near.

The fundamental problem of placing one part inside of another part, especially for closely fitted parts, has been taken for granted due to the ease with which a human can solve the problem. However, when applying robots to closely fitted assembly operations, jamming is a common problem and many current algorithms to improve parts alignment reduce the speed of the operation noticeably. While currently available RCC's are helpful, their range of accommodation is not large, they are not programmable, and they are not totally successful in preventing jamming due to misalignment. Current work on the generic "peg-in-a-hole" problem and the explicit geometry of jamming are likely to result in more effective and more efficient methods of assembling tight tolerance parts.

At this time, error handling methodologies for assembly sequences still require improvment. Not only should an error during assembly be detected, but recovery from the error should be attempted. Whether recovery consists of discarding the entire assembly and starting over or discarding a single part and trying again should be determined by the type of error and the part of the assembly sequence involved. At this time, most assembly robots, when an error is detected, simply stop and wait for human assistance.

Application Example

An interesting example of robotic assembly has been shown by SRI, using two coordinated robot arms, one with a force sensing wrist and the other with a hand mounted camera to assemble a part of a printer carriage. There are four types of parts involved in this assembly, a square shaft with four plastic rocker arms already attached, four plastic rockers that snap into the rocker arms and two sizes of roller shafts that snap into the rockers. The sequence begins with Arm number 1, the one with the camera attached, picking up the shaft/rocker arm assembly and placing

it in a specially designed support fixture. Arm 1 then acquires rockers
and places them in the correct location with respect to the rocker arms.
Arm number 2 is used for assembly because of its force sensing capability.
The force exerted by Arm 2 is monitored as it presses on the rocker, with
seating indicated by a rise in force followed by an abrupt drop as the
rocker snaps into place. While this is being done for all four rockers,
Arm 1 has placed the roller shafts in an aligning fixture. In the next
step, Arm 1 lifts the shaft/rocker arm/rocker assembly and turns it over,
placing it on top of the fixtured roller shafts. Arm 2 then pushes down
on each rocker until it snaps onto the roller shafts. Force sensing is
again used to sense completion of the snap fit.

5. Painting/Coating

There are a variety of ways to apply coating materials to objects
that will be included in this application. Painting can be performed
by techniques ranging from dipping to electrostatic spraying. While not
dealing with paint, thermally sprayed coating to produce a metallic surface
is included because the method of deposition is very similar to spray
painting. Of these methods, spray painting is the most common in manufacturing, and is the primary focus of this section.

Process Descriptions

Dip Coating:
- o lower work piece into coating material reservoir
- o lift out
- o allow to drain; spinning can be used to remove
     excess

Flow Coating:
- o pour paint over work piece
- o allow to drain

Paint Spraying:
- o atomize paint
  - air spraying uses the violent mixing of paint
    with high pressure air
  - airless spraying applies pressure directly to
    force paint thru nozzle
- o direct paint to target
  - air and airless spraying use residual momentum
    from the atomizing process
  - electrostatic spraying uses electrical attraction
    between the charged paint droplets and the oppositely
    charged target.

Thermal Sprayed Coating:
- o melt coating material
  - Flame spraying feeds coating into gas (e.g. propane)
    flame
  - Arc spraying feeds coating material through

                    an electrial arc
            o atomize molten coating material with c o m p r e s s e d
               air jet
            o direct spray to target

Process Considerations

    The goal of painting/coating applications is generally to achieve
an even, controlled thickness, coating on the target. Too thick of a
coating is wasteful and may cause problems due to excessive drying time
while too thin of a coating may defeat the purpose of the process. In
dip and pour coating, control of the viscosity of the coating material
and manipulation of the object during the draining (e.g., tilting, spinning)
are used to control the coating process. In spraying operations, coating
thickness and evenness are determined largely by the path and speed of
the spray gun with respect to the target. A successful coating requires
that the spray gun be moved smoothly along its path, maintaining a constant
distance and speed while following the contour of surface to be painted.
This path control problem becomes more complex when the object to be painted
is moving along an assembly line or requires manipulation, such as   door
opening in automotive applications, during the painting process.

    The painting of large assemblies requires mobility from the painter,
and increases the difficulty of producing an even coating due to the distance
over which the evenness must be maintained. Convoluted and partially
enclosed structures are especially difficult to paint, requiring that
the spray gun be moved into confined spaces and carefully manipulated
to provide even coverage of interior surfaces.

    The environment in which spray painting is performed is a harsh envi-
ronment since the process itself generates solvent vapors and a paint
mist that envelop the work area. Personnel working in the vicinity of
painting operations wear protective clothing and breathing gear to reduce
exposure to these hazards, but the price of protection is worker discomfort
and fatigue. Not only is this environment unhealthy, but the risk of
fire or explosion is severe. The accumulation of paint mist in confined
spaces can become severe enough to reduce visibility, and paint buildup
on surfaces and equipment requires frequent removal.

Basic Elements

    Mechanical - Dexterity is needed by the robot manipulator arm; five
degrees of freedom allows painting of three dimensional surfaces although
one or more additional degrees of freedom or mobility enhance the ability
of the robot to reach interior spaces. Actuators must be explosion proof
due to the flammable atmosphere; as a result, hydraulic actuators are
generally used.

    Large assemblies require robot manipulators with large work volumes
or some form of mobility such as a rail transport system. End effectors
usually consist of permanently mounted spray guns, although some models
allow a walk-through teaching handle to be attached.

Controller - The primary task of robotic controllers in spray painting applications is to provide smooth continuous path control along the surface being painted. Virtually all of the current robotic painters are programmed by the walk-through method in which a skilled painter physically moves the robot arm along the desired path while the robot records the required motions. The moving target situation can be programmed in this way with the constant speed of the target added to the learned path by the controller. Editing capability is highly desirable to allow corrections or adjustments of the program, since changes to walk-through taught programs generally require that the entire program be retaught from the beginning.

Sensing - Very little sensing is incorporated in painting robots; satisfactory results are achieved with good path control, carefully taught programming and reliable painting equipment.

Justifications

The improvment in consistency is an advantage of robots for all of the painting/coating applications. Once successfully programmed, a robot will turn out properly painted or coated pieces time after time, unaffected by fatigue, paint fumes or boredom. Not only does this consistency raise the quality of the process, it also reduces the waste of deposited material due to the precision of its path and spray stop and starts. Overspray can be minimized when programming, and the robot will repeat this savings every duty cycle.

Removing personnel from the spray painting environment not only protects their comfort and health but also reduces the ventilation requirements for the spraying area which are based on hazard to humans. The energy required to heat or air condition the fresh air brought in for the painter is substantial, and increasingly stringent standards on venting of contaminated air require more and more sophisticated and expensive cleaning of paint laden air before it can be released.

As in arc welding, the skill required of a human painter to produce high quality paint finishes quickly is the result of training and experience. Reducing the standards required for a painter expands the pool of available personnel, but may require slowing down the assembly line, increasing the number of units requiring correction or allowing quality control standards to slip. Robots offer a solution to the problem of conflicting demands for higher productivity and quality versus human aversion to working in unpleasant and unhealthy environments.

Current Technological Constraints

Current industrial painting robots incorporate little or no sensing. As a result, there is virtually no fault detection capability incorporated in the robot. Inspection of surfaces prior to coating for dents, gouges or contamination, if performed at all, is done by human inspection. Process monitoring is largely restricted to detection of gross failure of the painting equipment, such as a clogged paint nozzle. Current technology could be applied to provide constant monitoring of the spraying process, including average droplet size and velocity and paint spray density.

This type of monitoring would allow adaptive reaction to fluctuations in the paint supply, improving quality control and reducing the incidence of parts requiring repainting.

The walk-through programming method most commonly used for painting robots is cumbersome. While current users, whose painting is predominantlty large batch, consider this time consuming set up process acceptable, users who operate on smaller batches would find it uneconomical. Additionally, the very high cost of painting robots is a major barrier to the use of robots for small batches.

Application Examples

A manufacturer of molded plastic parts for the audio industry has installed a robot to do the final painting. The robot uses a 5 DOF articulated arm with 32K words of available memory, programmable with either continuous path or point to point motions. Parts handling is accomplished by means of a double conveyor system arranged in a "V" pattern on either side of the robot. This arrangement increases the through-put of the system. In addition to the advantages of removing workers from a dangerous environment and reducing the company's dependence on skilled labor, the manufacturer has seen a 140 percent increase in daily productivity as well as an eightfold reduction in defective parts.

General Motors has recently installed what they believe to be the most advanced painting system in use at their Doraville, Georgia, assembly plant. The painting system consists of one painting robot and one door-opening robot mounted on tracks on each side of an assembly line. All four machines operate under computer control. The path tracking is accomplished by operating a single robot in the teach mode and then mirror-imaging the taught path into the robot on the opposite side of the line. With this system GM claims to be able to paint all external surfaces plus interior surfaces such as station wagon tailgates, deck lids, pick-up truck beds, door hinges, and door openings.

6. Other Applications

6.1 Sealing/Bonding

Process Description

Although each bonding application will have its own specific consider-ations, there are several generic steps that must be performed in a bonding application. These include:

- o securing the work piece to a fixed, known position
- o applying the bonding material
- o aligning the work pieces
- o fixing the work pieces together

Sealing applications may involve two work pieces, or may involve simply covering a hole in one work piece.

Process Considerations

In sealing and bonding, one of the major application considerations is the speed at which the material sets up. Because of the short working time of most commercial bonding materials, successful applications require a well controlled and coordinated process scheme. The applicator speed must be adjusted to give a minimum application time, as well as be coordinated with the material pumping and flow rates to ensure that a consistent bead of material is maintained. The path must be well planned to cover an appropriate area with economy of both time and material. In addition, the applicator must be accurately controlled through this path, in each repetition to maintain economy.

Basic Elements

The basic robot requirements for sealing and bonding are similar to those necessary for paint spraying and arc welding. These include five or six degrees of freedom in the manipulator for dexterity and continuous path control capabilities in the controller. End effectors for sealing consist of specialized sealant applicators mounted directly on the end of the robot arm. Controllers for bonding robots usually have the ability to directly control the flow rate of the sealant through the nozzle. Vision systems and specialized air jet sensors have recently been used to detect breaks in the material bead, and the weight of the material container is monitored to prevent the supply pump from running dry.

Justifications

The high repeatability of a robot can significantly reduce wastage in this application since the robot, once programmed with an economical path, will follow this path more closely than a human worker. Robots can increase productivity by applying sealing or bonding material faster than a human, and by eliminating the fatigue that results from manipulating the heavy adhesive gun. A final incentive for the use of robots is the removal of human workers from an environment of very high temperatures and noxious fumes.

Current Technological Constraints

Because sealing applications are fairly straightforward, most problems encountered can be solved with proper planning techniques. One problem that has not been solved yet, however, is that of error handling. Although sensors can detect a missed section of bead, appropriate methods are not yet available for returning to repair the gap.

<u>Application Examples</u>

At a General Electric Company in Kentucky, a robot is being used to apply a foamed hot melt adhesive to seal perforations in refrigerator cases. Refrigerator cases are transported horizontally along a conveyor to the sealing station, where they are automatically tipped so that the holes to be covered are on an angle. The robot applies a metered amount of sealant above the hole and gravity pulls the sealant over the hole to seal the hole.

6.2 <u>Finishing</u>

Because the manufacturing definition of finishing processes differs from the general use definition, we give here a brief explanation of finishing as it applies to this report. In general use, finishing usually refers to painting or coating type operations, performed as the last step in the manufacturing process. In the strict manufacturing sense, however, finishing refers to a category of cleaning processes, such as trimming flash from castings, sanding, deburring, and polishing. For the purposes of this report, finishing will refer to the strict manufacturing definition.

<u>Process Description</u>

Finishing usually consists of one or more of the following operations:

        -trimming flash, by:
            o saw trimming
            o spark cutting
            o laser cutting
        -grinding flash
        -sanding
        -deburring
        -polishing

<u>Process Considerations</u>

The first consideration in the finishing process is the shape of the part to be processed. The part, usually a casting, will be "raw" in the sense of having unpredictable burrs and pieces of flashing in unknown positions. The scattered flashing and heavy weight of the work piece make manual handling potentially dangerous and automated handling difficult. Depending on the condition of the work piece, it may be necessary to perform a combination of finishing operations, with or without inspection between the process steps. The cleaning processes themselves must be considered both to prevent deformation of the work piece, and to minimize the production of irritants such as dust and grit.

<u>Basic Elements</u>

There are two generic modes in which finishing operations are performed.

In one mode, the finishing tool, such as a grinding wheel, is fixed in a permanent position. The robot picks up the work piece, orients it, and passes it over the tool in a prescribed path. In the second mode, the work piece is positioned in a jig, and the robot moves the finishing tool. Payload weight can determine the mode selected: it is usually desirable to have the robot hold the lighter of the two objects. If the robot holds the tool, it may either hold the finishing tool in a standard gripper, or have the tool mounted permanently on the robot arm.

Specialized controllers and sensors are important for finishing operations. The relative needs for each of these are interdependent. If there is no sensing involved, the robot needs a very accurate model of the finished part, stored in a database, to which it can refer during the finishing operation. On the other hand, an integrated system using vision to detect flaws and force sensing to guide the finishing tool, would not need such a detailed model. Current state of practice is to use a simple internal model of the part combined with low-level sensing, such as force sensing.

Justifications

Human workers in the finishing environment are exposed to a variety of hazards, including high noise levels, airborne dust and grit, and disintegrating grinding wheels. Robots can remove humans from these dangers, enhancing plant safety.

Current Technological Constraints

Although force sensing provides an adequate means of controlling an operation such as grinding, the robot must still be programmed to traverse the entire workpiece. Additional sensing such as vision could be used to guide the finishing tool to only those areas of the work piece that require cleaning.

Application Examples

A Swedish foundry has installed a two-robot system for grinding operations. The first robot carries a permanently mounted grinding wheel and is used to cut ingots. The robot is equipped with both force and torque sensing, the latter used to detect wheel wear. Work pieces are fixtured on a rotating, two-position work table. The second robot handles the ingots directly, passing them by several finishing machines composing a work cell.

A truck manufacturer is using a robotic system that finishes cast iron gear housings. In the first step of the process, the robot arm picks up an abrasive cut-off wheel driven by a hydraulic motor that is used to remove risers and external flash from the raw casting. In the second stage, the robot replaces the cut-off tool with a gripper that picks up the casting and moves it to a floor mounted grinder. Before grinding, a sensor on the robot arm locates the surface of the grinding wheel to set a reference that compensates for wear of the wheel. The casting is then moved against the wheel to remove the parting line along the outside diameter of the casting. Flash is removed from the inside of the casting by positioning the casting over a floor mounted impact tool fitted with a chisel. The final finishing step, deburring inside surfaces, is performed by moving the casting to a floor-mounted abrasive deburring machine. This deburring machine includes automatic wear compensation and programming to shut down the system in the event of tool failure. The robot operates unattended during the night shift, with tool replacement and maintenance being performed during the day shift.

6.3   Investment Casting

Process Description

Investment casting is based on single-use molds; a new mold must be formed for each casting. The molds are formed in the following way. First, a wax model of the part is formed. The model is then coated with a lubricating and releasing agent. The mold is then dipped into a ceramic slurry, and coated with sand. The slurry/sand mixture is allowed to dry, and the dipping is then repeated. After five or six coats, the mold is placed in a heating unit, usually a steam autoclave, to melt out the wax model. The hollow mold is then fired in a kiln and used for the metal casting.

Process Considerations

The most critical consideration for a successful investment casting is quality control of the ceramic shell. Consistent thickness of the individual coats of ceramic slurry will result in a more uniform and higher quality finished shell. The dipping, rotating, and swirling motions while the mold is in the slurry are all important factors in the final coat thickness, and must be carefully controlled. Other considerations in investment casting are the wide range of weights to be lifted from the beginning to the end of the coating cycles, and the time and temperature control necessary during the drying cycle.

Basic Elements

The basic robot elements necessary for investment casting operations are similar to those required for dip painting. In addition, it may be necessary to have a particularly robust manipulator to handle the weights involved; in some applications payloads can weigh hundreds of pounds by the end of the dipping process. A desirable, though not essential, robot element used in investment casting is a flexible, easily re-programmable controller. Easy re-programming allows for economical small batch jobs.

Justifications

A robot is well suited to performing the dipping operations for investment casting because the key to a quality shell is the consistency of the slurry coats. Once a successful pattern of dipping, rotating and swirling the mold in the slurry is programmed into the controller, the robot will repeat those motions exactly. Another factor favoring a robot to perform the dipping operations is fatigue. The heavy weights involved often cause fatigue in human workers who dip the mold, reducing consistency and productivity.

Current Technological Constraints

The high cost of current robotic systems is not always justified for those applications where very small batches are common.

Application Example

A manufacturer of marine outboard engines has been using robots in their investment casting applications since 1974. The implementation is straightforward and required a minimum of plant reorganization. At full manufacturing output, this company produces many different types of castings, ranging in weight from 1/4 to 8 1/2 lbs. Each type of casting requires six individual coating cycles, with specific dipping and swirling motions. The robot controller is responsible for cycling the slurry bath motors as well as the fluidizing bed air supply. This robot implementation has increased both output and casting quality.

6.4   Die Casting

Process Description

The die casting process, unlike the investment casting process, re-uses the mold in which the product is cast. This necessitates additional steps to maintain the quality of the mold. The process as a whole consists of:

- Preparing the die
  o clearing the mold of any obstructions
  o lubricating the mold

- Pouring the liquid
  o checking the temperature of the liquid metal
  o controlling the pour rate of the liquid

- Controlling the time and temperature of the cooling cycle

- Extracting the workpiece from the die

- Checking the mold for parts remaining in the die

Process Considerations

There are several variables which must be carefully controlled for a successful die casting. These include temperature, which must be controlled for molds to be predictable, and cooling time. There is a delicate balance between the metallurgical requirement for adequate cooling time and the economic need for short cycle times. Die cleaning and lubrication between cycles must be thorough and consistent to prolong die life and give higher quality castings. A final consideration in die casting is safety when handling molten metal.

Basic Elements

The basic robot requirements for die casting are similar to those involved in general material handling, i.e. average manipulator dexterity. In addition, there are several robot elements that are especially useful in the die casting environment. These include temperature protected end effectors, which are necessary when working at the high temperatures involved in die casting, and smooth path control. Although not necessary for simple work piece handling, smooth path control becomes useful in the delicate mold cleaning process.

Justifications

An important reason for choosing a robot to work in a die casting environment is the consistency of the robot. The high repeatability of the robot can reduce scrap by as much as 20%, thus increasing productivity and decreasing re-melt costs. Furthermore, consistent and accurate die cleaning can significantly increase the useful lifetime of the die. Removing humans from a hazardous environment and having the capability of integrating the casting and finishing operations are additional reasons to use robots.

Current Technological Constraints

Although robotic die casting is fairly straightforward, there are several sensing capabilities that would enhance the current state of practice. These include better detection of incomplete part removal from the die and better real-time temperature sensing and control during the cooling process.

Application Example

Du-Wel's casting plant in Dowagiac, MI., casts parts for a variety of users, including automotive and appliance manufacturers. One of their most successful robot applications consists of servicing two die casting machines. The robot loads one machine, turns 180°, unloads the other machine, sprays the die with lubricant, deposits the piece into a quench tank, reloads the machine, then turns back to the first machine.

6.5 Forging

Forging, although an important backbone of many manufacturing processes, is in fact a very simple operation.

Process Description

At its greatest level of complexity, forging consists of:

- o acquiring the work piece
- o placing the work piece in a furnace
- o transferring the heated part from the furnace to a forging press
- o cycling the press
- o removing and quenching the work piece
- o inspecting the work piece
- o depositing the work piece

Process Considerations

Although a simple process, forging does require careful control of several variables, namely timing and temperature. The pre-forge temperature of the work piece must be precisely controlled for consistently successful forging. This can be accomplished by altering the time that the work piece spends in the furnace, by altering the furnace temperature directly, or by a combination of both. After forging, the work piece may need to be quenched. Improper quenching times or temperatures could result in undesirable crystallization of the metal. The environment of dirt, smoke, noise and high temperatures typical in a foundry is an additional consideration that affects productivity.

Basic Elements

The basic robot elements necessary for forging applications are similar to those required for general material handling, i.e. average dexterity in the manipulator movements to acquire, orient, present and remove the workpiece from the furnace and press. Variations of robot elements that are used in forging applications include specialized end effectors, sensors and controllers. The end effectors used in forging must be heat-resistant. The high temperatures involved in forging can easily damage the hydraulic or electric systems of an unprotected end effector. Sensors that are used in forging have been developed to take advantage of the forging conditions. For example, infrared sensors are used to detect the position and status of a work piece based on its heat output. Robot controllers used in forging applications are usually modified so that they can communicate with their environment, e.g., the controller is equipped to sense and/or control the furnace temperature, or to cycle the presses.

Justifications

The harsh environment of the work place is probably the most important justification for using a robot in forging applications. Because of the heat, dirt, noise and smoke, a human may need to take as much as three to four hours of breaks during one production shift. A robot can usually run continuously, unhampered by the environment. In addition, the precise nature of the robot controller allows very accurate and repeatable timing and motion control. This increases the consistency and quality of the forged parts.

Current Technological Constraints

While current robot controllers are capable of real-time temperature sensing, they are not sufficiently sophisticated for adaptive control of timing and temperature.

Application Example

An aircraft engine manufacturer has successfully incorporated a robot into the upset forging process in the manufacture of jet engine airfoil blades. This application begins with the robot acquiring the raw part from a vibrating parts feeder/orienter. An infrared sensor is used to check that the feeder is in fact loaded. The robot loads the part into a standard rotating hearth furnace, coupled to the robot controller. The temperature of the furnace is sensed by thermocouple sensors which detect simple over or under threshold conditions, while the position of the table is controlled by a stepper motor. After the hot workpiece is removed, the robot controller causes the furnace door to close, checks to see if the part is in fact in the gripper (by means of another IR sensor), instructs the manipulator to load the part into the press, then cycles the press. After cycling the press, the controller signals the press to eject the part, checks to verify that there is no part in the press, and then repeats the entire process.

6.6 Plastic Molding

Process Description

As with die casting, the individual processes associated with plastic molding are simple. The plastic molding cycle consists of:

- o loading the plastic charge into the die mold
- o loading the die mold in the molding machine
- o cycling the molding machine
- o extracting the molded part
- o inspecting and finishing if necessary

Process Consederations

Plastic molding is similar to die casting in that it involves most of the same process considerations as die casting. Among the more important are: time and temperature control, consistent and accurate die cleaning and lubrication, balancing the need for adequate cooling time against the need for fast cycle times, and the harsh environment of the molding workstation. Specific to plastic molding, however, are the noxious fumes given off by the molten plastic, and the delicate handling requirements of the pliant plastic.

Basic Elements

The basic robot requirements for plastic molding applications are similar to those of general material handling. Useful robot element variations for plastic molding include specialized end effectors and controllers. To speed cycle times, the robot must handle the molded parts while they are still warm. The end effectors used for this handling must be able to manipulate the hot, compliant parts without deforming them. As in die casting, the robot controller must be interfaced with the peripherals that it will be controlling, such as the molding machine.

Justifications

The justifications for using robots in plastic molding are similar to those in die casting. These include increased quality due to the control, consistency and repeatability of the robot, and the removal of workers from the hazardous environment.

Current Technological Constraints

The major technological barriers to the increased use of robots in plastic molding involve sensing and control. Current sensing systems cannot detect small parts of the molded piece adhering to the die rapidly enough to avoid interfering with the cycle time. As a result, robotic systems either leave occasional remnants in the die, which ruins the next molded part, or clean the entire die each cycle, which reduces the lifetime of the die.

Application Example

An appliance manufacturer is using robots in the molding of vacuum cleaner parts. A pick-and-place robot removes two molded parts at a time from a dual cavity injection molder, using a specially designed twin gripper. The robot presents each part to a broach machine for sprue removal and then deposits the parts on a cooling conveyor. The elimination of an unpleasant and hazardous job was the primary motivation for installing the robot, but the increased productivity due to the robot allowed an investment payback of less than two years.

Appendix B

Industrial R&D Activities

In this appendix, we will briefly describe R&D programs of individual companies active in the field of robotics, including both producers and end-users. This list is of course limited by the availability of information concerning private companies.

Producers:

Westinghouse/Unimation

With the acquisition of Unimation, Westinghouse became one of the larger companies to manufacture robots both for sale and for in-house use. The majority of the research performed, however, remains largely centered on application development. One of the on-going projects concerns the development of an automated turbine blade inspection system in their Winston-Salem plant. Westinghouse is working closely with Carnegie Mellon University on the development of this software-intensive project. Several other projects include the automation of circuit board assembly in several different plants, and the development of a laser manipulator tube for the Nirop plant in Minneapolis. Past projects have included development of the APAS assembly system and several vision inspection systems.

Prab Robots

Prab robots has as one of its corporate philosophies the view that the simplest robot that can perform the job should be used for the job. For this reason, Prab tends to spend a good portion of its relatively small R&D budget not on new, leading edge technology but on adapting their robot lines for use in various established industries and applications, such as palletizing, parts transfer, and machine tool load/unload.

GCA

GCA's corporate strategy is inclined towards development of complete automated manufacturing systems. GCA has acquired several smaller companies and licensed appropriate technologies in an effort to create an immediate market presence in the field of flexible manufacturing, while at the same time integrating these technologies through the strong software design capability of their Industrial Systems Group. One of the most important projects involves the development of an advanced controller capable of complete integrated system control.

GE

At the Corporate Research Center in Schenectady, research projects

421

range from basic to developmental. On the basic research end, projects on vision, tactile sensing, and process planning are in progress. Application research centers on assembly, laser machining, and arc welding. The arc welding project has produced "Weldvision", a novel method of monitoring the weld puddle to control the arc welding process. Smaller projects scattered around the corporation include off-line programming development and local area networking.

GMF Robotics

While the majority of the basic research efforts of GMF are performed by Fanuc in Japan, GMF performs some application development work in the U.S. Highest on the list of priorities is research in vision systems. Other projects include work on quick-change robot components for a modular robot and development of application-oriented off-line programming.

IBM

While IBM originally concentrated their robotics efforts on application-oriented developmental work, recently they have shifted their emphasis to more fundamental issues necessary for the development of their own robot lines. Current work centers on simple 3-D and sophisticated 2-D vision, geometric modelling, various assembly issues including compliance, and intelligent software support. In the past, research efforts have included development of the process oriented programming language AML, as well as the assembly gantry robot, model 7565.

Automatix

Research and development at Automatix is focused in two directions, vision and control. Vision research, on which Automatix built its reputation, currently centers on developing an inexpensive, fast 2-D vision system as well as development of 3-D vision. The vision research has led Automatix into the control area, which currently includes sensory-based control for seam welding and control for a flexible manufacturing system.

Adept Technology

Much of the current work on process-oriented language and system design at Adept Technology is an extension of the software effort, specifically VAL-II, that was developed by Adept Technology when it was known as Unimation West. While this work is still important, other projects at Adept include development of a three-wheeled mobile cart on which a PUMA robot can be mounted, a six degree of freedom manipulator, a high speed servo control system, a new robot using direct drive actuators and several vision sensors.

Allen Bradley

Allen Bradley is a component manufacturer specializing in controllers.

In addition to research efforts aimed at increasing the sophistication of their existing controller lines, Allen Bradley currently is involved in the development of an advanced programming language based on the Pascal language. More developmental work is currently underway in the area of AC servo drives.

Lord Corporation

The Lord Corporation is a component manufacturer, specializing in tactile sensors and tactile sensor control algorithms. Researchers at the Lord Cprporation are working very closely with the tactile sensing laboratory at Case Western Reserve University. In addition to tactile sensing research, Lord Corporation has begun work on 2-D vision for inspection.

Fared

The Fared group of companies is composed of three robot producing firms. Fared Robot Systems has as its R&D thrust the development of an assembly robot to handle clean room applications. Robot Defense Systems is concentrating on an autonomous mobile robot for security. Farad Drilling Technologies has developed massive robots capable of lifting 5,000 lb. pipe sections and currently conducts R&D on controls for these robots.

Cincinnati Milacron

Cincinnati Milacron is one of the pioneer robot producers. They include vision sensing, control system architectures, programming languages and integrated manufacturing systems among their R&D programs. Specifically, current projects are concentrated on combining laser technology with robotics, and automating the production of structural components out of advanced composite materials.

Machine Intelligence Corporation

Machine Intelligence Corp. is developing work cells which incorporate robots, micromanipulators, lighting systems and machine vision systems for use in the semi-conductor and computer-peripheral industry. These fully integrated systems are designed to perform precision measurements in the micro-realm for in-process inspection and statistical quality control for fully automated production lines.

End-Users:

McDonnell Douglas

Robotic R&D efforts at McDonnell Douglas are concentrated on off-line programming and system control. McDonnell Douglas is assessing the capabilities of the programming language MCL for off-line programming, and the use of MCL in an actual production mode. In another effort, the language system RAPT (developed by researchers at Edinburgh University in Scotland) is being used for intelligent reasoning on a database of geometric models. In the past McDonnell Douglas has developed several modeling and simulation software packages, including one known as "Place".

Northrop

In addition to several classified robotics projects, Northrop is developing robotic capabilities for aircraft parts manufacture and inspection. Specifically, Northrop is studying automatic placement of carbon-impregnated fabric in the manufacturing process of airplane wings, automatic drilling of holes in wings, and visual inspection of material texture. For the visual inspection project, Northrop is studying the applicability of 3-D vision.

Hughes

Research at the Hughes Research Laboratories is concentrated on the development of an intelligent, autonomous system. Research issues include: knowledge-based systems, image analysis, navigation, goal monitoring and planning.

Fairchild

Fairchild robotics research is unique among end-user R&D in that it is concentrated almost exclusively on fundamental, basic research. Issues under consideration include: 3-D vision, specifically for use in IC inspection, intelligent systems for VLSI design using PROLOG, and knowledge representation.

General Dynamics

General Dynamics is relatively active in aerospace robotics R&D. Beginning in the 1970's with the Air Force ICAM project, General Dynamics has led a strong program in application specific development, most notably the wing drilling project. Currently, R&D work at General Dynamics is focused on system integration.

GMC

The largest robot user in the automotive industry, GM has several research and development projects in progress. The largest of these efforts is the development of their NC painter, a painting system project whose goal is to remove human workers from all aspects of the painting process. Other projects involve vision research based on a CAD modeling system, and some assembly work directed at complex engine subassemblies.

Appendix C

Not-for-Profit and Academic R&D Activities

This appendix is included to give the reader a more in-depth view of the size, scope and directions of NFP and academic R&D programs than was practical to list in the text.

<u>NFP</u>:

Although there are at least four non-profit research centers participating in robotics research, the two which sponsor the largest programs are the Stanford Research International Laboratory (SRI) and the Charles Stark Draper Laboratory.

SRI has been conducting research in robotics since the SHAKEY Artificial Intelligence Project, begun in the late 1960's. Current research at SRI covers a broad range of areas, with an emphasis on vision. SRI developed the first algorithms for binary image processing and continues to develop binary and gray scale vision for depth perception and parts recognition. Other projects include vision controlled arc welding, assembly, semi-automatic process planning, circuit board inspection, voice control and flexible grippers.

The Charles Stark Draper Laboratory has conducted research in a number of areas, such as real time simulation of the space shuttle's robot arm, 6-axis force/torque sensors and batch assembly processes. Research has also included accommodators for robot wrists which allow tight fitting parts with varying tolerances to be assembled without additional movement of the robot arm.

There are two other non-profit research centers active in robotics research, the Jet Propulsion Laboratory (JPL) of the California Institute of Technology and the Manufacturing Productivity Center of the Illinois Institute of Technology Research Institute. The robotics effort at JPL consists of 14 staff members and includes research in sensor-based control, teleoperators, multiple finger grippers and artificial intelligence. Efforts at the Manufacturing Productivity Center include research in sensors, controls, assembly, welding and material handling.

<u>Academic</u>:

Carnegie Mellon

At Carnegie Mellon University (CMU) there are presently 72 researchers devoted wholly or in part to robotics research. This makes CMU the largest robotics research center in the country. Funding for CMU comes from both government and industrial sources. In FY 82 CMU received a total of approximately $4 million for research. Over two million of this was from industrial sponsors, such as Westinghouse and Digital Equipment. The Office of Naval

Research (ONR) contributed about half a million dollars and the National Science Foundation (NSF) $375 thousand in FY83.

The key areas of research at CMU are sensing, programming, arm and gripper design, mobility and factory automation. One example of their sensory research is a proximity sensor which utilizes six infrared LEDs in a circular pattern and an anolog spot position detector chip. The sensor is used to determine surface characteristics and position to an accuracy of 0.1mm. CMU is also developing a direct drive manipulator, the DD Arm, which is driven by rare-earth magnet DC torque motors. These motors, due to their low operating speeds and low weights, are used not only as actuators but also as joints. This design eliminates transmission mechanisms, thus increasing efficiency and eliminating backlash problems.

Stanford University

Research efforts at Stanford have been, primarily, in force sensing, vision and programming languages. Their most significant advances have been in programming. Stanford has developed one of the most advanced programming languages, the Arm Language, or AL, as well as ACRONYM, which is used for robot programming, geometric modeling and reasoning in model based vision systems. Workers at Stanford have also developed a software package called SIMULATOR for use in off-line programming, which allows users to test programs prior to use.

Massachusetts Institute of Technology

Robotics Research at MIT is conducted in the Artificial Intelligence Laboratory and the Mechanical Engineering Department. At the Artifical Intelligence Laboratory, research is conducted primarily in mechanisms, computer controls and vision. Some of the work performed at the laboratory has included tendon-activated hands and their control. This research has produced lighter, nimbler hands for a PUMA robot. Other research is directed on a "robot skin," which detects pressure to discriminate between similar objects and to determine part orientation. Robot vision research at the AI Laboratory includes real time processing and 3D vision. The Mechanical Engineering Department at MIT carries out research in computer-controlled teleoperators for undersea work, drive systems, vision and prosthetics.

Although the three universities mentioned above sponsor the largest robotics programs, there are many others also active in robotics R&D. In the following, we will list these universities with their focused areas of research and include, wherever possible, the principal researchers, funding level and estimated staff.

New York University

Principal Researcher: Jack Schwartz
Funding: $1.25 - $1.5M per year from NAVSEA/ONR
Staff: 2 Faculties, 6-7 Graduate Students

O Development of Special Purpose Robot Language
O Software Algorithms: Obstacle Avoidance, Peg-in-Hole Assembly

Ohio State University

Principal Researcher: Robert McGhee
Funding: $1M per year from NSF and DARPA

O Leg Locomotion (Machine and Human)
O Controls
O Dynamics

Purdue University

Principal Researcher: L. Paul, B. Arash, S. Nof
Funding: $100K per year from NSF

O Vision
O Programming Languages
O Control Systems
O Plant Modeling

Rensselaer Polytechnic Institute

Principal Researcher: Leo Hanifin, C. W. LeMaistre
Funding: Industrial Association, NSF

O Computer Graphics Simulation of Robots and Layouts
O CAM Controllers
O Infrared, Sonar and Radar Sensors
O Gripper Design
O Robot Safety

University of Alabama

Principal Researcher: J. Hill, X. D. Zhang
Funding: $0.5M to 0.75M per year from the Army
Staff: 5 - 6 Faculty, 6 - 12 Graduate Students

O Manufacturing System Simulation
O Stereoscopic Vision

University of Cincinnati

Principal Researcher: Ronald Huston
Staff: 4 Professionals

O   Robot Arm Design
O   Kinematics and Dynamics of Robots
O   Vision

University of Florida

Principal Researcher: Del Tesar
Staff: 10+ Faculty, 20 Graduate Students
Funding: Approximately $1 Million (FY 82) from
          DOE, NSF, Army and the State of Florida

O   Robot Arm and Actuator Design
O   Computer-Based Teleoperators
O   Kinematics and Dynamics of Robot
O   Locomotion in Battlefield Conditions
O   Hierarchial Controls Using Force Feedback

University of Maryland

Principal Researcher: Azriel Rosenfeld
Funding: $1M+ per year

O   Vision and Image Interpretation
O   Real Time Programming Systems for Sensor and Control Interaction
O   Artificial Intelligence

University of Massachusetts

Principal Researcher: Kioch Masubuchi
Funding: $300K from NSF. Additional Funding from
          NAVSEA

O   Part Design for Automatic Assembly
O   Economic Application of Assembly Robots
O   Control of Welding Operations

University of Michigan

Principal Researcher: D. E. Atkins
Funding: $500K from State of Michigan. Additional
          Funding from Air Force, AFSC, AFOSR
Staff: 17 Faculty, 35 Graduate Students

O One of AFOSR's "centers of excellence"

O Programming Languages
O Vision
O Control Systems

University of Rhode Island

Principal Researcher: Robert Kelley
Funding: $200K (FY 82) from NSF
          $1M (FY 82) from Industry

O Vision Software
O Dexterous Gripper Designs
O Programming Languages
O Manufacturing System Design

University of Rochester

Funding: $200K from NSF and ONR

O Vision
O Computer Graphics Languages Called PADL for Storing 3D Shapes
   in Computer
O Automation of Manufacturing

University of Tennessee

Principal Researcher: Dr. Gonzales
Funding: $0.5 to 0.75M/Yr from NSF
Staff: 3 Faculty, 20 Graduate Students

O Integrated Sensory Research

In addition, there are a number of universities that have participated
in robotics research to a lesser extent. For completeness, they are listed
as follows:

Case Western Reserve University, Clemson University, Duke University,
George Washington University, Georgia Institute of Technology, Illinois
Institute of Technology, Lehigh University, Louisiana State University,
North Carolina State University, Northwestern University, Oregon State
University, Rice University, Texas A&M, University of Arizona, University
of Central Florida, University of Connecticut, University of Illinois
at Chicago, University of Minnesota, University of New Mexico, University
of Pennsylvania, University of Southern California, University of Tennesse,
University of Texas, University of Virginia, University of Utah, University
of Washington, University of Wisconsin, and Virginia Polytechnic Institute.

Appendix D
Bibliography

1. Proceedings

Casasent, D.P. (ED.), <u>Robotics and Industrial Inspection- Proceedings of SPIE</u>, SPIE-Int. Soc. for Opt. Eng., Vol. 360 (1983).

Lee, C.S.G. et. al. (Eds.), <u>Tutorial on Robotics</u>, IEEE Computer Society Press, Silver Spring, Md. (1983).

Salter, G.R. (Ed.), <u>Developments in Mechanised Automated and Robotic Welding</u>, The Welding Institute, Cambridge, England (1981).

Society of Manufacturing Engineers, <u>13th ISIR/ Robots 7</u>, Robotics International of SME, Vol. 1 (1983).

Society of Manufacturing Engineers, <u>13th ISIR/ Robots 7</u>, Robotics International of SME, Vol. 2 (1983).

IFS, Ltd. <u>Robots in the Automotive Industry-An International Conference</u>, IFS Publications, Ltd., Kempston, Bedford, England (1982).

Society of Manufacturing Engineers, <u>Applying Robotics in the Aerospace Industry</u>, Robotics International of SME, (1983).

Society of Manufacturing Engineers, <u>Robots 6</u>, Robotics International of SME (1982).

Shaw, M., Chairman, <u>Proceedings Robotic Intelligence and Productivity Conference</u>, Wayne State University, Detroit, MI., (1983).

2. General Surveys

Ando, S. and Goto, T., "Current Status and Future of Intelligent Industrial Robots," IEEE Trans. Ind. Electr. IE-30, 291 (1983).

Baranson, J., <u>Robots in Manufacturings</u>, Lomond Publications, Inc., Mt. Airy, Maryland (1983).

Bredin, H., "Unmanned Manufacturing," Mech. Eng. 104, 59 (1982).

Gevarter, W., <u>An Overview of Artificial Intelligence and Robotics, Vol. II-Robotics</u>, U.S. Department of Commerce, Washington, D.C. (1982).

Gibbons, J., "Exploratory Workshop on the Social Impacts of Robotics:

Summary and Issues.," Natl. Tech. Info. Svc. (NTIS), Springfield, Va. (1982).

Heginbotham, W.B., "Present Trends, Applications and Future Prospects for the Use of Industrial Robots," Proc. Inst. Mech. Eng. 195, 409 (1981).

Hogge, N. and Cutchin, J., "Competitive Position of U.S. Producers of Robotics in Domestic and World Markets," U.S. International Trade Comm., Washington, D.C. (1983).

Hunt, V., Industrial Robotics Handbook, Industrial Press Inc., New York, New York (1983).

Inaba, S., "An Experience and Effect of FMS in Machine Factory," IEEE Control Sys. Mag. 2, 3 (1982).

McCann, M., "Robot Showcase," Automot. Ind. 163, 15 (1983).

McElroy, J., "Market Realities Cool Robot Mania," Automot. Ind. 163, 12 (1983).

Mergler, H.W., "A Focused Bibliography on Robotics," IEEE Trans. Ind. Electr. IE-30, 178 (1983).

Sackett, P.J., Rathmill, K., "Manufacturing Plant for 1985-Developments and Justification," Proc. Instn. Mech. 'engrs. 196, 265 (1982).

Tanner, W.K. ed., Industrial Robots, Vol. 2/Applications, Robotics International of SME, Dearborn, Michigan (1981).

Walker, P.L., ed. "American Metal Market/Metalworking News Edition," Fairchild Publications, New York, New York (1984).

Technical Database Corp., 1983 Robotics Industry Directory, Conroe, Texas (1983).

Tech Tran Corp., Industrial Robots, a Summary and Forecast, Naperville, Illinois (1983).

Tech Tran Corp., Machine Vision Systems a Summary and FAREYOMT, Naperville, Illinois (1983).

Technical Insights, Inc., INDUSTRIAL ROBOTS...Key to Higher Productivity, Lower Costs, Fort Lee, New Jersey (1980).

3. Applications

Asano, K., et. al., "Multijoint Inspection Robot," IEEE Trans. Ind. Electr. IE-30, 277 (1983).

Bopp, T., "Robotic Finishing Applications: Polishing, Sanding, Grinding," 13th ISIR/Robots 7 (Proc) 1, 3-61 (1983).

Bowles, P. J., Garrett, L. W., "An Appliance Case History," Adhes. Age. 26, 26 (1983).

Curtin, F., "Automating Existing Facilities: GE Modernizes Dishwasher, Transportation Equipment" p. 32 (1983).

Duewke, N., "Robotics and Adhesives--an Overview," Adhes. Age 26, 11 (1983).

Gerelle, E., "Assembling Thermal Contacts With Robots," Assem. Autom. 254(1981).

Gustafsson, L., "Cleaning of Castings-A Typical Job for a Robot," 13th ISIR/ Robots 7 (Proc) 1, 8-24 (1983).

Hambright, R.N. et. al. Engineering Study and Analysis on the Feasibility of Modernizing the Main, Southwest Research Institute, San Antonio, Texas (1983).

Hartley, J., "Applications Diversify," The Industrial Robot 9, 56 (1982).

Hartley, J., "An experiment in Robot Assembly-Building Electric Motors," Assem. Autom. 1, 266 (1981).

Iron, G., and George, L., "Thermal Spray Robots," 13th ISIR/Robots 7 (Proc) 1, 8-34 (1983).

Jablonowski, J., "Robots That Weld," Am. Mach. 127, 113 (1983).

Kretch, S., "Robotic Animation," Mech. Eng. 104, 32 (1982).

Lambeth, D., "Robotic Fastener Installation in Aerospace Subassembly" 13th ISIR/Robots 7 (Proc) 1, 10-1 (1983).

Manimalethu, A., "Agricultural Robotic Application," 13th ISIR/ Robots 7 (Proc) 1, 10-76 (1983).

Molander. T., "Routing and Drilling With an Industrial Robot," 13th ISIR/ Robots 7 (Proc) 1, 3-37 (1983).

Mortensen, A., "Automatic Grinding,", 13th ISIR/ Robots 7 (Proc) 1, 8-1 (1983).

Mullins, P., "Automated Assembly Gains Ground," Autom. Ind. 163, 27 (1983).

Parker, J.K., et.al., "Robotic Fabric Handling for Automatic Garment Manufacturing," J. Eng. Ind. Trans. ASME 105, 21 (1983).

Ranky, P.G., "Increasing Productivity with Robots in Flexible

Manufacturing Systems," Ind. Robot 8, 234 (1981).

Reid-Green, K.S. et.al., "CAD/CAM at RCA Laboratories--Tools and Applications, Present and Future," RCA Eng. 28, 29 (1983).

Safai, S., Manufacturing Technology for Advanced Automated Plasma Spray Cell (APSC), Pratt and Whitney Aircraft, West Palm Beach, Florida (1983).

Smith, R.C. and Nitzan, D., "A Modular Programmable Assembly Station," 13th ISIR/Robots & (Proc) 1, 5-53 (1983).

Stoops, B., and Ferrier, P., "Merging Two Technologies: Robotics and Hot Melt Adhesives," Adhes. Age. 26, 22 (1983).

Toepperwein, L.L. et.al., ICAM Robotics Application Guide, Tech Report AFWAL-TR-80-4042 Vol. 2 (1980).

Wernli, R., "Robotics Undersea," Mech. Eng. 104, 24 (1982).

Wevelsiep, K., "Reading Robot Sorts Parcels," Schweitz. Tech., Z. 79, 6 (1982).

Wolke, R.C., "Integration of a Robotic Welding System with Existing Manufacturing Processes," Weld. J. 61, 23 (1982).

Wong, P.C., Hudson, P.R.W., "The Autralian Robotic Sheep Shearing Research and Development Programme," 13th ISIR/Robots 7 (Proc) 1, 10-56 (1983).

Yurevitch, E.I., et. al., "Expanding the Use of Robots for Assembly in the Soviet Union," Assem. Autom. 1, 259 (1981).

Japan Industrial Robot Association, The Specifications and Applications of Industrial Robots in Japan, Tokyo, Japan (1984).

4. Sensing

Allen, R., "Tactile Sensing, 3-D Vision, and More Precise Arm Movements Herald the Hardware Trends in Industrial Robots", Electron. Des. 31, 99 (1983).

Bergemann, H., "CCD--Image Sensors: New Eyes for Robots," Elektronik 32, 74 (1983).

Corby, N. R. J., "Machine Vision for Robotics," IEEE Trans. Ind. Electr. IE-30, 282 (1983).

Crosnier, J.J. "Grasping Systems with Tactile Sense Using Optical Fibres," Developments in Robotics 1983, p. 167 (1982).

Harmon, L., "Automated Tactile Sensing," Int. J. Rob. Res. 1, 3 (1982).

Kinoshita, G. et.al., "Development and Realisation of a Multi-Purpose Tactile Sensing Robot," Developments in Robotics 1983, p. 185 (1982).

Okada, T., "Development of an Optical Distance Sensor for Robots," Int. J. Rob. Res. 1, 3 (1982).

Presern, S. et. al., "Application of Two-Degrees-of-Freedom Tactile Sensor for Industrial Welding Robot," Developments in Robotics 1983, p. 177 (1983).

Pryor, T. and Pastorius, W., "Applications of Machine Vision to Parts Inspection and Machine Control in the Piece Part Manufac-turing Industries", 13th ISIR/ Robots 7 (P$_{roc}$) 1, 3-21 (1983).

Raibert, M., "Design and Implementation of a VLSI Tactile Sensing Computer," Int. J. Rob. Res. 1, 3 (1983).

Rovetta, A., "A New Robot with Voice, Hearing, Vision, Touch, Grasping, Controlled by One Microprocessor with Mechanical and Electronic Integrated Design", 13th ISIR/ RObots 7 (Proc) 1, 3-57 (1983).

Villers, P., "Recent Proliferation of Industrial Artificial Vision Applications," 13th ISIR/ Robots 7 (Proc) 1, 3-1 (1983).

5. Control

Bonney, M.C. et. al., "Verifying Robot Programs for Collision Free Tasks" Developments in Robotics 1983, p. 257 (1983).

Cook, G., "Robotic Arc Welding: Research in Sensory Feedback Control," IEEE Trans. Ind. Electr. IE-30, 252 (1983).

Mutter, R., "Effective Interfacing Through End Effectors," 13th ISIR/ Robots 7 (Proc) 1, 4-1 (1983).

Nagel, R., "Robots: Not Yet Smart Enough," IEEE Spectrum 20, 78 (1983).

Schwartz, J. and Sharir, "On the Piano Movers' Problem: V. The Case of a Rod Moving in Three-Dimensional Space Amidst Polyhedral Obstacles," New York University, Computer Science Technical Report No. 83 (1983).

Warnecke, H.J. et.al., "Simulation of Multi-Machine Service by Industrial-Robots," 13th ISIR/ Robots 7 (PRAY) 1, 2-10, (1983).

6. Manipulation

Orin, D.E., "Application of Robotics to Prosthetic Control," Ann. Bromed Eng., 8, 3-6 (1980).

Vukobratovic, M.V., "Engineering Concepts of Dynamics and Control of Robots and Manipulators," Adv. in Comput. Technol., 1, 212 (1980).

Vukobratovic, M.V., "Modeling and Synthesis of Control of a Manipulator for Mechnical Assembly," Tekh. Kibern. (USSR), 18, 44 (1980).

Takegaki, M., "An Adaptive Trajectory Control of Manipulators," Int. J. Control (GB), 34, 219 (1981).

Bailey, S., "Precise Positioning: A Matter of Sensors and Loop Dynamics," Control Eng, 27, 50 (1980).

Diaz, R., "Robotic Actuators: A Technology Assessment," Adv. in Comput. Technol., 1, 225 (1980).

Nerozzi, A., "Study and Experimentation of a Multi-Finger Gripper," Proc. Int. Symp on Ind. Robots, 10th, Int. Conf. on Ind. Robot Technol. 5th, Milan, Italy, 215 (1980).

Kelly, F., "Recent Advances in Robotics Research," SAE, prepr., no. 800383, 5 (1980).

7. R&D Activities

Atkins, D.E. , Volz, R.A., "Coordinated Research in Robotics and Integrated Manufacturing," Air Force, AFSC contract f49620-82-c-0089 (1983).

Brown, D. et.al., R&D Plan for Army Applications of AI/Robotics, SRI International, Menlo Park, California (1982).

Carlsson, J. and Selg, H., "Swedish Industries Experience with Robots," Ind. Robot. 9, 88 (1982).

Rooks, B., Developments in Robotics 1983, IFS Publications Ltd. Kempston, Bedford, England (1983).

Schlussell, K., "Robotics and Artificial Intelligence Across the Atlantic and Pacific," IEEE Trans. Ind. Electr. IE-30, 244 (1983).

8. Personal Communications

Following is the list of information sources that were collected during the course of study through a number of personal contacts in various forms. These referencesare cited here according to the following format:
Name, Information Content, Form of Communication, Date

Brady, M., Industrial R&D Activities on robotics, Notes, March 1984.

Brady, M., Robotic R&D Activities in foreign countries, Interview notes, April 1984.

Nagel, R., Background information on robotic industry and research community, Interview notes, February 1984.

Nagel, R., R&D taxonomy for robotics and comments on technological forecast, Interview notes, February 1984.

Carlisle, B., R&D trends and issues on robotics, Interview notes, March 1984.

Hartman, D., Information on robotic industry and government programs, March 1984.

Kelly, R., Research issues and technological forecast, January 1984.

Isler, W., DARPA research program on robotics, Program document, February 1984.

Haynes,L., NBS robotic program, Interview notes and program documents, March 1984.

Griswold, R., Robotic R&D sponsored by the Air Force Manufacturing Sciences Program, Program document, March 1984.

Everett, B., Information on the RAID data base, Data base output and briefing notes, March 1984.

Mc Glone, S. Summary of Army robotic efforts, March 1984.

Borase, F., ICAM Products Catalogue, March 1984.

Windsor, Summary of AFOSR-sponsored projects on robotics, February 1984.

Brasseau, G., Summary of NSF-sponsored projects on robotics, February 1984.

Appendix E

List of Contacts

During the course of this study, DHR approached many agencies and individuals to obtain information about their involvement in the robotic field. Our interaction with them took place in various forms such as telephone interviews, personal meetings or written communications. Below is a list of these individuals and their affiliations. Due to the limited space, we are obliged to neglect a number of contacts which are also very helpful but of a lesser significance.

1. <u>Air Force</u>

USAF Headquarters
Fiorino, Col. Tom
Room 4C283, Pentagon
Washington, D.C. 20330
202/697-1417

Air Force Office of Scientific Research
Electronic and Material Sciences
Windsor, LtC.
Bldg 410
Bolling AFB, Washington, D.C.

Air Force Human Research Laboratory
AFHRL/X
McCall, Maj.
Brooks AFB, TX 78235
512/536-3853

Aerospace Medical Division
AFSC/AMD/RDX
Beatty, Col. Dave
Brooks AFB, TX 78235
512/536-3406

Contract Management Division
AFSC/AFCMD/PD
Glover, Vern
Kirtland AFB, NM 87117
505/844-9656

Air Force Flight Test Center
AFSC/AFFTC/XRX
Johnston, Robert
Edwards AFB, CA 93523
805/277-3837

Air Force Geographics Laboratory
AFSC/AFGL/XOP
Posiadjo, Ron
Hanscom AFB, MA 01730
617/861-3606

Air Force Weapons Laboratory
AFSC/AFWL/PRP
Algermission, Robert
Kirtland AFB, NM 87117
505/844-9376

Arnold Engineering Development Center
AFSC/AEDC/DEM
Hartig, Maj.
Arnold AFS, TN 37389
615/455-2724

Ballistic Missile Office
AFSC/BMO/PMD
Roundtree, Maj. W.J.
Norton AFB, CA 92409
714/382-6014

Space and Missile Test Organization
AFSC/SAMTO/PM
Stevens, Ted
Vandenburg AFB, CA. 93437
805/866-1662

Air Force Logistics Command Headquarters
AFLC/MAXT
Hary, Lazlo
Wright Patterson AFB, OH 45433
513/257-7114

Air Force Logistics Command Headquarters
AFLC/MAXE
Head, Lawrence J.
Wright Patterson AFB, OH 45433
513/257-6163

Aerospace Medical Division
AFSC/AMD/RDT
Herrah, Col.
Brooks AFB, TX 78235
512/536-2091

Air Force Materials Laboratory
AFSC/ASD/AFWAL
Lee, Sylvester
Wright Patterson AFB, OH 45433
513/255-5151

Space Division
AFSC/SD/PDP
Black, Henry
Los Angeles AFS, CA 90009
213/643-0854

Manufacturing Technology Division
AFSC/ASD/ARWAL/MLTC
Schultz, William

Air Force Materials Laboratory
AFSC/ASD/AFWAL
Russo, Vince
Wright Patterson AFB, OH 45433
513/255-2738

Manufacturing Technology Division
AFSC/ASD/AFWAL/MLTC
Hitchcock, M.
Griswold, Roger
Wright Patterson AFB, OH 45433
513/255-7371

2. Navy, Army and DARPA

Naval Air Systems Command
Warren, Robert
Washington, D.C. 20350
202/692-2515

Naval Air Systems Command
Aircraft Division
Shumaker, R.
JP-1
Washington, D.C. 20350
202/692-7443

Naval Sea Systems Command
SEA-90M
Everett, LCDR Hobart
Washington, D.C. 20362
202/692-6465

NAVSEA/David Taylor Naval Ship Div.
Johnson, Ralph
Computation, Math and Logistics Dept.
Carderock, MD
301/227-1058

Naval Surface Weapons Center
Werneth, Russell
Silver Spring, MD 20910
202/394-3256

Naval Ocean Systems Center/NOSC
Harmon, Scott
Code 8322
San Diego, CA 92152
619/225-2083

Office of Naval Research
Information Sciences Program
Schneck, Paul
Code 433
703/696-4303

DARPA
Defense Sciences Office
Isler, William
1400 Wilson Blvd.
Arlington, VA
703/694-4750

DARCOM/DRCMT
Michel, Fred
50001 Eisenhower Ave.
Alexandria, VA 22333

DARCOM
Indusrty Base Engineering Activity
Mc Glone, Steve
DRXIB-MT
Rock Island, IL 61299
309/794-3682

Human Engineering Laboratory/Army Proving Ground
Shoemaker, C.
Aberdeen, MD
301/278-5871

3. NSF, NASA, NBS, Dept. of Commerce.

NASA HQ/Computer Science and Automation
Larsen, Ronald
600 Independence Ave. S.W.
Washington, D.C.
202/453-2747

NBS
Automated Manufacturing Research Facility
Albus, James
Haynes, Leonard
Gaithersburg, MD
301/921-2181

Dept. of Commerce
Kravalis, Hedija
International Trade Administration
Washington D.C. 20230
202/377-4257

4. Industry

Hughes Aircraft Company
Hartman, Dale
Woodtli, Ernest
Krag, Neils
Harger, Robert A.
Rogers, Victor
McMahon, Steve
P.O. Box 1042
El Segundo, CA 90245
213/414-6154

Metcut Research Associates
Merchant, Eugene
3980 Rosslyn Dr.
Cincinnati, OH 45209
513/271-5100

Robot Insider Newsletter
Thornton, Jack
11 East Adams St., Suite 1400
Chicago, IL 60603
312/663-3500

Adept Technology, Inc.
Carlisle, Brian
Mountain View, CA 94043
415/965-0557

Cincinnati Milacron
Mauser, Kenneth
Lebanon, OH 45036
513/932-4400

Boeing Commercial Airplane Co.
Developmental Manufacturing
Christner, Richard G.
Seattle, WA 98124
206/655-4550

GCA Corp.
Industrial Systems Group
Kirkpatrick, Lane
Naperville, IL 60566
312/369-2110

H.E. Buffum Co.
Buffam, Harvey
Anacortes, WA 98221
206/725-5040

Ceeris International, Inc.
D'Agostino, Salvatore
Old Lyme, CT 06371
203/434-8740

LTV Aerospace and Defense Company
Vought Aero Products Division
Gandhi, Chander
Dallas,_TX 75265
214/266_2680

5. Un1vers1ty

University of Massachusetts
Arbib, Michael
Amherst, MA
413/545-2743

George Washington University
Bock, Peter
Washington, D.C.
202/676-6083

Massachusetts Institute of Technology
Brady, Michael
545 Technology Square
Cambridge, MA 02139

University of Texas, Austin
Busch-Vishniac, Ilene
Austin, TX 78172
512/471-3038

Louisiana State University
Conners, R.W.
Baton Rouge, LA 70803
504/388-5532